DOMESTIC TECHNOLOGY

A Chronology of Developments

DOMESTIC TECHNOLOGY

A Chronology of Developments

NELL DU VALL

G.K. HALL & CO. • BOSTON, MASS.

Printed on permanent/durable acid-free paper
and bound in the United States of America

Library of Congress Cataloging-in-Publication Data

Du Vall, Nell
     Domestic technology: a chronology of developments / by Nell Du Vall.
        p. cm.
     Includes index.
     ISBN 0-8161-8913-7
     1. Home economics--History. 2. Domestic engineering--History.
I. Title.
TX15.D8 1988
640'.9--dc19   88-21110
                                                   CIP

Contents

Contents

Preface

When we think about early people, it is difficult for us to understand how they really lived. In trying to reconstruct a character of a remote era I could not describe what people did or what activities occupied the day. Archaeologists of my acquaintance readily admit that much of what is written about early people is conjecture. They use what is known of how people at a similar level of technological achievement live today in order to form a picture of how people in remote ages lived. Even when written records exist they may not describe the mosaic of daily life. Since I could not solve the problem directly, because no historical source contained the desired information, I decided to attack it indirectly by finding out what tools and techniques were available at that time and how they might have been used. It also made me think about how I spend my own time and the heavy reliance I place on the products of technology. Because of modern appliances, supermarkets, and my children, I am able to work and still maintain a home. I have time for hobbies and friends. In more primitive times, much of my day would have been devoted to gathering, preparing, and preserving food, to acquiring raw materials, and to making clothes or other goods necessary for survival.

Survival was the main activity--or was it? The more I sought answers, the more questions I had. The result led me to examine the books and articles on the lives of early people and the evolution of civilization. Each date I found took me further back into early history. What I learned was a revelation to me; perhaps it will be less so to the professional historian.

I did find that little is known about domestic life even where written records exist. Most histories and historical studies are based on written records that focus on major events. Only a few concern themselves with what people ate, how they ate, and what they did in daily life. Archaeologists do try to reconstruct the context of life from the very scattered and limited remains, usually garbage dumps or ruins. What we can find is usually what was thrown away: broken tools, utensils, or the debris of their creation. Even when written records exist, it is not an easy task, because most do not describe the minutiae of daily life, except indirectly. Written history focuses on events and major inventions, not the more mundane aspects, such as whether, or what kind of, potato mashers were used. The recent enthusiasm for collecting almost anything more than a few years old has changed this, but surviving tools are still scattered, and often their use is forgotten or misunderstood.

Encyclopedias, handbooks, and source guides do provide some information on household technology and methods, but precise dates and chronologies are difficult to find, and it is often hard to ascertain actual uses and practices. Future generations might be as perplexed as I have been over some of the highly innovative but unknown devices invented during the 19th century. Without a patent description or another author explaining the intricacies of an apple parer-slicer or what the cone-shaped clothes pounder for manipulating wet clothes was, I would not have been able to guess their use. It has profoundly increased my respect for the difficult task faced by the archaeologist.

It has been an exciting and interesting journey, sometimes frustrating, but never dull. I hope to convey some small part of that excitement and engender an appreciation for how early people developed certain tools and the steady evolution that has occurred even during the so-called Dark Ages. I have also tried to include events beyond the English-speaking world, but these are limited to those contained in English-language publications.

Modern effects of technology have been considered by numerous authors and, more recently, the impact of technology on the home and the homemaker has been extensively studied by Caroline Davidson in *A Woman's Work Is Never Done*[1] for the United Kingdom, Ruth Cowan in *More Work For Mother*[2] for the United States, and Ann Oakely in *Woman's Work*.[3] This review does not repeat those sociological studies, but provides a broad framework for looking at daily life from earliest times to the present. The list of technologies required to support the modern household has not been exhausted, but the ones mentioned highlight some of the impact of technology on the home and standards of living. Technology provides the fabrics worn, the foods eaten, the means used to prepare those foods, the tools used in their preparation, the water drunk, the heat for homes, and protection from the weather. The final chapter explores these effects further and draws together the various issues raised in this work. I have enjoyed this search and hope the reader will, too. Human inventiveness is striking in its variety and consistent evolution.

My sincere thanks are due to the technical reviewers who read the first draft manuscript, the special library community, and the excellent and cooperative support provided by many OCLC member libraries. Particular thanks is due to the long-suffering OCLC Library staff, to the libraries of Ohio State University, and to the Upper Arlington Public Library, and, finally, to Sandy Albanese.

1
Beginnings

The wind howled outside, confusing the falling snow with the drifting mounds until the landscape was a white, whirling blur. The windowpane felt cold, but inside the air was warm and there was no need for a sweater. The glow of the lamps added to the feeling of warmth, security, and comfort. It was pleasant to settle back and forget the blizzard outside. Suddenly, the lights flickered and went out. The house was quiet: the refrigerator was no longer clicking on and off, the furnace blower had ceased to whir, the radio, television, and the clocks had all stopped. Even the washing machine had stopped only half way through its automatic cycle. No water flowed from the tap; the well pump was electric.

This was exactly the situation faced during a recent winter storm in one community. Some families were forced to burn furniture to keep warm and to melt snow for water. Those of us living in Western societies take much for granted in our daily lives. The startling contrasts we experience when we visit a less developed country or when there is a sudden disruption such as an extended power failure, suddenly make us aware of exactly how dependent we have become on our water supply, power systems, and fuel sources, and even on the distribution system that makes food readily available. It is hard for us to understand primitive living conditions, because most of us have never experienced them. Even our camping activities are supported by the complex distribution and supply system we rely on for food and other necessities. Our homes and daily lives are dependent on an intricate web of technology. What are these technologies and how did they originate?

Domestic or household technology as used in this book includes all the tools, appliances, and services we use to maintain our homes and our daily life-styles. It encompasses the foods we eat, the tools and utensils we use, the clothes we buy, our homes, and almost everything in them. It also includes some of the underlying structures and services including water supplies, waste disposal, and power. It does not include the development of telecommunications technology--telegraph, telephone, radio, or television-- nor other mass media such as magazines, newspapers, or films. This exclusion is made not because these technologies are unimportant, or because they are not used within the home, but rather because they are less essential to survival and in order to keep this overview within reasonable bounds. Those aspects of communications development related to writing, papermaking, and printing have been included because so many of the products used in the home require use of these, as have some aspects of home medicines and child-rearing.

Most studies of domestic technology have dealt with the impact on women and their changing role in society. Today, with more two-career families and single heads of households, and with an aging population, women are no longer the sole users of domestic technologies. All of us use technology to some degree, whether it is using a microwave oven to heat a frozen meal, a stove or other appliance to heat water or make coffee, or an automatic laundry to wash clothes. We may not all use the same tools, but we are all affected in some way.

There has been no study encompassing the development of domestic technology from human beginnings to the present. A number of excellent studies for specific periods or specific functions do exist. They are cited in the bibliography at the end of this work.

We have come a long way from primitive man huddled over a smoky fire in a drafty cave eating half-raw meat. We will continue to see more changes as scientific and technological advances are applied to the mundane tasks that support and cushion our lives. As in the past, many of these will initially pass unnoticed; their importance will be clearer in retrospect. With our modern systems of patents, advertising, and recording information, they should be easier to trace. The computer revolution is just beginning, and already major changes are occurring in daily life and in the tools we use. More will come. These are treated briefly, since my intent is to provide an overview of how we have lived--from earliest times to the astronauts on the space shuttle.

Invention is a process of trial and error, and an eventual successful invention is as much a matter of luck and marketing as it is the best combination of features. For example, Edison and Swan both developed the electric light, but Edison's genius was in developing not the best light, but an electrical distribution system; Bell was not the sole inventor of the telephone, but he was able to establish a monopoly; James Watt improved, but did not invent, the steam engine; and among programming languages FORTRAN and COBOL are well entrenched, but not necessarily better than languages such as PL1. Failed inventions often paved the way for successful ones. Invention is an incremental process and may also depend on other supporting technologies being available to make manufacturing successful.

This is not a scholarly, in-depth study, but rather a short tour highlighting some of the endless inventiveness and adaptability of the human animal. The primary focus has been to develop a chronology of specific events including where and when they occurred and who was primarily responsible. Where possible both the successful invention or principle and those regarded as unsuccessful, but which were likely to have contributed to the former, are included. Many household machines and appliances owe their existence to industrial or business machines that were later also applied to home use. Consequently, machines such as road sweepers that led to carpet sweepers have been included.

Some sources indicated whether the dates given were for invention, patent registration, start of manufacture, or sale; others did not. Occasionally the patent for a device or process was issued before the item became broadly available; in other cases the patent was issued later. These differences meant that different sources may give different dates for the same event. For some items all three dates have been included.

Although the span of time and breadth of material covered are broad, this study is not complete. Locating and identifying pertinent sources has been difficult since indexes and catalogs are not sufficiently detailed to identify adequately all the relevant information. Earlier developments are not well documented, and much is still a matter of conjecture and speculation for the following reasons: lack of written documentation; lack of surviving artifacts; changes in language or usage; varying conditions in different locations; the different rates of evolution of various cultures and geographic areas; and the rewriting and reinterpreting of history in the light of more recent discoveries. Nonetheless, the overview presented here does provide a wealth of detail not available in any single source and should assist general readers, librarians, and scholars.

Each of the following chapters contains a brief explanatory text and extensive chronological tables listing major innovations and inventions documenting the gradual spread of domestic technology. Each event has been checked in several sources, but only one source is listed in the tables. Some references are suspect and are so indicated in the text. The *Dictionary of Inventions and Discoveries* was one of the least reliable sources used.[1] The *Book of Firsts* was one of the better sources, and everything that could be cross checked was verified as being accurate.[2] The *Browser's Book of Beginnings*[3] was published after most of the text had been written as was the

World Almanac Book of Inventions.[4] Wherever possible, events have been verified in multiple sources, particularly against the *Encyclopaedia Britannica*[5] and the *World Book.*[6] Sources used are a mixture of scholarly works, popular press, trade books, almanacs, and lists of trivia. Many items are only available in the popular press or company archives. In a number of instances, for modern events, because I knew an approximate date or company name, I was able to contact the firm or a trade association for further information. Until something becomes "historic" or of interest to the academic community, it is not generally found in technical or scholarly sources.

Each abbreviated source is listed in full by the abbreviated designation in the Sources. Notes for each chapter are included for information not tied to a specific event given in the tables and are found in the back under the appropriate chapter and note number. No separate references are given for events in the tables beyond the Sources.

Domestic technology has been difficult to trace. While archaeology is interested in how people lived, history records major events and little of how people live day to day. Diaries and personal records seldom fully describe how people made soap or did laundry, but often mention that such an activity took place. Few women had time, or in some cases the knowledge, to write. Those who did were often of the upper class and, while knowledgeable, were usually more concerned with writing on intellectual matters rather than the details of housework or household matters.

As has been suggested in the preface, the extent and use of practices detailed in manuals are questionable. Before the industrial revolution much of the available technology was domestic technology, but that is no longer the case, and copies of superseded encyclopedias that might contain such descriptions are rare. In most cases trying to trace when these types of inventions and events occurred has been difficult. Recent events were among the hardest to locate, since the historians have not yet recorded them, and a few, such as when a brand of toothpaste or soap were introduced, appear inconsequential.

The spread of innovation has been uneven. Not all segments of society have benefited to the same degree. Labor-saving devices benefited the middle class and well-to-do first, because they were usually better educated, more informed, and could afford technology. Ultimately, however, almost all of society was affected. For example, initially only the wealthy could afford sanitation, adequate food, and clothes. The advent of the industrial revolution led to major societal changes. At first the poor benefited from more jobs, then were hurt as their occupations were changed or eliminated; but ultimately they to enjoyed the benefits technology made possible; shelter, clothing, more food, water, and improved sanitation.

4

Standards of hygiene, nutrition, and dress differ between societies and even among segments of the same society. Some of the difference is due to education and background. Some of it is economically driven and some is value driven. There are significant differences between rural and urban poor. The urban poor often do not live more cheaply, only differently.

In the United States, two devices are found in almost every home, the telephone and the television. Television has led to rising expectations among all segments of society. Commercials and programs abound with the use of technology: microwave ovens, vacuum cleaners, automatic washers and dryers, and automatic coffee makers are used routinely. Food is plentiful and floors sparkle. The use of, and access to, technology becomes the expectation of every household. People feel they have a right to such conveniences.

Television has become the great homogenizing influence, a leveling force creating similar expectations among different societal groups. This has become a matter of considerable concern to many and especially to the lesser-developed nations where such leveling is viewed as a loss of cultural identity and values and is shared to some degree by minorities and ethnic groups in the United States.[7]

Transfer of innovation and use of technology may occur at different rates. Widespread use of electrical appliances by rural populations in the United States did not occur until the major hydroelectric projects made cheap sources of power available. A revolution occurred in daily life within a short span of time.[8]

In Britain the coin-in-the-slot gas and electric meters created a similar revolution. Users paid in advance for gas or electric power by inserting coins in the meters, which released a measured amount of gas or electric power. Until such devices were available, gas and electricity were not accessible to many of the poorer segments of society. Availability of these power sources at relatively low cost, pay-as-you-go, led to a revolution in lighting, heating, and food preparation. Candles and rushlights were replaced first with gas and then with electric lights. Fireplaces or smoky, dirty, wood or coal stoves were discarded for gas and electric appliances.

Such changes were profound. They affected time allocation: no more wicks to trim nor lamp globes to clean, no sooty stoves to polish nor ashes to dispose of. Light was available at the touch of a switch, enabling tasks to be undertaken when convenient, and not just in daylight hours. With more efficient stoves food could be fixed more rapidly. A fire did not have to be built. No wood had to be cut or coal hauled from storage. Tasks were performed differently and some were eliminated entirely.

Identifying and tracing such effects is neither straightforward nor easy. Drawing conclusions based on scanty or insufficient evidence may lead to confusing cause and effect. It is hoped, however, that this overview will trigger associations in the minds of others with more time and expertise to explore all the complex relationships and possible ramifications. This work focuses on a chronology of events that document the evolution and migration of many of the technologies used in or influencing the home.

Human society has evolved through at least three stages, although different groups have coexisted at the same time in different stages. These stages are: hunter-gatherer, pastoralist/agriculturalist, and urban dweller.

Hunter/gatherers lived a literal hand-to-mouth existence. Depending on climate and availability, much of their time was spent acquiring food sources, warmth, or shelter. Where the climate was mild and game plentiful, life was not overly rigorous, but outside a fairly narrow geographic band more adaptation and ingenuity were demanded. In colder regions, food was scarce in the winter season, and without adequate planning, survival became difficult. Tracking the seasons became an element of survival that was further emphasized in early agricultural societies and may have led to the development in such regions of monuments such as Stonehenge or Avebury.

Some historians have argued that the movement to pastoralism or agriculturalism occurred because of declining game resources, but if hunters were constantly worrying about the next meal, it is hard to see where the foresight and planning necessary to plant and harvest crops developed. More likely, in relatively stable areas with assured food supplies such as from a nearby lake or river, people had time to observe and to use their observations to develop agriculture.[9]

It is likely that the first harvests were from wild plants that reseeded themselves naturally. Consequently, it is difficult to identify exactly where and when deliberate cultivation began.

As people developed out of the hunter/gatherer stage more specialization occurred. Groups traded goods with other groups, particularly for scarce materials or when a special or high degree of skill was necessary. Examples of scarce materials included certain types of minerals, stones, and metals, fibers such as cotton or silk, wines, and some types of furs. Examples of skills included weaving, metal working, or glass making.

In a pastoral environment people still migrated with the animals, but with the emergence of agriculture, they became more stationary. Some seminomadic cultures planted crops during one season, harvested them, and then moved on to a warmer location for winter. Such migratory patterns were influenced by climate, available food resources, pressure from other similar groups, and the adaptability of the group.

Most of what is known about early agriculture is associated with the Middle East, Egypt, and to a degree, with China. In all of these areas agriculture, written language, and urban centers developed. A limited amount has been learned about early settlement in the Indus Valley and in the Americas. Much less is known about those two groups partly owing to the lack of written records and the condition of surviving artifacts.

Urbanization increases specialization with fewer people directly involved in the actual food production processes. Roman civilization is an example of this, as are the later city states of the Italian peninsula.

In modern society only a small proportion of the population, less than 2 percent in the United States, grows food. A larger segment handles various food processes or selling. We have grown increasingly remote from our sources of food, or even of an understanding of how they are produced. It is not surprising that a city child finds it difficult to believe milk comes from an animal and not from plastic jugs.

A hunter-gatherer needed few tools: a knife, a bow and arrow, a few scrapers, a lance or spear, perhaps a few bone needles. If he gathered grain, he may also have used a rock or two suitable for crushing or grinding grain. Pots and baskets were used, but too many possessions made it difficult to move easily and quickly.

As people became more settled they needed more tools and found it easier to accumulate and store them. Pottery increased in type and variety, while plows and hoes became essential for farming. Depending on the climate, they looked for and adapted various materials for clothing. Leather is hot and, unless well treated, tough and irritating to the skin. People learned early that various animal and vegetable fibers, could be twisted to form rope or yarn and plaited or woven to form mats and later fabric.

Gradually humans developed a host of new skills including metal working, textiles, and decorative arts. Surviving artifacts from primitive societies exhibit a fineness of detail and decoration that implies a love of beauty beyond mere utility. Artists and craftsmen became an accepted part of society.[10]

At first goods and services were exchanged by barter. Later, money was developed and served in place of barter agreements, and commerce evolved with other groups that had access to scarce resources or to products or skills not yet available in the home group. Trade, merchants, markets, and stores developed, and along with them increasingly sophisticated technology.

Urban centers were a concentration of people that individually could not grow or hunt enough food to support themselves. Places for extended storage of food such as root cellars were not readily available. Sanitation systems satisfactory for scattered dwellings or small groups were no longer adequate.

Many of these problems, and the sometimes creative solutions to them, are apparent in the development of some ancient cities and in medieval towns and cities. As a consequence of living in confined spaces, citizens had to work together to solve common problems. This critical mass both created and solved problems, albeit not always the same ones.

In different societies or environments people may choose different solutions to similar problems. One example of different solutions to a similar problem is demonstrated by the use and evolution of ovens in the United States and in England. In the Colonial United States sparse early settlements required that almost all homes had ovens or baking facilities,[11] while in the more densely settled United Kingdom with scarce wood resources more reliance was placed on the village baker.[12] Similarly, pay-as-you-go gas (1887) and electric meters (1901) became common in the United Kingdom, but were seldom used in the United States. The differences in the spread and features of stores in North America and in the United Kingdom are another example. Markets where people come to buy and sell a variety of goods remain common in Europe, but did not develop significantly in the United States where trading posts and general stores were more common. The recent development of flea markets in the United States is a return to forms common elsewhere, but is still a peripheral retailing mechanism. Most goods are sold through stores. Telemarketing, the direct sale of goods by telephone or television advertising, and catalog sales have grown dramatically, but have not replaced stores.

To help place the evolution of technologies more firmly in perspective, this overview begins with human evolution from hunters and gatherers to city dwellers, and is followed by a review of the individual technologies and their development and by a consideration of some aspects of their impact on daily life.

Three aspects of domestic technology are integral to survival. These are: food gathering, dress, and shelter. They are treated more extensively than some of the other technologies, which, while an important part of our daily life, are more peripheral to survival. These other technologies include lighting, food preparation, cleaning, sanitation, health and child care, and educational and general tools.

2

Food Origins and Production

A guaranteed wholesome supply of food is basic to our survival. The ability to produce enough food for consumption and to reserve seeds for planting and cultivation marked the beginning of agriculture as an alternative to hunting and gathering foods for subsistence. Although it is difficult for us to imagine the significance of the changes agriculture has introduced into human life and into the technology used for daily living, consider that a hunter-gatherer requires some fifteen square miles to support himself, but a farmer can support 5,000 people on produce from the same area of land.[1] These estimates depend on the sources of food available, the climate, and the means of food preservation, but indicate the dramatic shift agricultural technologies have caused in the organization of human societies.

Origins

Many of the early innovations are hard to trace. Some occurred simultaneously in different areas and others developed in one location and gradually spread to other areas.[2] Similarly, some foodstuffs grew or developed naturally in different parts of the world while others were transferred deliberately by humans and animals. The events described below identify when technologies and foodstuffs were first used in a culture and when they were transferred to other cultures. For convenience, foodstuffs have been divided into some broad groupings: cereals, fruits and vegetables, animals, beverages, and seasonings.

9

Figure 2.1. Overview of Food Sources

YEAR		
1987		
	Cloning techniques	
1975		
		Miracle rice
	Wheat hybrids	
1950		
	Corn hybrids	Crop dusting
1900		
		USDA, Morrill Act, Hatch Act
	Artificial fertilizers	
1800	Sugar beet refining	
	Steam power	
1600		
	Discovery of New World plants	
1200		Rigid horse collar
	Windmills	
400		
	Watermills	
		Iron
-1200		
	Bronze	
	Irrigation, wheel	Wine
-4400	Rice (China)	Potato (Peru)
	Domesticated pigs	Domesticated cattle
-10800	Farming begins	
		Sickles, celts (axes)
-23600		
-59200		
	Crude grinding stones	
-110400		
-212800		
-407600		
	Fire	

This chapter will explore the origins of the food we eat and how we grow, harvest, and process them. Two popular sources for these topics are *Food in History*[3] and *The Browser's Book of Beginnings*.[4] They are particularly helpful in tracing the origin and migration of foods. The *Encyclopedia Britannica* also has extensive articles on food and agriculture.

Scientists have tried to determine the origin, evolution, and migration patterns of food by studying the number of chromosomes in various species of plants and animals. Trade between different geographic areas was significant in the transfer and migration of foods, as were wars and colonial expansion. Alexander the Great, for example, discovered rice and sugar cane in India during the course of his wars of conquest (c. 327-326 B.C.). Similarly, the discovery of the New World led to significant transfers in a relatively short time. Crops and animals introduced from the Old World to the New included wheat, beans, broccoli, cabbage, coffee, grapes, olives, rice, oranges, sugar cane, bananas, apples, almonds, horses, pigs, and cows. Foodstuffs from the New World were also introduced to the Old and many became significant staples. These included potatoes, corn (maize), tomatoes, sweet potatoes, turkeys, eggplant, cacao, manioc, cashews, peanuts, chili peppers, avocados, pumpkins, and vanilla.

Cereals

Cereal grains are one of the essential components of the human diet. Hunter-gatherer societies used grains, as the surviving crude stones used for milling more than 75,000 years ago testify. Wheat was grown in the Near East by 7000 B.C. and barley was cultivated by 6000 B.C. Cereals were grown in the Mageburg-Cologne-Liege area about 3200 B.C.

Rice was cultivated at least 4,000 years ago in China, but spread slowly beyond Asia. It grew in Southeast Asia and in India and was discovered in India by Alexander during his conquests (327-326 B.C.). The Moors introduced rice to Spain during the 8th century A.D. and the Spaniards introduced it to Italy in the 15th century. Rice has never been as popular in the West as other grains, partly because it is not suitable for bread making.[5]
Corn originated in the Americas and is a popular crop in the United States and other countries. (In Europe, *corn* means any kind of seed grain, including barley and wheat; *maize* is used for Indian or New World corn.) Corn (maize) was grown in Mexico almost 7,000 years ago, but was unknown to Europeans until discovered by Columbus's crew in Cuba in 1492. It was taken to Europe and thence to Africa. Since corn is relatively easy to grow, it displaced other local crops and became a staple in African diets. Manioc (cassava root, from which tapioca is produced), another New World crop, was introduced to Africa and Asia by the Portuguese.

Figure 2.2. Cereal Grains

YEAR	WHEAT	RICE	CORN
1987			
1975			
	Gaines semidwarf	New rice strains, Int'l Rice Research Institute	96 percent of corn grown from hybrids
1950	Sears-wheat hybrids		
1900	Charles Sanders's Marquis wheat Mendel, Darwin McCormick Reaper		Corn hybrids
1800			
	Wheat cleaning machine		
1600		South Carolina	Maize cleaning machine
1200		Italy	Columbus discovered corn
		Spain	
400			
		Rice huller (China) Discovered by Alexander in India	
-1200			
-4400			
		China	
	Near East		Mexico
-10800			

Dependence on a single food source such as corn or manioc can cause nutritional deficiencies. New World diets built on corn and manioc also included tomatoes, peppers, and beans, which contain essential nutrients that these cereals lack. These supplements were not as readily available in Africa and other countries, however, and overdependence on corn, manioc, and potatoes, which can lead to malnutrition, continues to be a problem in some lesser-developed countries. An additional risk is that when crops fail, famine may occur. During the Irish Potato Blight in the 1840s, the Dingle Peninsula, lost one half of its population to starvation and migration.[6]

Vegetables, Nuts, and Fruits

Vegetables, nuts, and fruits were as important to early people as cereal grains and were used as sources of food, beverages, and oils.

Vegetables

The existence of early vegetables can be traced back to 6500 B.C., when chili pepper and squash were grown from seed in Tehuacan, Mexico. The history

12

of the cultivation and use of onions, potatoes, and cucumbers also reaches back several thousand years. Onions were depicted in Egyptian tomb decorations from 3200 B.C., and asparagus was grown there from 3000 B.C. Peas were grown in the Near East by the same time, while radishes and soybeans were cultivated in China and the potato was grown in Peru. Cucumbers were introduced to China from India in 140 B.C.

Figure 2.3. Selected Vegetables

YEAR	TOMATO	POTATO	AVOCADO
1987			
1975	Harvester		
1950		Harvester	

1900	A. Livingston's red variety
	U.S. Supreme
	   Court decision
	R. Johnson eating
	   demonstration
		Burbank potato
		
		Irish Blight
			California (from Mexico)
			George Cotton improved
			
			Florida (from West Indies)

1800	Grown by Jefferson	Grown by Irish immigrants in New Hampshire, Europe, Germany, England, Spain	
1600	Mexico, Peru	Columbia, Peru	
1200			
400			
-1200			
-4400	Peru	Peru	

The expansion into the New World in the 15th century led to the interchange of many foodstuffs across the continents. Most Old World vegetables were eventually introduced to the New World as settlement took place. Carrots were grown by Jamestown colonists in 1609, and Dutch farmers in Kalamazoo, Michigan, grew celery commercially by the 1870s. Broccoli was popular in Italy during Roman times and was brought to the United States near the end of the 19th century by Italian immigrants. The D'Arrigo brothers, California vegetable growers, grew broccoli in California and first shipped it east to Boston in 1923. Broccoli retains its flavor and texture when frozen and became popular as a vegetable in frozen TV dinners in the 1950s.

Just as Old World vegetables were transplanted to the New World, there was a sudden influx to Europe of new plants, vegetables, and fruits. Cieza de Leon, writing in 1553, mentioned potatoes grown in Peru and Colombia in 1538; by 1573 potatoes were recorded in the accounts of the Hospital de la Sangre. The sweet potato was introduced to England in 1565, and Sir Thomas Harriot brought Colombian potatoes to England in 1586. They were taken by the English shortly thereafter to Ireland.

The potato appeared in Austria in 1588 and was planted in Germany in 1621. In 1651 Frederick William I threatened to cut off the nose and ears of those unwilling to grow them, and his son Frederick William II made cultivation compulsory in order to alleviate food shortages in the latter part of the century.[7] By 1765 it had become a popular food in most of Europe. It was introduced into the United States by Irish immigrants to New Hampshire in 1719. The Irish Potato Blight in the 1840s caused over a million Irish to migrate to the United States.[8]

Nuts

Among the more difficult foods to trace are nuts. They are found in many locations and were eaten by humans from earliest times. Hazel nuts were, and are, found throughout Europe and are common in prehistoric sites. According to Friedlander, chestnuts were supposed to have been eaten by Xenophon's army during their retreat from Asia Minor in 401-399 B.C. Cashews were native to Central and South America and were taken by the Portuguese to India in the 16th century. Almonds, native to North Africa and Western Asia, were cultivated in the Mediterranean, and Friedlander notes they were called Greek Nuts by the Romans.[9] Poole records they were introduced in the 18th century to California by the Spanish missionaries. The trees thrived. By 1897 seventy-one California almond growers formed a cooperative association that became the California Almond Growers Exchange. One year later they grew 50,000 pounds of almonds, and by 1970 the total had reached 135 million pounds.

In the New World, pecans, native to North America (specifically found in Texas and Louisiana), were first cultivated in El Paso near the end of the 17th century and were carried further north by soldiers after the Civil War. Another New World food, the peanut, was used by the Aztecs and discovered in Haiti by Columbus. The Portuguese used peanuts as a cheap food for West African slaves awaiting shipment to the Americas and also carried the peanut to Asia. The slaves called peanuts *nuguba*, which evolved into *goober*. They brought the peanut to North America. Peanuts are a dietary staple in Asia and Africa and are often used in paste form as well as whole.[10]

Fruits

Among the earliest and the most important of early fruits were the olive and the grape. The olive was important for both its oil and the nutritional value of the fruit. The oil was used by the Egyptians for lubrication and as fuel for their lamps. Olive trees were cultivated in Syria and Crete by 4000 B.C. and introduced to the Italian peninsula in 600 B.C.

Grape cultivation began in the Caspian and Black Sea coastal regions around 4000 B.C. By 500 B.C. viticulture had been introduced to the Italian peninsula. Wine was important in trade with the provinces during the Roman empire and was traded with the Gauls for soap and woolens. The Romans introduced grape cultivation to southern France and Spain.

Other fruits such as the apple, orange, grapefruit, and tomato, like the grape, were valued both as foods and as beverages. Most early apples appear to be species of crab apple; the first apples similar to modern varieties were grown in the Caucasus from a cross breed of Asian and European crab apples. Carbonized apples have been found in Stone Age sites in Switzerland, while apples dating from 6500 B.C. have been found in Asia Minor. The first record of cultivation dates from approximately 1200 B.C. when Rameses II of Egypt ordered apple orchards to be planted along the Nile. Homer (ca. 850 B.C.) mentioned apple trees in the garden of King Alcinous, but it remains unclear if he meant the apple or the quince, which was also known as the Cretan Apple.

The Greeks and Romans considered apples luxury items. By the 2d century B.C. Marius Portius Cato (the Elder) in *De Agricultura* described grafting techniques for apple stock, and by 77 A.D. Pliny in his *Historia Naturalis* described thirty-six varieties of apples. Apples were mentioned in recipes in *The Forme of Cury* (1390) and in *Le managier de Paris* (1393).[11] Although some species of apple are native to the North American continent, most eating apples are hybrid varieties. Apple growing in the United States began from European seedlings in 1629, and the first apples were picked in 1639 on Beacon Hill, Boston.

By 2200 B.C. the Chinese had cultivated the orange, and mandarin oranges were available by 220 B.C. The Arabs carried oranges to Spain after the 7th century, and from there oranges were taken to France and, by 1300 A.D., to the Italian peninsula. In the 15th century Columbus carried Canary Island orange seeds to Hispaniola. By 1516 oranges had been introduced to Panama and were being grown in Florida by 1579. The orange was carried to Arizona in 1707-10 and to California by 1769 (the navel orange from Brazil was not introduced into California until 1873).

Of the other large, acidic fruits, the pineapple began in the New World and the grapefruit, while found there, may have had its roots in Southeast Asia. The pineapple was discovered on Guadaloupe Island by the Spaniards in 1493 and introduced to Italy in 1616 and to Hawaii around 1790. The grapefruit appeared in Bar Valos in 1755. Vietmeyer speculated that it may have evolved from the pummelo, which was introduced to the Caribbean region from Southeast Asia somewhat earlier.[12]

According to the *World Book*, the peach originated in China by 2000 B.C. and was cultivated there by 1000 B.C. Confucius immortalized the peach in his writings in the 5th century B.C. The peach had traveled to Persia by the 3d century, was imported from there to Rome by 1st century A.D., and was long thought to have been of Persian, not Chinese origin. The Spaniards carried it to the Americas in the 16th century and planted trees in Mexico and Florida. Large-scale commercial cultivation in the United States began around 1820.

Bananas originated in southern Asia, and the root was carried by various migrating peoples to the Pacific Islands. In 327 B.C. Alexander's troops discovered bananas in the Indus Valley. By the 2d century A.D. Chinese scholar Yang Fu extolled the banana in his *Record of Strange Things*. Arab traders carried bananas from India to the Middle East and northern Egypt and from there to Africa. The fruit was carried to the west coast of Africa, and the Portuguese took it from Guinea to the Canary Islands in 1492. From there a Spanish priest, Friar Tomas de Berlanga, took the banana to Santo Domingo in 1516. By 1854 Elder Dempster and Company was regularly importing bananas from the Canary Islands to the United Kingdom.

By 1850 ships brought the fruit to the United States, and in 1876 tinfoil-wrapped bananas were sold for ten cents at the Philadelphia Centennial Exposition. They remained somewhat of a novelty in the United States until the end of the century when Captain Lorenzo Dow Baker founded the United Fruit Company, which led to the export of bananas to the United States and to a growth in the banana business.[13]

Unlike the apple, orange, and banana, the tomato was a New World fruit and originated in the mountains of Peru. The Aztecs named the fruit *tomatl* and developed a larger, yellow variety.[14] The Spanish took it to Europe where the Italians call it *pomodoro* or "golden apple." European botanists identified it as a member of the poisonous Solanaceae family, which includes nightshade and black henbane. For a number of years the tomato was used primarily as an ornamental garden plant, and then as a treatment for skin diseases such as ring worm and athletes' foot. In 1781 Thomas Jefferson recorded *tomatas* in his garden journal at Monticello, Virginia.

By the late 18th century, however, tomatoes were grown as field crops in the Italian peninsula. They remained an ornamental plant in the United States until Robert Johnson, Salem, New Jersey, showed tomatoes were not poisonous by eating them in a public demonstration (ca. 1820). Alexander Livingston of Reynoldsburg, Ohio, developed a smooth, deep-red tomato in 1870. Reynoldsburg now labels itself as the tomato capital of the world.

By the later part of the 19th century tomatoes had become an important crop in the United States. Although imported vegetables were taxed, fruits were not. Eager to protect the domestic market from foreign competition, American growers argued that tomatoes were used as vegetables and should be taxed as vegetables. In 1893 the United States Supreme Court ruled that tomatoes were taxable, arguing that tomatoes were "a principal part of the repast and not . . . a dessert.[15]

Avocados, or alligator pears, are also a New World fruit and were eaten by the Aztecs who called them *ahuactal*, which translates as "tree testicle."[16] Avocados were introduced into Florida in 1863, but it was not until 1900 that horticulturalist George Cellon found a method to produce fruit of uniform size and quality. Avocado cultivation began in California in 1924 when Mexican avocados were introduced.

The boysenberry was developed by Rudolph Boysen by crossing varieties of blackberries, red raspberries, and the loganberry. The bushes were discovered around 1930 by Walter and Cordelia Knott who popularized the fruit and established Knott's Berry Farm.

Some fruits were also used to make beverages. These included herbal teas, coffee, wine, beer, cider, and mead.

Beverages

Early beverages included milk, beer, and wine. Milk was drunk in Eurasia by 6000 B.C., and milk drinking is illustrated in a frieze at Ur, southern Iraq, dating from 2900 B.C. Beer and wine are manufactured beverages and are discussed in the next chapter.

Use of coffee, tea, and chocolate as beverages spread more slowly than did the use of beer and wine. In fact, coffee was first used as a food. The raw beans are high in protein, which is lost when they are made into a beverage. African tribes used stone mortars to crush the ripe berries from wild coffee trees and mixed them with animal fat, which they rolled into balls. The balls were easy to carry and a compact source of both food and stimulant. Later, the Africans also used coffee as a beverage--including a coffee wine--but it was the Arabs who cultivated it and devised hot coffee.[17]

The first written mention of coffee dates to circa 1000 by Avicenna, an Arabian philosopher and physician, who recorded its medicinal function. By 1200 the Arabs were using the dried hulls of the beans, which they roasted to create a beverage; by the 16th century they were roasting and grinding the whole bean. When *qahveh khaneh*, coffeehouses, opened in Mecca, the doctors opposed the houses' sale of coffee because it competed with the doctors' lucrative prescription business. The more pragmatic government officials prevailed, however; perceiving a new source of tax revenue, they allowed the now taxable coffeehouses to continue operating.

Figure 2.4. Selected Beverages

YEAR	COFFEE	TEA	CHOCOLATE
1987			
1975			
	Freeze-dried products		
			Peanut M&M's
1950		Instant tea	
		Constant Comment	
	Instant coffee		
	(Nescafé)	Iced tea, tea bags	
1900			
			Hershey Bar
		Lipton tea	Milk chocolate
		Indian tea (England)	Cocoa
1800			Eating chocolate
	Brazil (from West		
	Indies)		
	U.S.		U.S., Germany
	France, England	Paris	Italy
1600			
	Venice		Spain
	Turkey		
1200			
	Arabia		
		Japan	
400			
-1200			
		China	
-4400			

Coffee reached continental Europe via Italy: it was brought to Venice from Turkey in 1580. Despite the objections of a number of priests, Pope Clement VIII (1592-1605) sampled the coffee and blessed it. With papal approval, coffee rapidly spread throughout the rest of Europe. It was introduced to Paris in 1643 and to Oxford sometime before 1650 by Conopios, a disciple of Cyrill, Partiarch of Constantinople, who had fled to Oxford after Cyrill was slain in 1637. Pasqua Rosee established the first London coffeehouse in 1652, and the city of Marseilles followed with its first coffeehouse in 1671. Coffee was available in Germany in 1670, offered by the elector of Brandenburg to his guests in 1675 and served at the first German coffeehouse opened in Hamburg by a London coffee merchant in 1678.[18]

As coffee usage spread, new ways were found to make and flavor it. Viennese coffee (strained so the grounds were not left in the coffee) was devised by Franz Georg Kolshitsky in 1683; he is also credited with popularizing the practice of adding milk and sugar.[19] The Buena Vista Café, San Francisco, added cream and Irish whiskey to make Irish Coffee in 1953.

Coffee reached the American Colonies with the Mayflower (1620); a coffee mortar and pestle were listed on the Mayflower's cargo list. A license to sell coffee was issued to Dorothy Jones of Boston in 1670. Despite these early traces of its presence in the colonies, coffee was not the drink of choice among colonists until the British tax on tea caused Americans to turn to coffee in protest (1773).

The Arabs maintained a stranglehold on the coffee trade until a group of Muslims visiting Mecca smuggled coffee beans out of the country to India. Dutch coffee spies also managed to steal some plants, which they cultivated in Java. Coffee growing spread throughout Southeast Asia to Ceylon, India, Sumatra, and Malaya. Unfortunately all were struck by the blight, *Hemeleia vastatrix* (more popularly called "coffee rust"), in the mid-19th century and the coffee plantations were wiped out. The fungus does not reproduce well at higher altitudes and can be controlled with copper compound sprays during the wet seasons. Unfortunately this solution was discovered too late to save the Southeast Asian coffee plantations.

Luckily for the world's coffee drinkers the Dutch botanical gardens had given plants freely to the rest of Europe's botanical gardens, including a 1714 gift to the Jardin des Plantes in Paris. Although jealously guarded, one of these valuable plants was stolen from the Jardin and later became the source of coffee throughout Latin America. In 1723 a French naval officer from Martinique, Gabriel Mathieu de Clieu tried to obtain a plant from the Jardin, but was turned down. He eventually managed to steal one and transported it, with some difficulty, to Martinique where it thrived.

During the early 18th century, the Dutch and French started coffee plantations in Guiana. A Brazilian officer stationed in the colony seduced the wife of the governor of French Guiana and persuaded her to give him a bouquet containing fertile coffee cuttings. With these precious cuttings cultivation began in Brazil in 1727. Together with the West Indies, Brazil provided a second New World source.[20]

Although coffee eventually became the drink of choice in the United States, the British are still noted as tea drinkers as are the Chinese and the Japanese. Tea drinking originated in China, perhaps as early as 2737 B.C., but it was not cultivated in China until around 350 A.D. By 450 Turkish tribesmen were bartering with Mongolians for tea and by 780 the Chinese government, seeking, as most governments do, new sources of revenue, had imposed a tax on tea. No one is certain when it reached Japan, but the Emperor Shomu gave powdered tea to one hundred Buddhists priests in 729, which inspired many of them to plant tea gardens. Among the group of priests was Gyoki, 658-749, who is credited with building forty-nine temples with attached tea gardens. By 800 tea was widely cultivated in Japan.

Europeans first heard about tea from the Portuguese Jesuit Gaspar da Cruz, who described it in 1560. Jan Hugo van Lin-Schooten, a Dutch navigator who published a journal of his travels in 1595, also mentioned tea. His volume was translated into English in 1598. The first tea was shipped to Europe from Java in 1610 and by 1615 the British East India Company agents were drinking tea and acquiring it. By 1636 tea appeared in Paris, by 1650 it had arrived in England, and by 1651 it was available for public sale in England. The Dutch settlers brought tea to the North American colonies by 1659, and it was licensed for sale in Massachusetts in 1690: it was a popular drink until the Boston Tea Party of 1773.

During this period (late 16th to mid-17th century), tea had also reached Russia by the overland route, brought there by two Cossacks, Ivan Petroff and Boornash Yalysheff, in 1567. By 1618 Czar Alexis had been presented with several chests of tea from the Chinese embassy, and tea soon became a popular beverage.

Most European and American tea was imported from China or Japan. In 1684 a German physician, Andreas Cleyer, grew tea in Java as an ornamental garden plant, but it was not until much later, as tea became popular, that European countries considered cultivating tea. In 1827, the Dutch government appointed Jacobius Isidorus Lodewijk Levien Jacobenson to start an Indonesian tea industry under the auspices of the colonial government in Indonesia. Jacobenson managed to obtain 7,000,000 tea seeds and fifteen workmen, but did not receive the permission of the Chinese to export the seeds. Lack of adequate attention to the cultivation of the seedlings provided a bitter tasting harvest. The experiment was a failure and the Dutch continued to import tea.

The British had been so myopically focused on Chinese tea that they failed to notice the possibility of cultivating Indian tea. C. A. Bruce, a former army officer, had been stationed in Assam after the Burma War. He recognized the wild Assam tea and began cultivating it. His first shipment reached London in 1838.[21] When John Peet introduced Assam tea seed from India to Java in 1878, it adapted well to the new, but similar climate. The mid-19th century devastation of the coffee plantations, contributed to the acceptance of tea as an essential replacement.

Thomas Lipton relied on the English concern for high quality tea in founding the Lipton Tea Company. His interest in improving the packaging and freshness of tea led him to establish his own tea plantations to provide choice teas to his customers. He began selling these teas in 1890, and won first prize at the Chicago World Fair in 1893.

Chocolate was a popular beverage of the Aztecs who introduced it to the Spaniards in the early 16th century. It was imported from Mexico to Spain in 1520 and to Italy in 1606; it was introduced to England in 1657. The first chocolate factory was established in Germany in 1756 and the first in the United States was established in 1765 by John Hannon in Massachusetts. Fifteen years later Hannon sold his factory to Dr. James Baker, who founded the Walter Baker Chocolate Company.

Almost three hundred years after chocolate was introduced to Europe, cocoa was first prepared by Coenraad van Houten of Amsterdam in 1828.

Until the 19th century, chocolate was used primarily in beverages or flavoring.

Animals

Although plants provided significant sources of foods and beverages, animals and seafood, both wild and domesticated, were also valued. In agriculturally based societies, animals had many uses including food. Hides, fiber, fur, and bone were useful for clothing production and as, or in the production of, tools. Animals also provided a major source of power and were used in the production of other foodstuffs.

Hunter-gatherers used meat from wild animals as an important source of protein. Domestication of animals occurred in Neolithic times after 10,000 B.C. Although most animals, except the donkey and the camel (which were considered too unpalatable to eat), were sources of food, some animals became too useful in other ways to remain food sources. Dogs, early in the history of their domestication, became the companions and servants of humans and were regarded as too valuable to eat except in extreme hunger or in the Far East.

Chickens and pigs were domesticated relatively early, shortly after the first crops were cultivated. Domestic pigs have been traced to 6750 B.C. in Iraq. Some historians have speculated that pigs may have been useful in helping to prepare soil for planting. Pigs, when rooting for food, tear up plant coverings and disturb the soil. Their droppings also act as fertilizer.[22] Europeans brought the pig to New World settlements. Domestic pigs were introduced to California by Spaniards in 1769, and the European wild boar was introduced into the Carmel Valley, California, in 1923.

Sheep were valuable sources of both food and wool. Some of the first domesticated breeds had little meat and were valued mainly for their wool. White, woolly sheep were herded in Sumer by 3000 B.C., and a fat-tailed sheep was bred in Egypt about the same time. The existence of these two breeds indicates that sheep must have been domesticated much earlier,[23] perhaps as early as 6500 B.C.

Domesticated horses were relative latecomers: it was not until about 2700 B.C. in the Ukraine that these animals were tamed. Used in warfare and to pull light loads, reliance on the horse for heavy field work did not become possible until 800 A.D. The reasons for this are discussed more fully later in this chapter. The Spaniards introduced the horse to the New World, which led to significant changes for some native cultures as illustrated by the Plains Indians use and adoption of the horse.[24]

Animal breeds can also disappear. By 1681 the dodo had been hunted to extinction, followed by English wild boars in 1683 and by the passenger pigeon early this century. By contrast some domesticated animals have also reverted to the wild, including the camel in Australia, the burro in the American Southwest, and the pig in Florida and California. Feral pigs and monkeys had become pests in certain areas of Florida by the 1980s, and California had problems with feral pigs soon after the domestic pig was introduced by the Spaniards in 1769.

Seasonings

People used herbs and spices as seasonings early in human existence. Their use was dictated by those available in the immediate area or their accessibility via trade. Salt was mined near the Caspian Sea in ancient times, although in coastal areas it was generally obtained by evaporating sea water. At one time part of a Roman soldier's pay was in the form of salt.[25]

The impetus for exploration during the 15th and 16th centuries was to seek less expensive means of acquiring spices and silks. The New World Columbus found was not the Spice Islands he was seeking, but it did yield new flavorings including chocolate, chili peppers, and vanilla. Sugar, for much of its history in the West, was treated as a spice and used in cooking meats and savories.

The human sweet tooth was heavily dependent on fruits and honey. While sugar was known and used in Asia, it did not reach the West until the 8th century and remained expensive until the 19th century.[26] Sugar making was known in India as early as 3000 B.C. and the *Athavaveda*, a sacred Hindu book from around 800 B.C., described a crown of sugarcane. Alexander's troops discovered it there during his conquests. By 710 A.D. it was cultivated in Egypt and introduced under Moorish domination to Spain during the 8th century. The Venetians imported sugar into Europe in 996, and the Italians maintained a stranglehold on the European sugar supply for some time.[27]

Sugar was scarce and costly partly owing to the limited growing areas. The introduction of sugarcane by Columbus to the West Indies helped increase the source of supply, but honey was the most common form of sweetener until the advent of sugar beet cultivation.

The cultivation and processing of the sugar beet made sugar the staple it is today. Andreas Marggraf first extracted sugar from sugar beets in 1747, but the first factory for processing was not established until 1801. Beets were a cheaper source of sugar since they could be grown easily in northern climes. This changed the entire economic base of the sugar industry and had a significant impact on trade.

Table 2.1 Food Origin and Migration

Dates B.C.	*Event*
75,000	Crude milling instruments used (EB).
15,000	Sickles, grinding stones, celts (axes) used (WB).
7000	Wheat cultivated in the Tigris-Euphrates Valley (BBB).
	Herds of sheep kept in Iraq (EB).
6750	Domesticated pigs kept in Iraq (EB).
6500	Apples used in Asia Minor (remains of carbonized apples found) (BBB).
	Chili peppers, cotton, and squash grown from seed in Tehuacan, Mexico (EB).
6000	Domesticated sheep kept in the Balkans (EB).
	Barley cultivated in the Tigris-Euphrates Valley (BBB).
	Domesticated cattle kept in Argessa (EB).
	Mortar and pestle used for grinding grain (WB).
	Milk drunk in Eurasia (BBB).
5510	First record of bees in tombs at Abydos, Egypt (DID).
5000	Wild strawberry seeds used at sites in Denmark, Switzerland, and England indicate collecting and eating of berries (BBB).
4000	Rice cultivation tools used in China (EB).
	Grapes grown in the Caspian and Black Sea coastal regions (BBB).
	Figs grown in Syria (BBB).
5000-3500	Mutant forms of maize planted in Tehuacan, Mexico (EB).
3500	Cakes made from coarse ground grain used in Swiss lake dwellings (EB).
	Olive trees cultivated in Syria and Crete (BBB).
3200-3000	Cereals grown in Magdeburg-Cologne-Liege area (MM).
3000	Radish cultivated in China (BBB).
	Pear used as a medicine in Sumer (BBB).
<3000	Sugarcane grown in India (BBB).
	Sugar-making techniques employed in India (EB).
	Potato cultivated in Peru (BBB).
	Asparagus cultivated in Egypt (RM).
	Soybeans grown in China (BBB).
	Peas grown in the Near East (BBB).
2900	Evidence of milk drinking depicted in frieze at Ur, southern Iraq (BBB).
3200-2780	Laborers depicted eating onions in Egyptian tomb decorations (EB).

2737	Tea drunk in China, recorded in the diary of Emperor Shen Nung (BBB).
2700	Rhubarb grown in the eastern Mediterranean lands and in Asia Minor (BBB).
2500-2000	Domesticated chickens kept in Babylon (TH).
2400	Cantaloupe grown in Iran (BBB).
2300	Bit and bridle introduced (EB).
2200	Orange cultivated in China (BBB).
2100	Pear seeds buried in Chinese tombs (BBB).
2000	Chinese milked animals (BBB).
	Watermelon grown in Central Africa (BBB).
1800	Plums grown in Babylon (BBB).
1200	Ramses II, Egypt, ordered apple orchards planted along the Nile (BBB).
10th c	Peach cultivated in China (USAIR).
647	Spinach grown in Iran (BBB).
580	Olive tree introduced to Italy (FR).
500	Carrot grown in Afghanistan (BBB).
	Artichoke cultivated in Central Mediterrean (BBB).
5th c	Confucius wrote of the peach (BBB).
500-450	Viticulture developed in Italy and Gaul (TH).
	Dams used in India (TH).
401-399	Chestnuts eaten by Xenophon's army in Asia Minor (SSVF).
327	Alexander's troops found the banana in the Indus Valley (SSVF).
326	Alexander found rice and sugarcane grown in India (EB).
325	Sugar first mentioned in western India by Nearchus (DID).
3d c	Peach reached Persia (BBB).
	Peaches cultivated in Persia (BBB).
220	Mandarin oranges grown in China (BBB).
200	Asparagus grown in the Eastern Mediterranean (BBB).
2d c	Marcus Portius Cato (the Elder) in *De Agricultura* described grafting of apple stock (PW).
140-86	Cucumber introduced from India to China (EB).
A.D.	
1st c	Romans imported peaches (BBB).
77	Pliny in *Historia Naturalis* described thirty-six varieties of apples (PW).
2d c	Chinese scholar Yang Fu extolled the banana in his *Record of Strange Things* (BW).
286	Sugar first mentioned in China (DID).
350	Tea cultivated in China (BCT).

>450	Turkish tribesmen barter for Mongolian tea (USAIR).
7th c	Arabs introduced banana cultivation into northern Egypt (BBB).
8th c	Moors introduced rice to Spain (WB).
>700	Arabs introduced orange to Spain (BBB).
710	Sugar grown in Egypt (TH).
729	Emperor Shomu, Japan, inspired one hundred Buddhist priests to plant tea with his gift of powdered tea (BCT).
658-749	Japanese Buddhist priest Gyoki planted tea gardens in temples he built (BCT).
780	Chinese taxed tea (BCT).
<800	A crown made of sugarcane described in the *Athavaveda*, a sacred Indian text (EB).
800	Japanese began tea cultivation (EB).
805	Dengyo Dashi, Japanese Buddhist student, returned from China with tea seed and planted it (USAIR).
996	Venice imported cane sugar from Egypt (TH).
1000	First use of coffee as a beverage recorded by Avicenna, Arabian philosopher and physician (1ST).
11th c	Lemons grown in Burma (BBB).
1002	Arabs introduced bitter orange to Sicily (BBB).
1200	Arabs made decoction of coffee from the dried hulls of the coffee beans (BCT).
1299	Sugar from Morocco to Durham, Great Britain (1ST).
14th c	Papaya discovered in the West Indies (BBB).
>1300	Arabs introduced the sweet orange to Italy (BBB).
1368	Charles V planted strawberry vines at the Louvre (BBB).
1394	One of early French cookbooks published, *La Menagier De Paris*, also contained apple recipes (EB).
15th c	Spanish introduced rice to Italy (WB).
1450	Coffee known and used as stimulant in Abyssinia (DID).
1468	Spaniards introduced rice to Italy (EB).
1492	Columbus's crew discovered maize in Cuba (EB).
	Portuguese found the banana on the coast of Guinea and took it to the Canary Islands (SSVF).
>1492	Columbus discovered the peanut in Haiti (USAIR).
1493	Columbus carried orange seeds from the Canary Islands to the West Indies (BBB).
	Pineapple discovered by the Spanish at Guadeloupe Island (BBB).

16th c	Cashews, native to Central and South America, taken by the Portuguese to India (SSVF).
	Spanish brought the peach to Mexico and Florida (USAIR).
	Arabs roasted whole coffee bean to make coffee (BCT).
1516	Oranges introduced to Panama (BBB).
	Spanish priest Friar Tomas de Berlanga took the banana from the Canary Islands to Santo Domingo (SSVF).
1517	Coffee imported to Europe (TH).
1519	Bernal Diaz reported Aztecs served plums to Spaniards (BBB).
1520	Chocolate imported to Spain from Mexico (TH).
	Vanilla imported to Spain from Mexico (RM).
1538	Cieza de Leon in 1553 in *Cornica De Peru* recorded potato cultivation in Colombia and Peru circa 1538 (WB).
1539	DeSota planted orange seedlings south of St. Augustine, Florida (BBB).
	Sir Anthony Ashley introduced cabbage into Great Britain (1ST).
1544	Tomato introduced to Italy from South America (EB).
1548	Guinea pepper plants grown in England (TH).
1553	Cieza de Leon, Seville, Spain, mentioned the potato in *Cornica De Peru* as grown in Colombia (WB).
1554	First recorded coffeehouse opened in Constantinople (1ST).
1556	Tobacco introduced into Europe from Brazil by Thevet Angolene, France (TH).
1558	Jean de Lery, minister, Fort Coligny, Rio de Janiero, mentioned sugar (DID).
1559	Venetian book described beverage drunk by Arab traders, thought to be tea (RM).
1560	Portuguese Jesuit Father Gasper da Cruz described tea (BCT).
1565	Sir John Hawkins introduced sweet potatoes and tobacco to England (TH).
1567	Ivan Petroff and Boornash Yalysheff, two Cossacks, brought word of tea to Russia (BCT).
1573	Potatoes itemized in accounts of Hospital de la Sangre (1ST).
1579	Spaniards were growing oranges south of St. Augstine, Florida (EB).
1580	Coffee imported to Venice from Turkey (TH).
1586	Sir Thomas Harriot brought potatoes to England from Colombia (1ST).

1595	Jan Hugo van Lin-Schooten, Dutch navigator, published journal of his travels describing tea customs (BCT).
1596	Tomatoes introduced into England (TH).
1598	Lin-Schooten's *Travels* published in English; contained first printed reference on tea in English (BCT).
1599	Anthony Sherley, English adventurer, mentioned coffee in an account of his expedition to Persia (BCT).
17th c	Limes discovered in Americas and were used by sailors to treat and prevent scurvy (BBB).
1600	East India Company obtained charter (USAIR).
>1600	China began cultivation of peanut from Malay Peninsula plants (USAIR).
1606	Chocolate imported to Italy (BBB).
1608	Cultivated strawberries first mentioned by Sir Hugh Plat (BBB).
1609	Chinese tea introduced to Europe (TH).
	Jamestown colonists grew carrots (HS).
1610	First shipments of Japanese and Chinese teas from Java to Europe (BCT).
1611	Jamestown colonists raised cattle brought from Europe (RM).
1615	British East India Company agents drank and acquired tea (BCT).
1616	Pineapple introduced to Italy (BBB).
1618	Czar Alexis presented with several chests of tea by the Chinese ambassador (BCT).
1620	Mayflower cargo list recorded a wooden mortar and pestle for grinding coffee powder (BCT).
1621	Potatoes planted for the first time in Germany (TH).
1622	Potatoes imported to Virginia (RM).
1629	Apple seeds and propagation wood arrived in America from England (RM).
1630	English colonist sampled popcorn at first Thanksgiving (WG).
1633	Thomas Johnson, London, exhibited bananas in his shop window (LST).
1636	Tea drunk in Paris for the first time (TH).
>1637	Conopios, disciple of Cyrill, patriarch of Constaninople, fled to Oxford and introduced coffee there (BCT).
1639	First apples grown in America from European stock picked at Beacon Hill, Boston (RM).
1643	Coffee drinking popular in Paris (TH).

1650	Tea first drunk in England (TH).
	Sir Richard Westin advocated the cultivation of turnips (TH).
	First coffeehouse in England opened in Oxford, England (TH).
1651	Thomas Garway held the first public sale of tea in England (1ST).
	Frederick William I threatened to cut off the noses and ears of peasants who did not grow potatoes (PB).
1652	First London coffeehouse opened by Pasqua Rosee (BCT).
1657	Chocolate drinking introduced in England (TH).
1659	Dutch settlers brought tea to U.S. (RM).
1660	French merchants in Marseilles drank coffee (BCT).
	Samuel Pepys wrote in his diary: "I did send for a cup of tea (a China drink) of which I never had" (BCT).
1662	Charles II and Queen Catherine, England, popularized tea drinking (USAIR).
>1664	Frederick William forced German peasants to plant potatoes to alleviate food shortages caused by war (RM).
1668	East India Company imported its first shipment of tea to England (USAIR).
	English Parliament forbade the importation of Dutch tea and granted a monopoly to the East India Company (USAIR).
1670	Coffee introduced into Germany (BCT).
	Dorothy Jones, Boston, granted a license to sell coffee (HS).
1671	First Marseilles coffeehouse opened (BCT).
1675	Elector of Brandenberg served coffee (BCT).
>1675	Rice cultivation introduced to South Carolina (WB)
1678	French botanist Marchand demonstrated to Academie des Sciences how to culture mushrooms (RD).
	First German coffeehouse opened in Hamburg (TH).
1679	London merchant opened a Hamburg coffeehouse (BCT).
>1680	Pecans cultivated in El Paso, Texas (SSVF).
1683	Wild boars became extinct in England (1ST).
	First coffeehouse in Vienna opened and Viennese coffee served (TH).
1684	Tea grown in Java as an ornamental shrub by German physician Andreas Cleyar (BCT).
1689	Coffee drinking reached America (BBB).
1690	Tea licensed for sale in Massachusetts (BCT).

18th c	Spanish missionaries introduced almond trees into California (NFFO).
1706	Twinings Tea and Coffee Merchants began selling tea (USAI).
1707-10	Oranges planted in Arizona (BBB).
1714	Dutch botanical garden gave coffee plant to Jardin des Plantes, Paris (BCT).
1719	Irish migrants began potato cultivation in New Hampshire (RM).
1720	Mustard first sold inLondon in paste form (1ST).
	Broccoli was introduced into England by Italians (USAIR).
1723	French naval officer, Gabriel Mathieu, of Martinique stole coffee seedling from Paris Jardin (BCT).
1727	Coffee first planted in Brazil (TH).
	Wife of French governor of French Guiana gave coffee seedlings to Brazilian officer (BCT).
1736	Coffee trees transplanted from Arabia to West Indies (DID).
1747	Beet sugar extracted from sugar beetroot by Andreas Marggraf, Berlin (TH).
1751	Sugarcane planted in Louisiana (WG).
1755	Grapefruit found in Barbados and may have been a mutant of pummelo from Southeast Asia (SCI).
1756	First chocolate factory in Germany established (TH).
1765	Potato widely used foodstuff in Europe (TH).
	Chocolate factory established in Massachusetts Bay Colony using cacao beans from the West Indies (BBB).
1769	Oranges planted in California (BBB).
	Spaniards introduced domestic pigs to California (CDIS).
1781	Thomas Jefferson listed *tomatas* in his garden journal at Monticello, Virginia (USAIR).
1790	Thomas Jefferson grew peanuts in Virginia (USAIR).
	Pineapples first planted in Hawaii (HS).
1791	Antonio Mendez made sugar from Louisiana cane (WG).
1801	First sugar beet factory was built by F. Achard, Silesia (DID).
>1820	Robert Johnson demonstrated the tomato was edible by eating one in public and surviving (USAIR).
	Large-scale peach cultivation began in the U.S. (USAIR).
1825	Jefferson Plum developed in Albany, New York (BBB).
>1826	C. A. Bruce cultivated wild Indian tea in India (BCT).
1827	Jacobius Jacobenson was appointed by the Dutch to start a Javanese tea industry (BCT).

1828   Coenraad van Houten, Amsterdam, prepared cocoa by extracting excess cocoa butter from cacao beans (1ST).

1830   Sugar beet factory built at Utling, Essex (1ST).

1832   Jacobenson brought seven million seeds and fifteen workmen from China to Java, but without official permission (BCT).

1836   Scientists discovered how to pollinate vanilla artificially, enabling cultivation outside Mexico (RM).

1838   First Indian tea from C. A. Bruce arrived in London (BCT).

1839   Indian tea was imported and sold in London (1ST).

1850   The banana was first brought to the U.S. (SSVF).

>1850   Large-scale cultivation of asparagus undertaken in the U.S. (RM).

1860   Goodenough introduced a machine for making horseshoes (DID).

1863   West Indian avocado introduced into Florida (USAI).

1869   Joseph Campbell and Abraham Anderson of Camden, New Jersey, successfully canned tomatoes (USAIR).

1870   Alexander Livingston, Reynoldsburg, Ohio, developed a smooth, red tomato (USAIR).

        Dutch farmers in Kalamazoo, Michigan, began growing celery for commercial consumption (HS).

1873   Washington Navel Orange, originally from Baia, Brazil, planted in California (BBB).

1876   Tinfoil wrapped bananas sold for ten cents at the American Centennial Exposition in Philadelphia (SSVF).

1878   John Peet introduced Assam tea seed from India to Java (BCT).

1884   Bananas regularly imported from Canary Islands to Great Britain by Elder Dempster & Company (1ST).

1890   Thomas Lipton began selling tea (TIME).

1893   Thomas Lipton entered his tea in Chicago's World Fair and received first prize (TIME).

        U.S. Supreme Court ruled that the tomato was a vegetable for import and taxation purposes (USAIR).

1897   Seventy-one almond growers formed a co-op, which later became the California Almond Growers Exchange (NFFO).

1899   Captain Lorenzo Dow Baker founded the United Fruit Company, which exported bananas to the U.S. (SSVF).

1900   Horticulturist George Cullen, Florida, developed an avocado strain that produced fruit of a uniform size (USAIR).

1923	European wild boar introduced into the Carmel Valley, California (CDIS).
1924	Mexican avocado introduced into California (USAIR).
1928	Broccoli introduced to the eastern U.S. from California by the D'Arrigo brothers, vegetable growers (HS).
<1930	Rudolph Boysen developed the boysenberry, a cross of the blackberry, red raspberry, and loganberry (HS).
1930	Poultry farmers, U.S., built special sheds for rearing young fowl at a lower cost (TT).
1936	First artificial insemination association established in Denmark, which led to improved cattle (TT).
1953	Irish coffee first served in the U.S. at the Buena Vista Café, San Francisco (HS).
1955	25,000 tons of almonds grown in California (NFFO).
1969	Cross-bred tick-resistant cattle bred for the tropics introduced (TT).
1970	135 million pounds of almonds grown in California (NFFO).

Production

How and when humans began using tools in agriculture is open to dispute. As archaeological investigations continue, new evidence changes views and old evidence is reinterpreted. We push back further in time the use of tools. Current data indicate that handmills for grinding grain were used by Stone Age peoples: crude stones for this purpose have been traced back 75,000 years ago. More sophisticated models appeared along with early agriculture by 6000 B.C.

This section highlights some of the food production and gathering tools and their evolution. Histories of agriculture such as Fussell's *Farming Techniques from Prehistoric to Modern Times*,[28] Partridge's *Farm Tools through the Ages*,[29] Blandford's *Old Farm Tools and Machinery*,[30] and Rossiter's *The Emergence of Agricultural Science*[31] provide more detailed analysis of the impact of these developments and describe some aspects not mentioned here. The broad subjects discussed are animal power, planting, harvesting, processing and milling, fertilizers and plant protection, crops and livestock, and sugar processing. Tools and metalworking are discussed more fully in chapter 10.

Prior to the advent of metalworking, tools were made from stone, bone, or wood. Durability, the ability to maintain a cutting edge, and strength were limiting factors for such tools. Early metal tools were also limited and stone tools continued to be used during the Bronze Age. Bronze, the first widely used metal, appeared in 3500 B.C. and was followed shortly thereafter in 3000 B.C. by the wheel. Irrigation appeared about the same time along with plowing, raking, and manuring in Egypt, while dams were built in India around 500 B.C. Gears and the ox-driven water wheel appeared about 200 B.C.

Figure 2.5. Agriculture

YEAR			
1987			
	Shellfish farming		
1975			
		Miracle rice	Tomato harvester
		Gaines, Mexican hybrid wheats	
1950	Improved transmissions		Potato harvester
	Combine harvester		Crop dusting
	Self-propelled combine	Corn hybrids	
1900	Gasoline tractor	Marquis wheat	
	Moldboard, disc plow	Fertilizers	USDA
	Self-propelled cultivator		McCormick reaper
1800			
	Steam power		
			Two-man plow
1600	Peruvian guano (Portugal)		
	Wheeled plow		
1200			
	Rigid horse collar		
400		Mills	
	Water wheel	Heavy plow	Trip hammer
	Dams (India)	Horseshoes	
-1200			
	Irrigation	Horse domesticated	Bit and bridle
	Plowing, raking, manuring (Egypt)		Wheel
-4400			
	Grinding stones	Bone sickle	
-10800	Farming began		
	Sickles, celts, grinding stones		
-23600			
-59200			
	Crude grinding stones		
-110400			

Animal Power

Animals have been used in many ways: as sources of food, apparel, tools, or power. As a power source animals have carried loads and pulled vehicles or agricultural tools. In many countries, donkeys and mules were used for carrying heavy loads and, in the East, elephants and camels were used. Oxen were used in the Middle East before 3000 B.C. and are still used in lesser-developed nations. Initially, horses were used primarily as a form of transportation rather than for agricultural purposes. Horses were unsuited to pulling heavy loads until the rigid horse collar was developed sometime in the 9th century, and oxen were used for plowing until that time. This device distributed the load more evenly and allowed the horse to use its powerful shoulder muscles without cutting off its breath as older harnesses did.[32]

Horses also required special feed, oats, that needed heavier soils. Although the use of horses greatly increased efficiency, it also led to growing additional crops for feed, and contributed to the development of the three crop rotation system (13th century). With this system, one field is left fallow, one is used for winter crops, and one is used for spring crops. Every fourth year the pattern for each field repeats.[33]

Outside Western Europe, other cultures and climatic conditions were less receptive to the horse. It was not used for agriculture in Africa, the Middle East, or Southeast Asia. Most plowing in these areas was done with oxen, water buffalo, mules, or people.[34]

Planting

The earliest known use of the plow has been traced to early civilizations in the Middle East and in Egypt. The plow, used to turn and loosen soil, eased the work in preparing fields and planting crops. These innovations led to more consistency and to greater crop yields. Usually a plow was pulled either by a man or by animal power. The carruca, a heavy Roman plow, was used in the 1st century in the foothills of the Italian Alps.

The plow continued to evolve, and significant improvements were made during the Middle Ages. The advent of the iron edged plow and of the heavy wheeled plow, in use by 990, changed agriculture significantly. They not only made plowing more efficient, but also made possible the cultivation of soils too heavy or difficult to cultivate with the lighter, older style plows.[35]

Advances in agricultural equipment increased during the 17th and 18th centuries. David Ramsey and Thomas Wildgoose patented a plowing machine in 1619, and in 1627 the two-man plow was patented. A team lead by William Parham patented a plowing engine in 1634, and by 1701 Jethro Tull of England had devised the seed drill. In 1830 Disney Stanyforth and Joseph Foljambe, also of England, patented the Dutch or the Rotherham plow, which was modified and further improved by James Small, who introduced the modified version to Scotland in the 1760s. Two other major enhancements by Robert Ramsone were the use of tempered cast iron for the plow in 1803 and a bolted plow in 1808 that made replacement and repair of the plow easier. In the U.S. in 1797, Charles Newbold developed a cast iron plow which was further modified by John Wood in 1819 when he used it in a plow with interchangeable parts. The steam engine, devised late in the

17th century and significantly improved in the late 18th century, was finally applied to the plow in 1837 when a steam plow was patented. John Deere constructed a plow with a steel moldboard in 1846 that broke up and turned the soil covering and burying residue as it tilled. A self-propelled cultivator was offered in 1855, and the disc plow (originally developed in 1847) was reintroduced to the United States in 1893.

Steam power was adapted to plows and harvesting equipment, but was overtaken early in the 20th century by smaller, more flexible gasoline powered equipment. The gasoline engine was first used with a plow in the 19th century with the first gasoline tractor built in 1889. By 1912 steam and horses were being replaced by gasoline engines. The plow rapidly became just another attachment to hook to a tractor.

Gasoline engines, smaller and lighter than steam engines, provided more flexibility in design. Henry Ford had developed a tractor by 1915 and began production of the Fordson tractor in 1917. By providing an inexpensive, reliable tractor, he contributed to the mechanization of the family farm. The development of an efficient power take-off mechanism in 1918 increased the versatility of the tractor and placed more control with the operator. Transmissions were further improved in 1954.[36] Tractors are now used for a variety of tasks. Small, relatively inexpensive models are used for mowing lawns or working small garden plots while their larger brothers are used in agribusinesses.

Harvesting

After the soil is prepared and the seeds planted, the crop must be harvested before it can be used for food. Early harvesting tools were simple. Sickles first appeared about 15,000 B.C. and a bone sickle fitted with bits of flint was found in Syria dating from 6000 B.C.

Major mechanization of harvesting equipment did not occur until the 17th century. A threshing machine was invented in 1636 by John van Berg and a horse drawn thresher appeared in 1722. A machine for hulling barley and pepper was invented in 1677, a sifting machine for grain processing appeared in 1686, a maize cleaning machine appeared in 1715, and a wheat cleaning machine in 1725.

Developments continued steadily throughout the 18th and 19th centuries. Of major importance was the reaping machine patented by Cyrus McCormick in 1834, to which March added a conveyer belt in 1851. In 1910 the Holt Company of California introduced the first self-propelled, gasoline-driven combine harvester, and by 1924 harvesters were being used for soybeans. Allis-Chalmers introduced the "baby" combine harvester in 1936. Engines and combine harvesters continued to be improved and applied to more varieties of crops.

Other types of harvesters were developed for potatoes, sugarcane, and tomatoes. A potato harvester was introduced in 1941, but still required subsequent hand sorting. A new sugar harvester was demonstrated by Massey Ferguson to Australian farmers in 1958, but in parts of Florida sugarcane is still harvested by hand because soft, wet soil prevents the use of heavy equipment.[37]

The tomato presented a number of problems for mechanical harvesting. These problems were largely resolved during the 1960s, and several different machines were made available as a result of machines developed initially at the universities of California and Michigan. New species of tomatoes also made harvesting by machine easier. Farm labor groups in California have brought law suits protesting the reduction in farm harvesting jobs caused by these improvements.[38]

Today farm equipment for planting and harvesting represent a major financial investment for farmers, but also reduces the number of people required for planting and harvesting crops. One example of the increased productivity is the dramatic drop in the percentage of population involved in agriculture in the United States. Currently less than 2.2 percent of the U.S. population are farmers, whereas as recently as 1960 it was 8.7 percent and in 1820 72 percent of the United States population were employed in agriculture. The ratios are declining world wide, but it should be noted that in some countries as many as fifty percent of the population are engaged in farming.[39] Furthermore, although United States farming today employs less than 3 percent of the work force, it generates food supplies beyond our capacity to consume and yet people still go hungry in some areas. Food gathering is only one of the areas affecting our daily life, and although vital to our survival, it is not by itself sufficient to ensure it.

Processing and Milling

Among the first applications of mechanization were the processing and milling of grain. The Chinese used a crank handled winnowing fan in 40 B.C. By 290 A.D. the Chinese were using a trip hammer mechanism for hulling rice. In Europe manual tools for separating the chaff from the grain were still heavily used until the late Middle Ages. St. Jerome mentions the jointed flail around 400.

Although windmills were known to the Greeks and Romans, and Vitruvius described in detail the vertical post windmill in 22 B.C., they were mainly curiosities and not used for heavy work. The first indication of windmills used regularly for work is a reference to a Persian windmill builder in 644. Watermills, however, were used by the Romans and were in common use by 700 A.D. in Europe.

Use of mills reduced the labor required to process grain into flour. It would take forty men to grind by hand what one Roman watermill could grind in the same time period.[40] The quantity would depend on the size and speed of the mill, but significant gains were possible, which made both mills and watercourses suitable for driving the mills valuable resources. Numerous lawsuits involving the building of dams that affected the operation of other watermills on the same river provide one indication of the importance of mills. One such suit dragged on for fifty years.[41] Too, lords and clergy who controlled mills forced nearby peasants to use and pay for the services of the mill, sometimes using force to ensure compliance. One abbot confiscated all hand mills to ensure peasants had no alternative to using his mill.[42]

During Roman times, mills were used primarily for grinding grain rather than other tasks because manpower was readily available and there was less need for labor-saving devices. During the Middle Ages, however, labor was in short supply and mills were adapted to a variety of uses including cloth processing, beer making, and iron manufacture. Gimpel's *The Medieval Machine* describes a variety of such applications.[43] Mills were also a major power source in the northeastern United States early in U.S. history, and led, during the 19th century, to heavy industrialization.

Fertilizers and Plant Protection

The use of fertilizers evolved along with agricultural practices. The earliest fertilizers were natural waste products. Shards of pottery found in Iron Age fields have led to speculation that garbage and other waste products were taken from middens and spread in the fields as fertilizers.[44] Beans and other nitrogen fixing crops also aided in conditioning fields.

Fertilizer application began in a more concentrated way in the 19th century. In 1804 Nicolas Theodore de Saussaur discovered that saltpeter promoted the growth of cereals, and by 1830 nitrates were being shipped from Peru and Chile to Europe for use as crop fertilizers. Investigation of other suitable fertilizing substances continued, and in 1839 John Lawes applied superphosphate of lime to turnip cultivation. He began manufacture of this artificial fertilizer in 1843. Other substances used were slag and bones. Slag may result either from volcanic activity or as a by-product of metal production. Its use as fertilizer provided a profitable means of waste disposal for steel producers. Basic slag was used as a phosphorus source for fertilizer in 1861 and bones were used in 1880. In 1909, Fritz Haber discovered a nitrogen fertilizer, that was manufactured in 1913. Haber's ammonia process was used for the manufacture of both fertilizers and explosives.

Home gardeners found the Fison Gro-bags, first marketed in England in 1974, ideal for patio or small gardens. The plastic bags, filled with a soilless growing medium, conserve moisture and protect plants from weeds, but are not heavily used at present in commercial agriculture.

The scientific protection of plants and crops from disease and insects is a relatively recent phenomenon. While ground tobacco plant was used by the French in 1763 to control aphids, most serious pest control efforts began in the 19th century. In the 1840s lime and sulfur were used on vine powdery mildew and "Bordeaux mixture," a combination of copper sulphate and lime, was used in the 1880s to combat downy mildew on grape vines in the wine-growing regions of France. Most of the compounds available were relatively weak, and large-scale applications of more effective chemicals followed the growth of the chemical industries in the 20th century and availability of equipment suitable for applying pest control chemicals to large areas.

In fact, large-scale use of agricultural fumigants did not begin until 1920.[45] The Ohio Agricultural Experiment Station initiated aerial crop dusting in 1921 using powdered arsenate of lead against leaf caterpillars on catalpa trees, and the Huff Daland Dusters, Inc., of Georgia offered the first commercial service in 1925 using calcium arsenate against the cotton boll weevil. DDT, discovered in 1874 by Othmar Zeidler of Strasbourg, was not used until 1943 in the United States, although it was first used as an insecticide in Switzerland in 1939. In 1945 the herbicide 2,4-D was patented, followed by other herbicides in the 1950s and 1960s.

Some insecticides and herbicides were too strong or had undesirable side effects, causing concern among environmentalists and others. *Silent Spring*, a commercial book by Rachel Carson (1962), traced the negative effects of such pesticides and led to the banning of DDT from the United States in 1969. More recently, the Environmental Protection Agency banned the use of ethylene dibromide (used as a fumigant for grain and citrus storage) in September 1984, and in 1985 banned two herbicides containing dioxin, Silvex, and 2,4,5-T. The latter was one of the chemicals used in Agent Orange (widely used during the Vietnam War as a defoliant), and was also used to control weeds in forests and rangelands and along roadways.

The herbicides that kill weeds efficiently can also harm crops, so some herbicide manufacturers are seeking ways to make herbicides more selective or to develop crops that are more resistant to existing herbicides. Monsanto Company, in 1985, announced work with a genetic engineering firm to develop herbicide resistant crops, and American Cyanimid announced the development of a herbicide-resistant corn hybrid.[46]

Crops

The development of hybrid crops during the beginning of the 20th century led to the "Green Revolution" of the 1960s that greatly improved crop yields with new, disease-resistant strains. These improvements benefited both the lesser developed countries and agriculturally rich countries such as the United States by making more food available at a lower cost. Much of the success in developing new crop strains derived from the pioneering work of the Austrian monk Gregor Mendel, who in 1866 set forth his theories on cross fertilization and genetic traits. Mendel's work was not applied immediately, but was rediscovered forty years later in 1900 by researchers working independently in Austria, Germany, and Holland.[47] Together with the work on self-fertilization published by Charles Darwin in 1876, Mendel's publication laid the basis for developing hybrid strains of plants.

Unaware of Mendel's or Darwin's work, farmers had experimented with crossing various strains and had been successful in developing some cross-bred Indian corns.[48] Similarly, seed growers also experimented with cross fertilization, and Luther Burbank produced his hybrid potato in 1873. However, as scientists became aware of Darwin's and Mendel's work, they were able to develop new strains more quickly, and maverick attempts at hybridization were replaced by more controlled experiments focused on increasing crop yields.

Corn hybrids were produced in 1921 at the Connecticut Agricultural Experiment Station and offered commercially in 1926. Their impact on crop yields was dramatic: in 1933 only one percent of the corn was grown from hybrid seed and yields were 23 bushels/acre, but by 1960, 96 percent of the corn grown was from hybrids and yields were 72 bushels/acre.[49]

Wheat also was the subject of scientific investigation and improvement. Wheat growers owe much to Sir Charles Saunders who produced the Canadian Marquis wheat and released it to farmers in 1900. The work of Ernest Robert Sears in 1954 led to new, improved wheat hybrids in the 1960s. Gaines, a strong, semi-dwarf variety, was introduced in 1962, and by 1965 a similarly developed Mexican strain was being used worldwide.

The Rockefeller Foundation, after its successful program with Mexican wheat, also established the International Rice Research Institute in the Philippines in 1962. The Institute developed a new strain of miracle rice in 1964. This rice doubled yields and was responsive to fertilizer, and early maturing in some areas allowed an extra crop per year.

The new hybrids grew more rapidly, produced larger yields, and were bred for increased resistance to disease. The world food supply has been substantially increased, as has our choice of foods.

Livestock

Animal husbandry and breeding have also changed significantly. Changes in the food supply and methods of rearing livestock reflect a trend to increase the efficiency and yield of animals and their products. Animals have been developed for specific qualities such as leaner, larger pigs, higher milk yield for dairy cattle or more meat for beef cattle, heavier wool coats for sheep, and larger, meatier turkeys and chickens. Some animals such as sheep and turkeys would find survival difficult or impossible without human care. The sheep's heavy, high quality wool fleece would probably kill the animal from heat in the summer unless sheared. The turkey has not increased its intelligence or survival abilities with its greater bulk.

Agricultural activities were seasonal and those related to livestock were heavily influenced by climate and geography. In northern areas, winter fodder was in short supply and many herds and flocks were culled in the fall. Where possible meat was dried, salted, smoked, or otherwise preserved. Culling reduced the pressure on available grain stocks and reduced the competition between people and their animals for scarce resources.

At one time the laying season for chickens was restricted, and most did not lay eggs during the winter season. Modern improvements include battery cages for hens, introduced in 1925, which promoted increased egg production. Special sheds for intensively rearing young fowl were introduced in 1930 enabling more birds to be raised at a lower cost. New methods have reduced the time a young fowl requires to reach an adequate weight from three months to seven weeks. Improved feeds, new vitamins, hybrid breeds, and a controlled environment have made the term "spring chicken" obsolete.

Improvements related to cattle include the portable milking shed introduced in 1927, artificial insemination (the first association to control the process and maintain breeding histories was established in 1936), and the development of cross-bred tick-resistant cattle for the tropics in 1969.

Fish were a dietary staple among a number of hunter-gatherer societies,[50] and during the Middle Ages fish ponds were kept and used to supply fresh fish. Encouraged by religious preferences for fish, gentile housewives also practiced aquaculture. More recently, with an emphasis on healthy diets, fish has again become popular, and there is a new emphasis on aquaculture. Methods to increase the successful spawning of shellfish has made shellfish farming practical. The methods have been applied to both oysters and abalone. Catfish have become a "cash crop" and are also becoming more popular.[51]

Sugar Processing

As noted earlier, honey was the most common sweetener, and until the middle of the 19th century sugar was rare and expensive. The pivotal role the Italians played in controlling trade was exercised in many areas, including sugar and cloth manufacture. The Italians held a virtual monopoly on European sugar refining and most early developments occurred there. A three roller wooden sugar mill was invented by Pietro Speciale of Sicily in 1449, using water or ox for power.

The introduction by the Spaniards of sugarcane to the West Indies provided another source of raw ingredients. Refineries were gradually established in other parts of Europe. The first German refinery was built at Augsburg in 1573. A more durable, iron-clad roller for sugar processing was introduced by George Sitwell in 1693 and a woolen filter was added in 1700.

Mechanization of sugar processing followed the development of steam power sources and their successful application in the textile industry. The first steam driven sugar mill was patented in 1802. Hydraulic pressure was introduced in 1859 by Jeremiah Howard. Mechanization occurred about the same time sugar beet cultivation began on a large scale. The combination of the two led to cheaper sugar that most people could afford.

Agricultural Education and Research

Also important to improved food production processes was the agricultural educational and research infrastructure developed in the United States within the last hundred and thirty years. Among key legislation were the bills establishing the Department of Agriculture (USDA) (1862), the land grant colleges (1862, 1890), the Agricultural Experiment Stations (1887), and the Cooperative Extension Service (1914). These institutions promoted the development and application of scientific techniques to food production, food technologies, nutrition, and the use of technology in the home.

The two Morrill acts of 1862 and 1890 established funding for the land grant colleges by providing an initial allocation to each state of 30,000 acres for each senator and representative in Congress. The lands were to be sold and the proceeds used to create colleges for the agricultural and mechanical arts. The 1890 Act provided for continued federal support to the colleges. Home economics was included in the land grant colleges providing a formal educational basis for future research, teaching, and application. This inclusion of home economics led to advances not only within the home, but also to knowledge useful in industrial and scientific applications.

The Agricultural Experiment Stations received similar federal support with the passage of the Hatch Act in 1887. The first such station, established in 1875 at Wesleyan University in Middletown, Connecticut, had been supported by private and state funds. W. O. Atwater, the first director of the USDA Office of Experiment Stations, included nutrition studies on the research agenda for the stations. The many publications of the experiment stations in areas related to food, nutrition, and home technology are ample evidence of how important their activities have been to those engaged in food production, food preparation, farming, and in maintaining a better standard of living. A new Hatch Act, passed in 1955, extended the original act and incorporated relevant subsequent legislation.

The Cooperative Extension Service coordinates programs of informal education in farming and homemaking. Extension work began in Massachusetts in 1854 when a farmers institute convened to discuss common problems. When the USDA was established in 1862, the federal government and the state agricultural colleges began a cooperative program to send trained workers into farming districts to assist with education and provide information to farmers. The Smith-Lever Act, passed in 1914, provided for federal, state, and county funds to support cooperative extension work. The 4-H clubs for young people, which evolved from groups established in the early 1900s, are part of the cooperative extension activities. Lastly, the extension service provided for county agents and home demonstration agents to make available the scientific information in agriculture and home economics to the general population, particularly to inhabitants of rural areas.

The educational, research, and communication activities of the USDA, the land grant colleges, the experiment stations, and the county agents have promoted the use and application of scientific techniques to food sciences and homemaking. They have done much to shape the modern way of life that so many in the United States enjoy and have established an example other countries have sought to emulate.

Table 2.2. Food Production

Dates	Event
B.C.	
75,000	Crude milling instruments used (EB).
15,000	Sickles, grinding stones, celts (axes) used (WB).
10,000	Farming began (WB).
6000	Sickle of bone set with flints used in Palestine (DID).
	Mortar and pestle used for grinding grain (WB).
3500-3000	Plowing, raking, and manuring techniques used in Egypt (TH).
3000	Irrigation used in the Middle East (WB).
2700	Horse domesticated in the Ukraine (EB).
800-700	Sledges with heavy rollers used (TH).
	Horse-drawn chariots used (TH).
300-400	Nailed horseshoes left at Maiden Newton, Dorset (DID).
40	Crank handled winnowing fan used in China (DID).
22-11	Vertical (post) windmill described by Vitruvius, a Roman (DBE).
A.D.	
1st c	Romans imported peaches from Persia (BBB).

1	Carruca, heavy plow, used in the foothills of the Italian Alps (DBE).
290	Chinese used trip hammer mechanism for hulling rice (MM).
>400	St. Jerome mentioned the jointed flail (DID).
>600	Horizontal windmill used (DID).
644	Earliest reference to windmills was to Abu Lu'lu'a, Persian windmill builder (EDE).
700	Waterwheels used in Europe (TH).
>700	Three field crop rotation used (MM).
800	Rigid horse collar adopted in Europe increased the load a horse could pull (MM).
800-900	Horseshoes used in Yenesei Region, Siberia (DID).
886-911	Horseshoes mentioned by Leo IV of Byzantium (DID).
>890	Horse for plowing in Norway (MM).
>900	Horizontal windmill used in Arab countries (DID).
900-1000	Horizontal windmill used in Low countries (DID).
987	Watermill used for the manufacture of beer at Saint-Sauveur, Montreuil-sur-mer (MM).
>990	Heavy wheeled plow used (MM).
1170	Earliest recorded tide mill used in Britain, Woodbridge, Suffolk (DID).
1180	Post-type windmill in Normandy mentioned by Leopold de Lisle (DBE).
1185	Windmill used at Weedley, York (DBE).
1300	Alum discovered at Rocca, Syria (DID).
1449	Three roller wooden sugar mill invented by Pietro Speciale, Sicily, water or oxen driven (DID).
16th c	Mango discovered growing in Southeast Asia (BBB).
1523	Use of the wheeled plow noted by Fitzherbert (DID).
1573	First German cane-sugar refinery began operation at Augsburg (TH).
1602	Peruvian guano used as fertilizer in Portugal (DID).
1619	David Ramsey and Thomas Wildgoose patented a plowing machine (DID).
1627	William Brouncher, John Aprice, and William Parham invented a two-man plow (DID).
1634	William Parham, John Prewitt, Ambrose Prewitt, and Thomas Derney invented an engine for a plow (DID).
1636	John van Berg invented a threshing machine (DID).
1677	Edward Melthorpe and Charles Milson invented a machine for hulling barley and pepper (DID).

1686	John Finch, John Newcomb, and James Butler invented a wire screen sifting machine for grain processing (DID).
1693	Iron-clad roller sugar mill introduced by George Sitwell (DID).
>1700	Wool cloth filter for sugar refining first mentioned (DID).
1701	Jethro Tull devised the seed drill (1ST).
1715	Thomas Martin invented a machine for cleaning maize (DID).
1722	Horse-drawn threshing machine devised by Due Quet, France (DID).
1725	George Woodroffe invented a machine for cleaning wheat (DID).
1731	Thomas Ryley and John Beaumont invented a fodder (chaff) cutting machine (DID). Michael Menzies invented a threshing machine with rotary flails driven by water wheel (DID).
1737	Meiffen invented a horse-drawn threshing machine (DID).
1747	Beet sugar extracted from sugar beetroot by Andreas Marggraf, Berlin (TH).
1751	Sugarcane planted in Louisiana (WG).
1754	John Smeaton invented the triangular sugar mill with three rollers (DID).
1762	De Malassagny, France, invented a threshing machine (DID).
1765	John Milne designed a sifting machine driven by wind or water for grain processing (DID).
1770	James Edgill invented a spiral-knifed fodder cutting machine (DID). James Sharp invented a bean-spitting mill and winnowing machine (DID).
1773	John Fleming patented the Wallerer Wheel for sugar mills (DID).
1775	George Robinson designed a man- or horse-operated dressing machine for grain (DID).
1784	Andrew Meikle, Scottish millwright, invented the threshing machine (TH).
1787	James Cooke described the modern-type chaff cutter (DID).
1791	Antonio Mendez made sugar from Louisiana cane (WG).
1797	Charles Newbold, Burlington County, New Jersey, farmer, developed a cast iron plow (FFF).
1799	Joseph Boyce invented a machine for cutting wheat and corn (DID).

1800	Thomas Wigful, Norfolk, invented the post threshing machine (DID).
	Robert Mean invented a machine for cutting standing corn and grass (DID).
1801	First sugar beet factory was built by F. Achard, Silesia (DID).
1802	Richard Trevithick and Vivian, Cornwall, patented a steam-driven roller sugar mill (DID).
1803	Robert Ransome, England, developed the tempered cast iron plow (OFTAM).
1804	Nicolas Theodore de Saussaur, France, discovered saltpeter promoted the growth of cereals (DID).
1805	Steam "cane-engine" built by James Cook of Glasgow (DID).
1808	Robert Ransome, England, patented the bolted plow, removable parts made replacement easier (OFTAM).
1811	Salmon described a reaping machine with clippers and delivery system (DID).
1812	John Lorain, Phillsburg, Pennsylvania, farmer, successfully crossed flint and dent varieties of Indian corn (EB).
1819	John Jethro Wood, Poplar Ridge, New York, developed a plow with interchangeable parts using Newbold's 1797 cast iron plow (FFF).
1830	Nitrates were first shipped from Peru and Chile (DID).
	Matthew Boulton and James Watt designed a sugar mill (DID).
	Sugar beet factory built at Utling, Essex (1ST).
1833	Obed Hussey patented a reaper (six months before Cyrus McCormick patented his reaper) (MITU).
1834	Cyrus McCormick patented his reaping machine (TH).
1837	John Upton patented the first steam plow (DID).
1839	Artificial fertilizer, superphosphate of lime, used by John Bennet Lawes for cultivation of turnips (1ST).
1840	Roller sugar mill was patented by James Robinson (DID).
	Peruvian guano was first used in England (DID).
1843	Peter Spence patented a method of manufacturing alum by treating burning shale and iron pyrites with acid (DID).
	John Bennet Lawes manufactured artificial fertilizer at Deptford Creek, London (1ST).
1846	John Deere constructed a plow with a steel moldboard (TH).
1847	Disc plow developed (EB).
1851	Marsh incorporated a conveyer belt in a reaping machine (DID).

1852   Potash found at Strassfurt (DID).
1854   Farmers Institute, Massachusetts, provided first example of later cooperative extension service work (WB).
1855   Self-propelled rotary cultivator built by Robert Romaine, Montreal (1ST).
1859   Hydraulic pressure into sugar mill by Jeremiah Howard (DID).
1861   Basic slag first used as a source of phosphorus for fertilizer (DID).
1862   Morrill Land Grant Act granted every state 30,000 acres for each senator and representative (WB).
        U.S. Department of Agriculture established (WB).
1866   Gregor Mendel published his work on plant genetics in Austrian journal (EB).
1873   Luther Burbank developed the Burbank potato (EB).
1874   DDT first synthesized by Othmar Zeidler, Strasbourg (EB).
1875   First Agriculture Experiment Station was established at Wesleyan University, Middletown, Connecticut (WB).
1876   Charles Darwin published the results of his studies on cross and self-fertilization of plants (EB).
1878   Appleby invented a knotting device for a reaping machine (DID).
1880   Bones used as fertilizer (DID).
1880-90 Bordeaux mixture, copper sulphate and lime, used on vines to combat downy mildew (DBE).
1887   Hatch Act provided financial support for the agricultural experiment stations (WB).
1889   First gasoline powered agriculture tractor built by Charter Engine Company, Chicago, Illinois (1ST).
1890   Disc plow invented (DID).
        Second Morril Act provided further funding for land grant colleges (WB).
1893   Disc plow reintroduced to the U.S. (see 1847) (EB).
1899   First Lake Placid Conference on Home Economics held at Lake Placid, New York (led to American Home Economics Association) (ER).
1900   Scientists in Netherlands, Germany, and Austria discovered the importance of Mendel's work (EB).
        Canadian hybrid Marquis wheat released to farmers (EB).
1901   Will B. Otwell, Macoupin County, Illinois, founded the Boys Corn Club (EAA).
1902   O. J. Kern, Winnebago County, Illinois, founded the Boys Experimental Club (EAA).

1903   O. J. Kern, Winnebago County, Illinois, founded the Girls
        Home Culture Club (EAA).
1904   O. H. Benson, later with the U.S. Department of
        Agriculture, used the three leaf clover which later
        became the 4-H four leaf clover sign (EAA).
1908   First gasoline engine tractor with crawler tracks produced
        by Holt, California; better traction, larger tool (TT).
        George H. Beal discovered that self-fertilization weakened,
        but cross-breeding restored, plant vigor (EB).
1909   Nitrogen fertilizer discovered by Fritz Haber, Karlsruhe,
        Germany (1ST).
        American Home Economics Association established (EOA).
1910   Holt, California firm, developed first self-propelled
        gasoline-powered combine harvester (TT).
1913   Nitrogen fertilizer manufactured by Badische Anilin and
        Soda Fabrik, Germany (1ST).
        U.S. Imported Meat Act (MMGT).
1914   Ammonia produced at Oppau, Germany, on large scale
        using Haber process for fertilizers and explosives (TT).
        Smith-Lever Act provided for federal, state, and county
        funding of the cooperative extension service (WB).
1915   Ford developed a farm tractor (TH).
1917   Ford began production of cheap, durable frameless tractors,
        increasing farm mechanization (TT).
1918   Donald E. Jones, Connecticut, used double cross breeding
        to produce stronger plants with desired features (EB).
        Efficient power take-off mechanisms developed for tractors
        (EB).
1919   U.S. Horse-Meat Act (MMGT).
1920   Agricultural fumigants first used (DID).
1921   Aerial crop dusting initiated by Ohio Agricultural
        Experiment Station; first plane flown by J. Macready
        (1ST).
        Connecticut Agricultural Experiment Station sold the first
        commercially produced corn hybrids (EB).
1924   Soybeans harvested for the first time by combine harvester
        (TT).
        Insecticide first used (TH).
        Henry A. Wallace (later secretary of agriculture and U.S.
        vice-president) sold cross-bred corn (EB).
1925   California agriculturists conducted earliest successful
        hydroponics experiments (TT).

First commercial crop dusting service offered by Huff
Daland Dusters Incorporated, Georgia (1ST).

Battery cages for higher egg production by laying hens
introduced, U.S. (TT).

1926 First factory ships for freezing and processing fish used by
Britain and France (TT).

First seed company devoted to the commercial production
of hybrid corn established (EB).

1927 A. J. Hosier, British farmer, invented the milking bail or
portable milking shed (TT).

1930 Corn hybrids produced in U.S. by cross-breeding (TT).

1931 Diesel caterpillar tractor manufactured by Caterpillar
Tractor Co., Peoria, Illinois (1ST).

1936 Inexpensive "baby" combine harvester marketed by Allis-
Chalmers (TT).

1939 DDT used as an insecticide by Dr. Wiesmann to combat
Colorado Beetle outbreak in Switzerland (1ST).

Rotolactor, milking machine, shown at the New York
World's Fair (CDIS).

1941 Aerosol insecticides developed by L. D. Goddhue and W. N.
Sullivan, U.S. (TT).

Packham potato harvester sped up crop production, but
potatoes still had to be sorted by hand (TT).

1943 DDT used in the United States (TT).

1945 Herbicide 2,4-D patented in the U.S. (TT).

1946 Fisherman used ultrasonic equipment to detect shoals of
fish (TT).

1954 Hybrid wheats made possible by work of Ernest Robert
Sears, wheat geneticist, U.S. (TT).

Tractor transmissions improved (EB).

1955 New Hatch Act incorporated and extended earlier
legislation for continued funding of agriculture
experiment station (WB).

1957 Quick-freeze plant for trawlers developed by Scottish
research scientists (TT).

U.S. Poultry Inspection Act (MMGT).

1958 Massey Ferguson demonstrated new labor-saving sugarcane
harvester to Australian farmers (TT).

B. A. Stout and S. K. Ries, Michigan State University,
devised crude tomato harvester (KM).

1959 Stout and Ries improved their tomato harvesting machine
(KM).

	University of California staff tested a tomato harvester (KM).
1960	Stout and Ries tested a further improved model of their tomato harvester in Florida and Michigan (KM).
	Blackwell Manufacturing Company built and sold the University of California tomato harvester (KM).
1961	Mexican and Colombian wheat crossed with Japanese varieties produced a Mexican cross bred wheat (EB).
1962	Rachel Carson's *Silent Spring* alerted the public to dangers of DDT and other chemical pesticides (EB).
	Gaines, first semidwarf wheat, produced (EB).
	International Rice Research Institute was established by the Rockefeller Foundation at Los Banos, Philippines (EB).
1964	New strain of miracle rice grown at Philippines International Rice Research Institute (TT).
1965	Mexican cross-bred wheat used internationally (EB).
1967	U.S. Wholesome Meat Act (MMGT).
1968	U.S. Wholesale Poultry Products Act (MMGT).
1969	DDT banned in the United States (1ST).
1970	U.S. Egg Products Inspection Law (MMGT).
1974	Fison's Gro-Bags, compost-filled sacks for growing fruit and vegetables, marketed in England (TT).
1984	Environmental Pollution Agency, U.S., banned the use of fumigant ethylene dibromide (WJ).
1985	Experiments reported on methods to increase successful spawning of shellfish making farming practical (WJ).
	Environmental Protection Agency, U.S., banned the use of Silvex and 2,4,5-T herbicides (WJ).
	American Cyanimid introduced Combat, a new, more effective roach poison (FORT).
	American Cyanimid announced new corn hybrid resistant to chemical weed killers (WJ).
1986	Plant Genetic Systems, Belgian biotechnology company, developed tobacco strain that kills insects (BUSWK).

This chapter has given an encapsulated overview of the discovery, use, and migration of various foodstuffs and the evolution of agricultural technology from the primitive plows to the sophisticated equipment used by today's agribusinesses. Having identified foods suitable for eating and devised the means to grow or to cultivate them, people were then faced with problems of how to store these foods until needed. The next chapter will explore some of the various means used for storing, preserving, and further packaging or processing foods.

3

Food Preservation and Processing

Most foodstuffs, unless properly prepared and stored, will deteriorate rapidly. In Western society, we have ready access to a large variety of foods, both in season and out, due to our preservation technologies and rapid and efficient distribution systems for goods and products. Technology makes it possible to transport, store, and make available a fragile, seasonal fruit like raspberries at any season of the year. This was not so for early people, who had to rely on fresh food or those few items that they learned to dry, preserve, and store for extended periods of time. This is still the case in many parts of the world.

Rotten meat or spoiled grain are not only unappetizing, but they can also make people sick. Grain spoiled by a fungus like rye ergot can cause illness and even odder effects similar to those of LSD. Rye ergot has been cited as one possible cause for the strange behavior of those accused of witchcraft during the Salem Witch Trials.[1] On the other hand, some side effects of partial spoilage may be beneficial and even provide new foods: grain fermentation processes create wine and beer, as well as leavened bread.

The knowledge of how to preserve foods successfully from deterioration and to store them for extended periods of time has been crucial to human survival. The technological developments that arose from seeking better preservation and storage methods have been significant contributors to our present standard of living. This chapter will explore the technologies of food preservation and manufacture, and the development of effective distribution systems, including stores and restaurants. Particularly useful sources include Kinder, Green, and Harris's *Meal Management*[2] and Caroline Davidson's *A Woman's Work Is Never Done*.[3]

Preservation

Storage and preservation of foods have been important to humans from earliest times. Curing and drying are two basic methods of preserving food. For example, animal meat would rot rapidly unless salted, smoked, or properly dried. Although there is no real evidence to show exactly when and how people began preserving meats, we do know that Palestinians were drying and salting fish (1000 B.C.) and Iron Age farmers in Britain (700 B.C.-500 A.D.) were salting meats to preserve them.[4] In the Middle Ages, salted herrings were prepared in continental Europe ca. 1260 and were first cured in England in 1359.

Figure 3.1. Preservation

YEAR			
1987		Puff drying	
1975			
	Frost-free refrigerators	Freeze-dried coffee	Aseptic packaging
	Domestic freezers		Irradiation of vegetables
1950			
			Dried soups
	Frozen foods		
1900	Domestic refrigerators	Freeze drying	Irradiation (pork)
	Refrigeration	Canned soups, milk	
		Metal cans	Meat extract
1800		Canning process	
	Refrigeration experiments		Dehydrator (France)
1600			
1200			Dried milk (Tartars)
400			
	Ice storage (China)		Dried fish (Palestine)
-1200		Pickling (Mesopotamia)	
-4400			
-10800			Dried grains
-23600			

Early peoples also dried grains, fruits, and vegetables; samples of dried grains have been found on Stone Age sites. Typically, grains were dried and stored in jars or in pits. Vegetables and fruits were either dried or stored in pits or cool locations such as caves.

Pickling and fermentation techniques developed early: pickles had appeared in Mesopotamia as early as 2000 B.C. Pickling originally used a salt brine to soak unwashed cucumbers (if washed first, the bacteria required to trigger the pickling process are absent).   During the 20th century John Etchells of North Carolina State University isolated the active agents in the pickling process and had two yeasts named in his honor. His yeast cultures have enabled picklers to develop more sanitary procedures.[5]

The Egyptians, in addition to using pickling, also developed the fermentation processes used in making both beer and bread. At first wine was stored in pottery containers, later versions of which included the Greek and Roman amphora. Wooden barrels were used in 500 B.C. for storing wines. The wood of these barrels aided in seasoning and preserving the wine.

Techniques now in common use include drying, freeze drying, canning, refrigeration, and freezing. New methods not as widely used in the United States include irradiation and ultra high temperature processing (UHT), although UHT foods are readily available in European stores and markets.

Drying

Early people dried foods using the sun or heat from fires. Most grain was sun dried before storage. In the Near East, foods such as dates, figs, and grapes were buried in hot sand to hasten drying. Fruits, vegetables, fish, and meat were among foodstuffs commonly preserved by drying. The first major improvement in drying foods occurred in 1795 when the French developed a hot-air dehydrator for vegetables.

Drying, salting, and curing cause changes in the texture or flavor of foods, particularly meats. During the 19th century several scientists sought ways to improve these processes, and much of the effort was focused on creating meat extract. Nicolas-François Appert (1750-1841), the father of canning, is also credited with devising a bouillon tablet used for flavoring.[6] In 1838 Baron von Liebig reduced flesh food to 15 percent of its bulk weight by drying, and in 1847 he perfected a process for producing meat extract. A dried meat biscuit was practical by 1850.

South America had many areas well suited to cattle raising, and as herds grew export to the European market was sought. This created problems for meat transportation, storage, and preservation. Initially, beef was salted or dried. Refrigerated shipping was not available until the last quarter of the 19th century. Consequently, the techniques, developed by Liebeg for creating meat extract, provided another way of providing a useful, easily transported product from South American beef. In 1864 meat extract was manufactured in Frey Bentos, Uruguay, and Oxo (bouillon flavoring) cubes were produced there in 1899. It was not until 1921 that bouillon cubes were imported from Europe to the United States.

Once bouillon cubes had been developed, it was only an enhancement of the technique to produced dried soups. The Lipton Tea Company began selling packaged dried soups with its chicken noodle soup in 1940. Its onion soup mix was popularized by Arthur Godfrey in 1954 for use in making California Dip by combining the dried soup mix with sour cream. Although a number of manufacturers make dried soups, Lipton's has dominated dried soups in the same way Campbell's has dominated canned soups. In an effort to break Lipton's virtual monopoly on the dried soup market, Campbell's introduced a dry soup in the 1960s marketed under the Comin' Home brand. By 1985 Campbell's was challenging Lipton's with dry soups sold under the Campbell name, but Lipton's still retains the dominant share of that market.

Beverages such as milk, coffee, and orange juice were also dried and used as concentrates. Marco Polo (ca. 1270) described dried milk used by the Tartars when they were traveling, but dried milk products apparently were not made in the West until the 19th century. A series of individuals helped develop the processes needed for the successful commercial production of dried milk. Parmentier (France) produced a dried milk powder in 1805. Nicolas Appert devised a dried milk tablet, but it was not until 1855 that F. S. Grimwade obtained his British patent and carried out industrial tests. Nearly fifty years later large-scale production began.[7]

Work continued on dried milk processes with Samuel Percy patenting his process in the United States in 1972. Other individuals experimenting with various processes and equipment were, in the United States, Campbell, Just, MacLachlan, Merrell, Gere, and Gray; in Denmark, Wimmer; and, in England, Hatmaker.[8]

Early milk powders did not mix evenly, and, unless the user was very careful, the final liquid often had lumps of undissolved milk solids. This changed with the work of D. D. Peebles, who patented variations on an instant milk process in 1936, 1955, and 1958. In 1946 the Instant Milk Company patented an agglomeration process that improved dried skim milk.

Like milk, a number of inventors experimented with soluble coffee before a satisfactory product was derived. Satori Kato sold the first powdered, instant coffee at the Pan American Exposition, Buffalo, New York, in 1901, and G. Washington developed a soluble coffee in 1906. Nescafé, the first commercially successful instant coffee, was developed in 1937 and patented in 1938 by Nestlé's after eight years of research. Nescafé remained the major instant coffee until the development and commercialization of freeze-dried coffee in the 1960s. Dried breakfast drinks were rounded out with the availability of instant orange juice in powdered form in 1962.

Freeze-drying brought a significant improvement to dried foods. In freeze-drying foods are frozen and the liquid in the form of ice is removed under vacuum conditions leaving the solid matter. Freeze-dried foods retain their shape and more flavor than with the traditional heat-processes used in dehydration. Freeze-drying was invented by J. A. d'Arsonval and F. Bordas in 1906 for preserving biological materials, and was independently discovered by

Shackwell in the United States in 1909; however, the technique was not applied to foods until the 1940s. A Swiss laboratory did freeze-dry coffee in 1934, but they never put their method into commercial use. In 1946 and 1947 E. W. Flosdorff demonstrated the use of freeze-drying for such products as coffee, orange juice, and meat. The process was adapted for preserving Texas shrimp in 1955, and General Foods used it to produce Maxim freeze-dried coffee in 1964, but did not market the coffee nationally until 1968. Nestlé responded to this threat to its dominance in the instant coffee market by introducing Taster's Choice coffee in 1966.

Since then freeze-drying has also been used for meats, vegetables, packaged entrees, and soups, and for camping foods. It has been popularized by NASA as part of the astronauts' menu in the space program.

Another technique that could rival freeze-drying for some foods is puff drying. In 1986, John Sullivan of the USDA adapted the puff-drying process originally developed for blueberries so that it could be applied to mushrooms as well. Food is partially dried in a conventional oven and then processed under high pressure steam. When released from the steam processing, the food expands rapidly, and further drying and later rehydration can be completed more quickly. The process is cheaper than freeze-drying and preserves the nutrients otherwise lost in canning processes.

Canning

Canning is a term applied to foods preserved in glass jars or in metal cans. War and difficult food production conditions during the late 18th and early 19th centuries prompted the French government to seek better food preservation techniques. Napoleon's troops had suffered from a diet of putrid meat and poor quality food during most of their campaigns since the available food could not be stored or transported long distances unless first salted, smoked, or dried. The French government announced a 12,000 franc prize for a useful method of preserving foods, which prompted Nicolas-François Appert to develop the canning techniques he later called "appertizing".[9]

The California Gold Rush and the United States Civil War also contributed to the development and spread of the canning industry in the United States. At first canned foods were expensive and limited in variety, but by the 1880s household management books were recommending the addition of the can opener to kitchen tools, and by World War I they were a common fixture in many U.S. kitchens.[10]

The first successful canning process used glass jars. Spallanzani first suggested preserving food by means of hermetic sealing in 1765, but it was 1795 before Nicolas-François Appert, a Parisian confectioner, designed the first preserving jar and 1804 when he began to apply his technique for canning food commercially.

In 1810 Peter Durand, England, patented the concept of using vessels of metal or tin for food preservation, and in 1812 the Donkin & Hall cannery, using Durand's patents, was established at Bermondsey. By 1819 Joseph Colin, Nantes, had produced canned sardines, and in 1925 canned salmon appeared.

In the United States William Underwood established a canning plant in Boston in 1820. His bookkeeper is credited with coining the term "can" as an abbreviation for canister in 1839. Thomas Kensett, a New York canner, was issued a United States patent for tin plated cans in 1825. Tomatoes and corn were canned in 1848, but general use and availability of canned foods took longer. By 1880 the first canned fruits and meats were appearing in stores.

Milk, in addition to preservation by drying, was also canned. A number of different individuals experimented with various processes. Appert in 1809 had reduced milk to one third its original volume and sealed it in a container, but the most successful processes were those devised by Gail Borden and Swiss immigrant John Meyenberg. Canned milk first became available in 1856 when Borden patented condensed milk, and by 1858 he was advertising it for sale. Borden applied a partial vacuum to remove moisture and obtained a more flavorful product than by open boiling as used in Appert's process. A sterilization process for evaporated milk was patented in 1884 and 1887 by John Meyenberg, St. Louis, Missouri. Condensed milk contains sugar while evaporated milk does not.

In addition to canned milk, other prepared baby food was available by the late 19th century. Mellin's Food and Murdock's Liquid Food (1875) were two brands being marketed as supplements for infants. Canned baby food appeared when Harold H. Clapp introduced Clapp's Baby Food in 1922; he was followed in 1927 by the Fremont Canning Company (renamed Gerber in 1941). Gerber led the change to marketing baby food in grocery stores instead of in drug stores where other baby foods were marketed.

Canning as a method for home preservation of foods was made practical with the development of the screw-threaded Mason jar in 1858, with rubber gaskets produced for such jars by B. F. Goodrich in 1871, and with the improved rubber-gasket sealing lid developed by Alexander Kerr in 1902. Goodrich also manufactured rubber bottle stoppers in 1871. The two-piece Economy lid, available in 1914, made the reuse of the metal canning lids practical and reduced problems associated with deterioration of the sealing gasket. This type of jar and lid are still used today.

Significant developments continued in commercial canning processes, including Isaac Salmon's 1861 discovery that the addition of calcium chloride to the water bath used to sterilize canned food would raise the boiling point of water to 240 degrees, and thus reduced the sterilization time from four or five hours to 25 or 40 minutes. As a result, canning production increased from 2,500 cans per day to 20,000.[11] In 1874 the pressure steam retort was introduced, once again reducing the sterilization time. Further advances included the double-seamed, no-solder can introduced in 1898 by George Cobb, Fairport, New York, and the addition of liquid to canned products in 1927 by the American Can Company, reducing spoilage and loss of vitamins.

A number of familiar names appeared in the latter half of the 19th century, including Van Camp, Campbell, and Heinz. Gilbert Van Camp began producing canned goods for sale in 1861, offered Heinz Catsup in 1876, and in 1891 he introduced Heinz Baked Beans in tomato sauce. Burnham and Morrill, Portland, Maine, offered canned sweet corn in 1867 and introduced their B&M baked beans in 1875. Campbell's soups were first sold in 1873, and by 1897 John Dorrance of Campbell's had developed a process for condensing soups. Campbell's new soup won a prize at the Paris Exposition in 1900, and Campbell added the medallion to their label. Their red and white label was adopted in 1898.

Competition in canned foods has continued. Campbell and Heinz became competitors when Heinz entered the soup market with its Great American Soups in 1967. Campbell retaliated with its Chunky soups in 1969, ultimately forcing Heinz to withdraw. Campbell entered the Italian food market with Prego Spaghetti Sauce in 1982. In 1984 the Progresso Quality Foods Company mounted a challenge to Campbell's dominance, the outcome of which is still undecided.

Cans have also been produced that not only preserve foods, but also heat or cool the food they contain. In 1941 Heinz produced a self-heating can for lifeboat and battlefield use. It was not produced for consumer use because of high production costs. More recently, in 1977, Pozel, a German firm, developed a process that causes an exothermic reaction when the container is opened. Their process was developed for beverages sold from vending machines. Similarly, a process for cooling canned beverages instantly was perfected in 1976 by Chill Car Industries.

Traditional canning requires exposing food to high temperatures for periods sufficient to kill harmful organisms in the can and the food. Such exposure often affects the taste, texture, and appearance of the food. Research efforts have been seeking methods for reducing these effects. The American Can Company developed a plastic can with a metal lid that was lighter in weight, stacked well, and required a shorter period of high heat for preservation. Plastic trays have also been used which, because of their flatness as opposed to the circular structure of a can, are even quicker to process.[12]

Another process similar to canning is aseptic packaging using hydrogen peroxide to disinfect the package and more heat for a shorter period of time than is used in traditional canning to process the food. This technique was used in Europe in 1965 for extended unrefrigerated storage of milk and has been applied extensively in the United States for fruit drinks and juices in cardboard or foil packages.[13]

Given a choice between canned foods and fresh foods, consumers are increasingly choosing fresh. Although canned foods such as tuna and some prepared foods are still staples in many homes, the popularity of canned fruits and vegetables is suffering from the increased competition of fresh produce. Improved transportation systems delivering fresh fruits and vegetables quickly and at relatively low cost have dramatically increased their availability and attractiveness. Emphasis on nutrition and health are also affecting consumer preferences. The reduced number of jobs in the canning industry, particularly in California, is one indication of this shift in consumer buying patterns.[14]

Canning has made a significant contribution to the availability of seasonal food throughout the year. It has provided convenient and economic preservation of foods and made them easier to store and transport. It has significantly improved the variety and nutritional quality of many people's diets. Although new techniques such as freeze-drying and aseptic packing are gaining in popularity, canning is still important for many foods.

Refrigeration

Although the refrigerator is a relatively recent invention (mid-19th century), ice has been used for much of recorded history. The Egyptians and the Indians both used evaporative techniques to make ice despite the fact that the air temperature never reached freezing. Such ice was created by setting earthenware pans of water in straw beds. The effect of nocturnal radiation coupled with rapid evaporation of the water caused ice to form. The Chinese were cutting and storing ice as early as 1000 B.C. Both the Chinese and the Romans knew how to preserve ice by insulating it.[15] The ice was packed in snow, wrapped with insulating materials, and stored in a cool place often underground. Ice was the only means of refrigeration until the late 19th century, and these techniques were used even in the early 20th century.

The next improvement was described by Blasius Villafranca in 1550 when he recorded the use of saltpeter in water for cooling wine. He claimed this method was also used by the Romans. A series of experiments were carried out in the 18th century, and in 1748 Dr. William Cullen produced liquid refrigeration with a vacuum hand pump. Sir John Leslie, in 1810, used a vessel with sulfuric acid and a hand pump to produce one and a half pounds of ice.

More complex machines for refrigeration were soon to follow. Thomas Moore coined the term "refrigerator" for an improved ice box he patented in 1803. In 1805 the American inventor Oliver Evans designed a machine for making ice, but he never built it. Jacob Perkins, originally from Massachusetts, had been a friend of Evans and may have been inspired by Evans's ideas when he built the first gas refrigerator in 1834.[16] Another American, John Gorrie of Florida, built a machine for making ice and cooling air in 1842. Although similar to Evans's original proposal, Gorrie's invention received the first United States patent for a mechanical refrigerator in 1851. Unfortunately, he died in 1855 without having obtained funding to manufacture his machine.

Commercial refrigeration was developed separately by Alexander Twining in the United States, and by James Harrison in Australia in 1850. Advances were fairly rapid. In 1857 Harrison used the 1834 Perkins process to transport meat by ship, and in 1859 Twining produced the first commercial ice, the same year Ferdinand Carré of France used ammonia in a refrigeration system.   Machines devised by Carré were used by the Confederate States to produce ice when the Civil War cut off Northern ice supplies.[17] Carl von Linde contributed to refrigeration development by designing a scientifically based ammonia compression system in 1876.

The need to refrigerate meat in storage or during transport was a significant spur to the development of refrigeration processes. A meat refrigeration plant was established in Sydney in 1861, and in 1879 the first Australian frozen meat was sold in London. South Americans saw the transport of frozen meat as a more effective solution than turning their meat into meat extract. Frozen meat was first shipped from Argentina to Europe in 1877. Like Argentina, the cattle raising and processing regions of the United States also faced problems of how to move meat, once butchered, to its place of sale or use. The first refrigerated railway cars were built for Chicago meatpacker Gustavus Smith in 1877.

Most of the progress toward domestic-use refrigerators occurred in the beginning of the 20th century. The first household refrigerator patent was issued to Albert T. Marshall in 1899, but the first domestic-use refrigerator was not manufactured until 1913. The Kelvinator appeared in 1918, followed in 1919 by the Frigidaire. The absorption-type Electrolux refrigerator, patented by Swedes in 1920, was not perfected until 1929. It was introduced to Swedish and American markets in 1931. The name Electrolux was used in Sweden, but Servel was used in the United States because the name Electrolux was already in use.[18] General Electric began manufacturing a hermetically sealed refrigerator in the United States in 1926.

Improved safety resulted from the adoption of Freon, a nontoxic, low-boiling liquid, for refrigeration and aerosols. Introduced in 1931, Freon reduced the hazards caused by escaping refrigerants, but more recent research has questioned its possible affects on the destruction of the earth's ozone layer. No definitive evidence has been found either condemning or exonerating Freon, and it is still used in many commercial and domestic refrigerator systems.

Refrigerators were expensive and not fully reliable when first introduced. Not all areas had electrical service, and the cost of running a refrigerator was higher than buying ice for an icebox. Too, the iceboxes of the 1920s and 1930s were considerably improved over 19th-century models and thirty cents of ice would last a week. As refrigerators improved and low-cost electrical power became more generally available, the icebox gradually disappeared.

By 1939 General Electric had introduced a dual temperature refrigerator capable of storing frozen foods for a limited time, but the true domestic freezer, capable of freezing fresh food, was first offered in the United States in 1955. The front-opening freezer appeared in 1961, and by 1965 the frost-free refrigerator was available, ending the messy task of periodically defrosting the refrigerator.

While frozen foods were available during the depression years, it was not until after World War II, with the spread of electrical services and with the availability of improved refrigerators, that frozen foods became a major source of food for many households. Once commercial freezers were available, however, the preparation and marketing of frozen foods became practical. Clarence Birdseye had patented his process for quick freezing food in 1923, but his first frozen foods were not introduced until 1930. In 1939 Bird's Eye offered a limited range of precooked frozen foods to householders, including chicken fricassee and a minute steak. Other food processors quickly followed with creamed chicken, beef stew, and roast turkey with dressing. In 1945 a United States airline began serving precooked frozen dinners to passengers; nine years later, in 1954, the TV dinner was offered to consumers. Since domestic freezers had become available by 1955, householders were able to store frozen foods until needed. By 1974 even pizza could be had in frozen form.

The first premium frozen foods were offered in 1958 by Stouffer, but they were not complete dinners. Premium frozen dinners premiered in 1980 when Armour introduced Dinner Classics, which were followed in 1982 with Campbell's highly successful Le Menu line of frozen entrées. These premium lines led rapidly to a number of imitators, but Armour and Campbell have maintained strong market positions. Le Menu entrées were marketed on sturdy plastic plates suitable for use in a microwave or a conventional oven and were one of the first frozen dinners packaged for dual use. The increased popularity of the microwave also influenced others; lower priced rivals used cardboard containers and Swanson, which had used aluminum trays for its frozen dinners since first offering them in 1954, finally switched to a dual-use, plastic tray in 1987.

Irradiation

Many people think of irradiation as a new technique; however, early efforts to use radiation for food preservation occurred in the beginning of this century. The first United States patent for food irradiation was granted in 1905, but it was not until 1921 that X rays were tested on pork as a treatment for trichinosis. Practical use stems from the work of Bernard E. Proctor, Massachusetts Institute of Technology, who demonstrated a process for the irradiation of foods in 1943. Typically, food is exposed briefly to a cesium or cobalt source that kills insects, larvae, and germs, and retards ripening without leaving any residual radiation in the food. Irradiated foods retain the taste and texture qualities of fresh foods, and irradiated meat products do not require refrigeration.[19]

By 1953 the United States Army had established a test program for irradiated foods, and by 1963 the United States Food and Drug Administration (FDA) had approved the use of irradiation for potatoes, wheat, and wheat flour. The Soviets, in 1958, were the first to use irradiation to treat commercial products. Irradiation has also been used for foods used in space and for hospital patients requiring germ-free conditions.

The FDA approved irradiation for use in preserving some forty spices in 1983. With the bans on certain agricultural fumigants and the recent concerns over various other additives and processes such as nitrites and sulfites, it is not unlikely that irradiation will be used more.

Irradiation is not only effective in killing germs and preserving foods for extended shelf life, but it is also energy efficient. Estimates indicate it could save three quarters of the cost of conventional canning, freezing, and chemical methods and up to half the cost of fumigants. Because of concerns over possible effects of radiation contamination, the technique has yet to be used widely despite almost forty years of experience showing it to be a safe, effective means of food preservation.[20] There are over one hundred irradiation plants throughout the world with thirty in the United States Among the United States plants only three are used for commercially processed food.[21] Twenty-five foreign countries irradiate a variety of foods.

Table 3.1. Food Preservation

Dates	Event
B.C.	
500,000	Fire used (HEP).
6500	Europeans mined salt at Hallstein and Hallstatt, Austria (BBB).
6000	Mortar and pestle used for grinding grain (WB).
2000	"Cold house" timber-lined ice pit used in Ur (FR).
1000	Palestinians salted and dried fish (FR).

	Chinese cut and stored ice (WB).
500-400	Wooden barrels used to store wine (TH).
>390	Theophrastus described vinegar (used with lead for pigments) (RM).
328	Alexander the Great found the Indians mining five kinds of salt (BBB).

A.D.

1260	William Beukelszoon, Biervleit, Holland, first cured herring by gutting and salting (DID).
1550	Blasius Villafranca's *Methodus Refrigerandi*, Rome, described saltpeter-in-water process for cooling (1ST).
1748	Dr. Cullen used a vacuum hand pump to produce liquid refrigeration (DBE).
1765	Lazzaro Spallanzani, Italy, suggested preserving food by means of hermetic sealing (TH).
1787	Ice first produced by artificial means (DID).
1795	Nicolas-François Appert, France, designed preserving jar for food (TH).
	French developed hot air dehydrator for vegetables (EB).
1803	Thomas Moore, Baltimore, Maryland, applied the term "refrigerator" to his improved ice box (FFF).
1804	Nicolas-François Appert, France, began commercial application of his technique for canning food (1ST).
1805	Oliver Evans, U.S., proposed machine for creating ice (WB).
	Antonine-Augustin Parmentier, France, first produced powdered milk (WABI).
1806	Frederic Tudor, Boston, shipped ice to Martinique, later to Charleston, New Orleans, and India (WJ).
1809	Appert, France, reduced milk to one third its original volume and sealed it in a container (HH).
1810	Leslie combined a vessel containing sulphuric acid with an air pump to produce 1 1/2 pounds of ice (EDE).
	Peter Durand patented the idea of using vessels of metal or tin for preserving food and licensed it to Donkin & Hall (1ST).
1812	Donkin & Hall established a cannery at Bermondsey (1ST).
1819	Joseph Colin, Nantes, canned sardines (1ST).
1820	William Underwood, Boston, established a U.S. canning factory (WB).
1825	Canned salmon was produced by John Moir, Aberdeen, Scotland (1ST).
	Thomas Kensett, New York canner, patented tin-plated cans in the U.S. (WB).

1834   Gas refrigerator developed by Jacob Perkins, Massachusetts (WB).

1838   Baron von Liebeg, Germany, reduced flesh food to 15 percent of bulk weight by desiccation (DID).

1839   William Underwood's bookkeeper abbreviated the word "cannister" to "can" (1ST).

1840   Canned lobster was produced by Tristram Halliday, St. John, New Brunswick, Canada (1ST).

1841   Canned anchovies produced at Deammen, Norway (1ST).

1842   Dirchoff, Russia, invented a milk powder (DID).

H. Benjamin froze food by immersion in a mixture of ice and brine (DID).

Bevan invented the vacuum chamber for food preservation (DID).

Dr. John Gorrie, U.S., invented mechanical refrigeration (DID).

1847   Meat extract prepared by Justus von Liebig, Royal Pharmacy, Munich (1ST).

1848   Canned tomatoes produced by Harrison W. Crosby, steward at Lafayette College, Easton, Pennsylvania (1ST).

First canned sweet corn was sold by Nathan Winslow, Portland, Maine (1ST).

Morton Salt Company, Chicago, was founded (RM).

1850   Dried meat biscuit invented (DID).

Australian James Harrison and American Alexander Twining each developed commercial refrigeration (1ST).

Alexander Twining patented ice-making equipment (EB).

1851   M. Mason, France, invented a method for drying vegetables (DID).

Lord Gorrie obtained the first U.S. patent for a mechanical refrigerator (EB).

1852   Lord Kelvin and Professor Rankine proposed open-cycle refrigeration (DID).

1855   Grunwald, Germany, carried out industrial tests to produce powdered milk (WABI).

1856   Gail Borden patented condensed, canned milk (RM).

1857   James Harrison, Geelong, Australia, used the Perkins refrigeration process to transport meat by ship (DID).

1858   Condensed milk advertised by Gail Borden, Burrville, Connecticut (1ST).

John Mason invented the threaded top Mason jar used for home canning (WB).

1859 First commercial ice produced by Alexander Twinning, Cleveland (EB).

Ferdinand Carre, France, used ammonia in refrigerator (EB).

1861 Meat refrigeration plant established by Thomas Mort, Darling Harbour, Sydney, N.S.W. (1ST).

Isaac Salmon discovered the addition of calcium chloride to canning water bath reduced canning time (RM).

Gilbert Van Camp, Indianapolis, began canning food products for sale (RM).

1863 First canned peaches produced by Cutting Company, California (1ST).

1864 Louis Pasteur developed a pasteurization process for improved wine preservation (TH).

Canned asparagus produced by William Hudson, Hunter's Point, Long Island (1ST).

Meat extract manufactured at Frey Bentos, Uruguay (1ST).

1865 Thaddeus Lowe invented an ice machine (TH).

1867 Burnham and Morrill Company, Portland, Maine, canned sweet corn (HS).

1869 Joseph Campbell and Abraham Anderson of Camden, New Jersey, successfully canned tomatoes (USAIR).

1871 B. F. Goodrich, Akron, Ohio, manufactured rubber preserving rings (EOOET).

1872 Samuel R. Percy, New York, patented dried milk in the U.S. (FFF).

1873 Campbell's soups offered for sale (RM).

1874 H. Solomon introduced a pressure-cooking method for canning food that reduced sterilization time (TH).

1875 B&M canned baked beans produced by Burnham and Morrill, Portland, Maine (1ST).

Mellin's Food and Murdock's Liquid Food sold as infant or invalid supplement (HG).

1876 Carl von Linde introduced the first successful refrigerator using ammonia in the compression system (EB).

1877 Railway refrigerator cars built for Chicago meatpacker Gustavus Smith (1ST).

First frozen meat shipped from Argentina to Europe (1ST).

1879 Australian frozen meat sold in London (TH).

1880 Baked beans in tomato sauce first sold in U.S. (1ST).

First canned fruits and meats available in stores (TH).

1891	Van Camp's, Indianapolis, canned its first baked beans with tomato sauce (1ST).
1892	First canned pineapple sold (TH).
1897	Dr. John Dorrance, Campbell's, invented a process for condensing soup (RC).
1898	George Cobb, Fairport, New York, developed the double seamed, solderless can (RM).
	Campbell's Soup adopted the red and white label (UNITD).
1899	First U.S. household refrigerator machine patent was granted to Albert T. Marshall, Brockton, Massachusetts (FFF).
	Oxo (flavoring) cubes produced at Fray Bentos, Uruguay (1ST).
1900	Campbell's Condensed Soup won a prize at the Paris Exposition (UNITD).
1901	Satori Kato, Chicago, sold powdered, instant coffee at the Pan American Exposition, Buffalo, New York (ADBE).
	American Can Company founded (RM).
	Campbell (U.S.) and Wimmer (Denmark) dried concentrated milk on trays to make powdered milk (HH).
1902	Alexander Kerr perfected the Economy lid with a rubber sealing gasket fixed to the lid for home canning (WB).
	Just patented a drum milk drier that was later improved by Hatmaker (HH).
1905	Carl Linde patented the ammonia compression refrigerator (DID).
	First U.S. patent issued for using ionizing radiation to kill food bacteria (ACTB).
	MacLachlan developed spray processing for faster drying of milk, skim milk, eggs, or blood (HH).
1906	G. Washington, U.S., developed refined soluble coffee (DBE).
	Freezedrying invented by J. A. d'Arsonval and F. Bordas, Paris, for preserving biological materials (TT).
1907	I. C. Merrel, I. S. Merrel, and Gere patented improvements to milk spraying process for making dried milk (HH).
1909	Shackwell, St. Louis, Missouri, independently discovered freeze drying (WABI).
1910	Glass-lined railway milk tanker used by Whiting Milk Company on the Boston & Maine Railroad (1ST).
1913	First domestic electric refrigerator, Domelre, manufactured in Chicago (1ST).

C. E. Grey and Jensen patented further improvements to milk spray drying process (HH).

1914 Two-piece Economy lid with a separate screw band invented, a boon to home canners (WB).

1916 Mechanical home refrigerator sold for $900 (ACTB).

1918 Kelvinator refrigerator designed by Nathaniel Wales, U.S., marketed (WABI).

1919 Canned grapefruit introduced by Yankee Products Ltd., Puerto Rico (1ST).
Frigidaire refrigerator marketed in U.S. (WABI).

1920 C. Munters and B. von Platen, Sweden, patented a silent refrigerator (WABI).

1921 Bouillon cubes imported from Europe to U.S. (TT).
X rays used to kill trichinosis eggs in pork (ACTB).

1922 Canned baby food manufactured by Harold H. Clapp, Rochester, New York (1ST).

1923 Clarence Birdseye patented his process for quick freezing foods (HS).

1926 Gas refrigerator introduced to the American market by Electrolux, Evansville, Indiana (FFF).
B. von Platen and George Munters, Stockholm, Sweden, patented the absorption refrigerator in U.S. (FFF).
General Electric manufactured hermetically sealed household refrigerator (WABI).
First factory ships for freezing and processing fish used by Britain and France (TT).

1927 First silent electric refrigerator manufactured by Electrolux, Britain (TT).
American Can Co. discovered canning with liquid reduced spoilage and vitamin loss (RM).
Continental Can Company founded (RM).
Fremont Canning Co., later Gerber, began marketing canned baby food (ACTB).

1928 Machine for boning and cleaning kippers first used, Fleetwood, England (TH).

1929 C. Munters and B. von Platen devised a condensor for their refrigerator, the Electrolux (WABI).

1930 Clarence Birdseye, Springfield, Massachusetts, sold individually packaged products of frozen foods (1ST).

1931 Kinetic Chemical Company produced Freon, nontoxic, low-boiling liquid for use in refrigeration and aerosol sprays (TT).
Electrolux began refrigerator production in Sweden (WABI).

	Servel (Swedish Electrolux) began refrigerator production in U.S. (WABI).
1934	Refrigeration process for meat cargoes developed (TH). Swiss lab developed freeze-dried coffee (WABI).
1936	D. D. Peebles patented a process to reprocess dried milk for better reconstitutability (HH).
1938	Nestlé's introduced Nescafé, first commercially successful instant coffee (1ST).
1939	Bird's Eye introduced precooked frozen foods (1ST). General Electric introduced dual temperature refrigerator with frozen foods in separate compartment (WABI).
1940	Freeze drying adapted for food preservation (TT). Lipton's Chicken Noodle Soup (dried soup mix) introduced (TIME).
1941	Heinz developed a self-heating can for lifeboat and wartime use, but high production costs made it impractical for commercial use (TT).
1943	Bernard E. Proctor, Massachusetts Institute of Technology, demonstrated the use of irradiation to preserve foods (IEEE).
1945	U.S. airline began serving precooked frozen dinners (WABI).
1946	Instant Milk Company, U.S., patented agglomeration process improving dried skim milk (TT).
1946-1947	E. W. Flosdorff, U.S., demonstrated the use of freeze-drying techniques for foods, including coffee, orange juice, and meat (WABI).
1948	Instant tea marketed in the U.S. (WB).
1953	U.S. Army began test program to irradiate precooked foods (IEEE).
1954	First frozen TV dinners introduced (TT). Arthur Godfrey popularized California Dip made with Lipton's Onion Soup and sour cream (TIME).
1955	First domestic deep freezer capable of freezeing fresh food sold in the U.S. (TT). Freeze drying used to preserve Texas shrimp and Maryland crab (WABI). D. D. Peebles registered another patent for reprocessing milk for reconstitutability (HH).
1957	Quick-freeze plant for trawlers developed by Scottish research scientists (TT).
1958	Soviets used irradiation for commercial food products (ACTB).

D. D. Peebles registered another patent relating to dried milk reconstitutability (HH).

Stouffer's introduced the first premium-priced frozen foods (BUSWK).

1961   Domestic use front-opening freezers sold (TT).

Green Giant sold boil-in-bag frozen vegetables (PM).

1962   Instant orange juice available in powdered form (TT).

1964   First freeze-dried coffee marketed by General Foods (FFF).

U.S. Food and Drug Administration approved irradiation for potatoes, wheat, and wheat flour preservation (IEEE).

1965   Long-life milk packaged for nonrefrigerated storage (TT).

Frost-free refrigerators sold in the U.S. (CREPT).

1966   Nestlé introduced Taster's Choice freeze-dried coffee (HS).

1967   H. J. Heinz introduced Great American soups, but they were later withdrawn from the market (BW).

1968   Freeze-dried coffee marketed nationally in the U.S. (FFF).

1969   Campbell's introduced Chunky soups (BW).

1974   Frozen pizza first marketed (RD).

1976   Chill Car Industries perfected a process for instantly cooling a canned beverage (WABI).

1977   Pozel, German firm, developed an instant heating process for vending machine containers (WABI).

1980   U.S. Department of Agriculture assumed responsibility for former U.S. Army food irradiation program (IEEE).

Armour introduced gourmet frozen dinners, Dinner Classics (CDIS).

1982   Campbell Soup Co. launched Le Menu frozen foods (WJ).

Admiral introduced the Entertainer, premium-priced refrigerator with built-in wine rack (FORB).

1983   U.S. Food and Drug Administration approved general use of irradiation for preserving spices (IEEE).

1984   General Electric introduced electronically controlled refrigerator with sensor to detect open door (BUSWK).

Campbell Soup Company launched Chunky New England Clam Chowder (WJ).

Campbell Soup Company launched Great Start Breakfasts (WJ).

1985   Campbell's introduced dried soup mixes for the second time, but the first time under Campbell label (WJ).

1986   John Sullivan, U.S. Department of Agriculture, adapted puff drying techniques used for blueberries to mushrooms (WJ).

Food Processing, Manufacture, and Packaging

A number of different techniques are used in food processing, manufacture, and packaging in addition to those discussed above under preservation. This section reviews baked goods, cereals, dairy foods and dairy substitutes, meat, beverages, seasonings, candy, snacks, and packaging.

Figure 3.2. Development of Selected Foods

YEAR	BAKED GOODS	CHEESES, FATS, AND MEATS	ICE CREAM DAIRY
1987			
	Toaster strudel	Butter-flavored Crisco	
1975	Granola bars		
		Fluffo shortening	
1950			
			Dannon yogurt
	Sliced bread	Kraft macaroni and cheese	Eskimo pie
1900	Oreos	Crisco, Kraft cheese	Ice cream cone
	Doughnuts	Hamburgers, hot dogs	Ice cream soda
		Cheese, butter factory	
1800		Pineapple cheese	
		Camembert cheese	
	Croissants	Cheddar cheese	
1600			
			Ice cream (Italy)
1200			
		Brie, Mariolles cheeses	
400			
	Bakers (Rome)		
-1200			
	Bread (Egypt)		
-4400		Cheese (Mesopotamia)	

Baked Goods

Bread has long been a dietary staple in Western societies. The Egyptians had developed fermentation processes for bread by 2600 B.C. Only wheat, and to a lesser extent rye, flours have enough gluten to hold the gas produced by fermentation and provide leavening. Other bread grains such as corn, millet, barley, and buckwheat lack this property and, while used for flat breads in Asia, Africa, India, and Latin America, are not suited for leavened breads.[22]

In addition to developing leavening agents, the Egyptians also developed enclosed ovens to ensure the bread was cooked throughout.[23] In Colonial America, where ovens were not available, an overturned cast iron pot or a Dutch oven was used. Hot coals were often placed on or around the overturned pot to achieve the desired cooking conditions.[24] In the American

Southwest, Pueblo Indians used adobe ovens, many of which are still in use today.

Until the 20th century, most people, with a few notable exceptions, made their own bread. Commercial bakers existed in Roman times around 170 B.C. and the remains of a bakery was found in the Pompeii ruins, 79 A.D. After the fall of Rome, bread in Europe again reverted to home production. While bakeries existed during the Middle Ages, home baking was still the preferred method, although poor housewives often took their dough to the village baker for baking.[25] At the turn of the century in the United States 95 percent of bread was baked at home, but by 1950 95 percent of United States bread was made by bakeries.[26] The General Baking Company was founded in 1911 to offer white bread, and Wonder Bread, introduced in 1930, was the first packaged, sliced bread.

Home baking was simplified by new products offered in the mid-19th century. Self-rising flour was first marketed in England in 1845, baking powder was marketed ca. 1850, and Arm & Hammer Baking Soda was introduced in 1867. In the 20th century, baking mixes for cakes and cookies and refrigerated dough for a variety of baked products were introduced. Prepared cake mixes were first marketed by General Mills and Pillsbury in 1949. During the 1940s through 1960s manufacturers found that women baking at home still wanted to feel they were doing more than just opening a box and adding water, so most mixes required the addition of eggs.[27] However, as more single heads of households and working women place a higher premium on their time, different demands on manufacturers are leading to changes. Some cake mixes come complete with a pan and frosting mix, and only require the addition of water before mixing and baking. For those who feel adding water and baking is too much work, there are a large variety of packaged baked goods. One of these, Sara Lee baked products, was introduced by Charles Lubin in 1949.

A variety of commercially produced specialty bakery products became available in the late 19th and early 20th centuries, including doughnuts and cream cookies. Ring doughnuts were introduced in 1847. In 1892, James Mitchell developed a dough-extruding machine for cookies that was used in 1895 by the Kennedy Biscuits Works, Cambridge, Massachusetts, to make Fig Newton cookies. In 1908 the Hydrox Cream Cookies were marketed in the United States by the Sunshine Baking Company, followed in 1911 by Oreos from Nabisco. Oreos went on to become one of the most popular brands of chocolate cream cookies. In 1975 Nabisco introduced Double Stuf Oreos, followed in 1984 by mint-flavored Oreos.

Commercially manufactured cookies are usually fairly crisp, while home-baked cookies, at least initially, are soft and chewy. Researchers at Battelle Laboratories, Columbus, Ohio, discovered that cookies made with invert sugar would remain soft after baking instead of becoming hard.[28]

Invert sugar is a mixture of the sugars glucose and fructose found in honey and fruits, and because of its ability to prevent the crystallization of cane sugar, is used to maintain moisture content in confections, other food products, and medicine. Soft, chewy homemadelike cookies appeared in 1983 when Procter & Gamble introduced Duncan Hines cookies with homemade taste and texture. Nabisco entered the soft cookie market shortly after. Procter & Gamble filed suit for patent infringement, but Nabisco contended that the P&G patent was invalid because the two-dough method of making cookies was based on conventional technology and was not patentable. Nabisco's contention was based on a traditional Mennonite two-dough cookie recipe for riggelvake.[29] The suit was still pending in late 1987.

Cookie specialty stores were given an impetus when Wally Amos opened his first Famous Amos Cookie Store on Sunset Boulevard, Hollywood, California, in 1975. He was quickly followed by a number of imitators including Mrs. Field, David's Cookies, and Cheryl's Cookies. As competition became stiffer, some shops closed while others have diversified to carrying other foods such as ice cream and muffins and gift items.

Crackers were originally called biscuits. The appellation *cracker* was first used in 1801 when Josiah Bent opened his bakery in Milton, Massachusetts. He made a crisp biscuit that "cracked" when eaten. Graham crackers were devised around 1830 as a health food by the Reverend Sylvester W. Graham. Irish cream crackers were manufactured by William Jacob of Dublin in 1885. Nabisco began making Ritz crackers in 1934 and by 1937 were producing more than 29 million/day. By 1987 production had reached 60 million/day.

In the 20th century, some breads and bread companies evolved from meeting the health needs of special groups. Energen rolls, a bread substitute for diabetics, were sold in 1920 in London. Margaret Rudkin, who devised a wholesome bread to improve the health of her asthmatic son, initiated large-scale production with her popular Pepperidge Farms Bakery in 1937. The Rudkins sold Pepperidge Farms to the Campbell Soup Company in 1961, and Campbell's diversified its products to include frozen baked goods, cookies, and fruit juices. The latter effort was unsuccessful and the fruit juices were withdrawn.

With the growing interest in health and healthy foods by consumers, manufacturers responded by mass marketing new items such as Granola bars and fruit bars. Granola bars were introduced by General Mills in 1975 and followed in 1982 by Quaker with Chewy Granola Bars. Quaker also added a chocolate coating to create Granola Dipps. The chocolate coating significantly increases the caloric content of what is supposed to be a health food, but to many consumers they taste better. New brands and new variations of these chocolate-coated bars continue to appear on supermarket shelves.

The croissant was devised in 1863 to celebrate the victory of the Viennese over the Ottoman Turks. The buttery pastry migrated to France and became so popular it is now thought of as French. By the 1920s local bakeries in the United States made them, but popular success came in the 1970s when commercial bakeries began making and selling them. In 1985, La Vie de France Corporation, Virginia, introduced a low-cholesterol version using margarine instead of butter. Croissants are now marketed in supermarkets, bakeries, and speciality shops, and are widely available.

Other baked goods include cakes, cheese cakes, and pastries. Work on solving the technical challenges of a pastry product suitable for toaster heating began in 1980, and Pillsbury introduced its successful Toaster Strudel nationally in 1985. A recent fad has been the introduction of large size, fresh-baked cinnamon rolls. The first shop specializing in these gooey concoctions opened at the Ward Parkway Shopping Center, Kansas City, in January 1985 and has been followed by a number of imitators.

Cereals

The history of cereals, like that of canned products, is studded with familiar names and brands. The first ready-to-eat breakfast cereal, Shredded Wheat, was produced in 1893 by Henry D. Perky of Denver, Colorado. The first flaked cereal, Granose Flakes, followed in 1895, also prepared from wheat by Dr. John Kellogg of Battle Creek, Michigan. Kellogg's Corn Flakes appeared in 1898. Charles W. Post, while a patient at the Battle Creek Sanitarium, developed Postum (1894), an early cereal-based substitute for coffee, and Grape-Nuts cereal (1895). Originally these cereals were regarded as health foods and did not come into general use until 1906.

A number of now familiar cereal brands appeared in the early 20th century. The Quaker Oats Company was formed in 1901 to package and sell rolled oats. Wheaties, the "Breakfast of Champions" first marketed in 1924, was an accidental discovery when some cooked bran cereal fell on the hot stove. When it dried, it was tasty and flaky. Rice Krispies were launched by Kellogg in 1937.

By the 1970s nutrition conscious consumers demanded more from cereal manufacturers than presweetened cereals. By 1972 most of the major food companies--Post, Kellogg, Quaker Oats, and General Mills--responded by offering "natural" cereals lower in sugar content and higher in fiber. Although Nabisco and Quaker had offered natural cereals from the beginning, and Post and Kellogg originally offered their first cereals as "health foods," these companies now emphasized the nutritional benefits of cereal eating instead of promoting only the taste. Among some of these new, more healthful cereals were Special K, Fruit 'N' Fiber, and Nutri-Grain.

Dairy Products, Dairy Substitutes, and Oils

Dairy products discussed here include cheese, ice cream, frozen custard, yogurt, and butter. Margarine, shortening, and selected vegetable oils are reviewed briefly as well.

Cheese was made as early as 4000 B.C. in Mesopotamia. Egyptian and Chaldean cheese molds from 2000 B.C. have been found. Brie cheese, increasingly popular in the United States, was being made in France in the 9th century A.D. and was served at the court of Charlemagne. By 960 A.D. Mariolles cheese was being made at the Abbey of Thierache, France. Camembert was first produced in the 1780s near Vimoutiers, Normandy.

Many cheeses take their name from the area or region where they are made: the first Cheddar cheese was made in 1666 in Cheddar, England; the first Stilton cheese was served at the Bell Hotel, Stilton, England, in 1730; Port du Salut was manufactured in the early 1800s by Trappist monks of Notre Dame de Port du Salut, Entrammes, France. Other cheeses are named for their appearance or shape: pineapple cheese, named for the pattern of criss-cross marks on the cheese, was first produced in 1808 and patented in 1810; American Brick cheese, which took its name from its shape, was developed by John Jossi, Wisconsin, in 1877.

Early cheese-making was done by farmers' wives or by householders. The first cheese factory was established at Bern, Switzerland, in 1815, and the first large United States cheese factory was established by Jesse William at Rome, New York, in 1851 to produce Cheddar cheese. Now over 37 percent of United States cheese is made in Wisconsin.[30]

What constitutes a "real" cheese has been a topic of some dispute of late, particularly in regard to the cheese and cheese substitute toppings on refrigerated and frozen pizzas. The Filled Cheese Act of 1896 was one of the first U.S. attempts to deal with what is a cheese. By 1904 Kraft processed cheese was being made in Chicago. Some cheese lovers dispute the use of the name cheese for processed cheese products since they also contain other substances including vegetables dyes and coloring, but almost all noncream colored cheeses now contain coloring or dyes. Kraft, Schieber, and Borden are the largest manufacturers of processed cheese. Kraft began manufacturing and marketing Kraft macaroni and cheese in 1937. By 1982, the Milk Marketing Board of the U.K. had developed a cheese substitute.

One of the more popular dairy foods, ice cream, was invented (or rediscovered) in 1300 by Bernardo Buontalenti of Italy. The Chinese had made a frozen dessert of rice and milk much earlier, and Marco Polo is credited with having brought word of it to Italy. Wealthy Romans were also fond of ices or snow with flavoring. Café Procope, Paris, served ice cream to the general populace in 1670, and by 1677 it was a popular dessert in Paris. It was recorded in the accounts of the English king James II in 1686. Ice cream was also popular among the wealthy and in urban areas in Colonial America, and purchases were recorded in George Washington's accounts. The first United States ice cream company was established in 1786 by Hall in Philadelphia, but ice cream remained expensive until the development of a process to mass produce ice cream by James Fussell in 1851.

Several different ice cream concoctions were devised in the late 19th and early 20th centuries. The ice cream soda was invented in 1874 by Robert Green, Philadelphia. The sundae appeared in the 1890s in response to local laws prohibiting the sale of soda water on Sunday which also prevented the sale of ice cream soda. Ice cream with syrup, minus the soda water, was sold instead. The concoction, available only on Sundays, was called a sundae.

In addition to the soda and sundae, other forms of ice cream included the cone, the ice cream sandwich, the Eskimo Pie, and ice cream on a stick. The ice cream cone was first made just after the turn of the century. Italo Marcioni patented a mold for the ice cream cone in 1903 and he is credited with creating the idea. The ice cream cone was reintroduced in another form and popularized in 1904 at the World's Fair in St. Louis where folded waffles were used for cones. The ice cream sandwich was created in 1905 and the first Eskimo Pie appeared in 1921, followed soon after by Good Humor, with ice cream on sticks which was sold from trucks. Wrapped ice cream brickettes were introduced in 1922.

Critical to the taste and texture of ice cream are the conditions promoting rapid freezing. The size and rate of formation of ice crystals determine the texture of the resultant frozen mixture. If large crystals form, the mixture will be grainy instead of smooth and creamy. Methods of rapid freezing, coupled with the controlled addition of air, promote the most desirable texture. Liquid air, nitrogen, and carbon dioxide are used to speed rapid freezing. Carbon dioxide was first used in the manufacture of ice cream in 1924.

Frozen custard was invented in 1919 by a Coney Island vendor who added egg yolks to ice cream. Frozen custards are normally kept at 23°F instead of the normal 10°F used for ice cream. A popular chilled dessert between the world wars, it has gone into a decline because of its perishability.

During the 1980s there was an increased emphasis on health and nutrition, and a number of foods have become popular including yogurt. Twenty-six percent of Americans now eat yogurt regularly.[31] Dannon Yogurt was first marketed in 1942. In 1981 Dannon was acquired by the BSN Groupe, France, and Dannon began marketing a French-style yogurt with nuts and fruit in 1983. Other products aimed at the same market are tofu-based products like Jofu and Kefir, similar to yogurt, but with less sugar.[32] Tofutti and frozen yogurt have also become popular.

Butter was developed early: the Greeks and Romans used it as a treatment for burns and other skin ailments rather than as a food. In other geographic areas it was used as hair dressing.[33] The Hindu Vedas, Indian religious texts written before 1200 B.C., mention the use of butter as a food.

Butter remained largely a cottage or home product until the late 19th century,[34] even though the first butter factory was established in New York in 1856. Commercial production was aided by the cream separator, patented in 1876 by Swede Gustave de Laval. The New American Cream Separator, derived from Laval's design, was manufactured in the United States by Wahlin of New York in 1896. The mass production and use of the cream separator made possible the continuous production process for butter. From around 1900 until World War II, dairy farmers took cream for butter to centralized creameries for processing into butter. Since World War II and the use of improved processing equipment, farmers have sent whole milk to the creameries.

Margarine, like the canning process, was encouraged by the French government sponsorship of a competition to develop a substitute for butter to be used by the French Navy and the less prosperous citizens. Margarine was patented in 1869 and the first margarine factory established in 1871. Its name derives from the Greek word *margarites*, meaning pearls, and referred to its appearance during one stage of its processing. The flavor and texture of margarine were improved in 1903 through the use of hardened vegetable oils. By 1944 the British began to use sunflower oil in margarine. Many households were forced by shortages and by rationing to use margarine during World War II. It was sold in blocks that were colored after purchase by the addition of packets of dye by the householder. It was unpleasant tasting and much heavier than today's low-fat margarines.

Before the advent of shortening, lard, butter, or oils were used for cooking. Whereas lard and butter spoil if not properly stored, shortening, made from hydrogenated oils, can be stored at room temperature for long periods of time. Hydrogenation processes, adding hydrogen to vegetable oils, which makes them stable in solid form, were developed near the end of the 19th century and provided a ready use for all the cottonseed oil obtained as a by-product of cotton growing. In 1912, Paul Sabatier, a French chemist, discovered that nickel could be used as a catalyst in the hydrogenation process, making it faster and less costly.

Commercial shortenings became widely available in the United States early in the 20th century and have become a household staple. Crisco Shortening was first marketed in 1911. A yellow shortening, Fluffo, was introduced in 1954, but withdrawn soon after the introduction of butter-flavored Crisco in 1983. With the emphasis on low-cholestoral diets, there has been a greater interest in polyunsaturated oils such as sunflower and safflower oils as well as, more recently, in monosaturated oils such as olive oil.

Meat

As cities developed so did the need for butchers, meat processors, packers, and later markets and stores. Until the development of commercial refrigeration in the mid-19th century, most animal slaughtering was done in the winter.[35] Meat had to be smoked, processed, or dried until the advent of canning and later of refrigeration. The first United States meat-packing plant was founded in 1641 by William Pynchon of Springfield, Massachusetts, to pack pork in salt for use in West Indies plantations. The Union Stockyards opened in Chicago in 1865, and the Armour meat packing factory opened there in 1867. In 1906 the United States Pure Food and Drug Act was triggered in part by Upton Sinclair's *The Jungle*, which exposed the unsanitary conditions in the Chicago stockyards. The developments related to the preservation and transportation of frozen meat are described in the section Refrigeration.

Sausages have been made since ancient times and were popular in the Middle Ages, partly because some types of sausage could be kept longer than fresh meat. Modern consumers continue to eat sausages primarily because of the taste rather than the need for preservation. While there are disagreements as to the attribution of the American hot dog, the frankfurter is credited to a Frankfurt butchers' guild that devised the sausage in 1852. These sausages were sold at Coney Island in 1871 as "dachshund sausages" without buns. The term "hot dog" was coined in 1901 by Hearst sport cartoonist, T. A. Dorgan, who could not spell dachshund. At the 1904 St. Louis Louisiana Purchase Exposition hot dogs were first sold with buns.

Another meat product, Spam, was developed from two canned meat products, Hormel Spiced Ham and Hormel Spiced Luncheon Meat, first sold in 1927 in large cans to butchers, who cut and re-packaged the meat for householders. The popularity of the products led to imitators and to a change in marketing by Hormel. Jay Hormel packaged his spiced pork-shoulder mixture in twelve-ounce cans, changed its name to Spam, and began selling it to consumers in 1937. Spam has a seven-year shelf life, and also provided a use for the heretofore limited usefulness of pork shoulder. It acquired a bad reputation during World War II when any canned ground-pork mixture was called Spam. However, by 1954 it had rebounded and held 41 percent of the canned luncheon meat market and currently claims to hold 75 percent.[36]

The hamburger sandwich was an American invention first sold at Louis' Lunch, a small New Haven, Connecticut restaurant, in 1900 and popularized at the Louisiana Purchase Exposition in 1904. One explanation for the name relates to Jewish immigrants in the 1880s who made patties of shaved dried kosher beef and onion. The beef they used came from Hamburg. Harry H. Holly developed a hamburger patty machine in 1937 and introduced his

Ultimate System patty machine in 1985. His newest machine interweaves the strands of meat leaving air-spaces that trap meat juices and produce a more flavorful and tender patty when cooked.

Although restaurants are discussed more fully later in this chapter, it is worth noting here the tremendous impact the McDonald's chain has had on the popularity and availability of the hamburger. McDonald's did not establish the first chain of hamburger restaurants, but they have been the most successful. (White Castle restaurants, established in 1921, also offered a menu consisting mainly of hamburgers.) The first McDonald's was opened in Pasadena, California, in 1940 by the McDonald brothers, and they developed a self-service restaurant in 1948. The familiar chain was established in 1955 and claims to have sold more hamburgers than any other restaurant. Despite their success selling plain hamburgers, McDonald's has continuously expanded its menu to include chicken, fish, and a variety of different hamburgers. In 1985 McDonald's introduced the McDLT, devised by Will May at a Texas franchise. The McDLT was really a packaging innovation, not a new hamburger: the hot patty was separated from the lettuce and tomato topping so that the lettuce stayed fresh and crisp.

Beverages

Although water supplies were subject to less pollution in earlier times, droughts and difficult terrain have always necessitated, and taste preferences have encouraged, substitutes for water in the human diet. People drank the milk of domestic animals before 6000 B.C. Beer and wine were both in use by 4000 B.C., and a variety of other alcoholic and nonalcoholic beverages have been developed over the years.

Beer was largely a home-made product until recent times. In addition to home production, it was made in monasteries and in local breweries, but it was not until the 19th century that large-scale commercial production and distribution became dominant. Until then each locality had its own type of beer, although beer was imported from Flanders to Winchester, England, in 1400.

Beer lovers have long argued the merits of the best form of storage and containers for beer. Lagers were developed when monks discovered that storing beer for several months in cool caves improved the flavor. Beer was first bottled by Alexander Nowell, dean of St. Paul's, London, in 1586. Easy-carry cans for beer were not introduced until 1935. In 1985, Brassiere du Pecheur, a French company, developed a concentrate for making alcoholic or nonalcoholic beer, to which water was added. Concentrates have been used for over thirty years, but beer made from them has been regarded as inferior in taste.

Figure 3.3. Selected Beverages

YEAR		
1987		New/Classic Coke
	Plastic cans	Nutrasweet, Diet Coke
		Caffeine-free drinks
1975	Push-through tabs	
	Tab-top cans	
	Aluminum cans	Diet sodas, Sprite
1950		
	Canned beer	7-up (Lithiated Lemon)
1900	Milk cartons	Pepsi, Canada Dry
	Milk bottles	Coke, Dr. Pepper
	Wine pasteurization	Hires Herb Tea (Root Beer)
1800	Brandy from wine	Flavored soft drinks
		Carbonated drinks
	Brandy from potatoes	
1600		
	Bottled beer	
1200		
400		
	Arrack (distilled liquor, China)	
-1200	Hops in beer (Egypt)	
-4400	Beer and wine	
	Milk	
-10800		

Wines have long been popular and were a major trade good of both the Greeks and the Romans. The unfermented grape juice could only be kept for a limited time, but once fermented into wine, it could be stored for extended periods and transported long distances.

Recent developments affecting wine include its pasteurization, changes in agricultural methods to combat disease, and new wine drinks. Pasteur developed his sterilization process originally for wine in 1864. Wine growing in Europe was threatened in the 1880s by downy mildew, which was treated with copper sulfate, the first agricultural fumigant. The spread of the American root louse to Europe threatened all European vines in the 1870s through 1900. The eventual solution was to graft the European varieties onto

American root stock that was resistant to the louse. Recent additions to wine beverages include sparkling wines such as Cold Duck and wine coolers. Champagne, a sparkling wine, was produced using natural fermentation processes in France near the end of the 17th century. The modern sparkling wines are produced through carbonation, not natural fermentation. Wine coolers, introduced in 1981, have become a popular beverage. Such coolers combine fruit juice and other flavors to produce a flavored wine drink with a lower alcohol content.

Strong liquors are generally created by distillation, and the use and development of these liquors paralleled the development of distillation processes. In 800 B.C. the Chinese developed the first liquor, arrack, distilled from beer. European liquors were developed somewhat later. The Italians produced brandy from grapes around 1000 A.D., using primitive techniques. Improved distillation methods were introduced by the Arabs in the 10th century. By 1748 brandy was being distilled from potatoes, and Edouard Adam invented an apparatus for distilling brandy from wine in 1801. Brandy has been made from wine, apples, potatoes, plums, raspberries, strawberries, peaches, cherries, and blackberries. Peaches were also used in Peachtree Schnapps introduced in 1985.

Originally developed to imitate naturally carbonated waters found in natural springs, soft drinks have been in use for some 200 years. Artificial mineral water was prepared by Dr. W. Brownrigg with carbonic acid gas in 1741, and soda water was made by Richard Bewley in 1767. Manufacture of soda water began in 1785 in London. The first fruit-flavored carbonated soft drinks were made by Townsend Speakman, Philadelphia, in 1807, and Elias Durand commercially bottled soda water in Philadelphia in 1835.

Charles Hires, a Philadelphia pharmacist, investigated a New Jersey herb tea made from sarsaparilla root in 1870. It had no medical value, but Hires liked the taste and in 1876 began selling it in a dry powder as Hire's Herb Tea. He changed it to a liquid concentrate in 1880 and to a liquid bottled form by 1893. It was called Hires Herb Tea until Dr. Russell Conwell suggested it would sell better as Hires Root Beer.

Other popular soft drinks include colas such as Coca-Cola, Pepsi-Cola, and Dr. Pepper; ginger ale; and lemon soda. Coca-Cola was invented by Dr. James Pemberton of Atlanta, Georgia, in 1886, and Asa G. Chandler bought rights to the syrup recipe in 1892. The Coca-Cola trademark was registered in 1893, and it was first bottled by Joseph Biedenham, Vicksburg, Mississippi, in 1894. Dr. Pepper was also developed in 1886 and Pepsi Cola dates from 1898. Canada Dry Ginger Ale appeared in 1904 and the English soft drink Lithiated Lemon was promoted as a mixer drink, 7-up, in the United States in 1933. It was not until 1961 that Coca-Cola introduced Sprite, its lemon-lime drink, to compete with 7-up.

A new formulation for Coke was introduced in 1985, but the old formulation was soon reintroduced as Classic Coke after the strong protest of loyal Coke drinkers. Cherry Coke also debuted in 1985 to compete with Dr. Pepper. There was some industry speculation, after the collapse of the vanilla bean market following the introduction of the new Coke, that vanilla had been a key ingredient of the closely guarded formula of old Coke.[37]

Diet soft drinks began with NoCal, developed by Kirsch Beverages of New York in 1952, and were popularized ten years later by Diet Rite Cola, quickly followed by Tab and Diet Pepsi in 1963. Diet Coke finally appeared in 1982, and its formula was switched from saccharin as a sweetener to Nutrasweet (aspartame) in 1983.

The health-conscious 1970s and 1980s brought a revival of interest in sodas from natural springs or using natural flavors. Perrier was the in drink of the 1970s. The American Natural Beverage Company, Brooklyn, New York, was established in 1978 to market natural sodas made with sparkling water and natural fruit flavors. In 1985, Canada Dry added citrus flavors to their seltzer water, and Perrier did the same. Pepsi and Coke began marketing new drinks containing small amounts of real fruit juices. Despite the increased numbers of these alternatives to artificially flavored soft drinks, natural sodas currently account for 1 percent of the $26 billion soft drink industry.[38]

Caffeine-free soft drinks also became popular in the 1980s in response to reports on the effects of caffeine on health. Seven-up touted its lack of caffeine, forcing Coke and Pepsi to consider caffeine-free drinks. Pepsi introduced Pepsi Free in 1982 and Coke followed in 1983 with Caffeine-Free Coke.

In order to offset eroding market share, partly owing to the increasing popularity of soft drinks, research is being pursued for new ways to use milk. The Dairy Research Branch of the United Dairy Industry Associates has developed a method for carbonating milk and anticipates introduction of the new beverage in 1989 as a healthful alternative to colas.[39]

Iced tea was invented by Richard Blechynden, an Englishman, at the 1904 Louisiana Purchase Exposition, St. Louis, Missouri. The weather was hot and few fairgoers wanted hot tea. The iced drink became an instant seller. More recently, Constant Comment Tea, tea mixed with a blend of spices and orange peel, was first sold in 1945. By 1986 Lipton had also begun to offer herbal and spiced teas. Instant tea appeared in 1948 some ten years later than instant coffee.

Fruit juices have also been popular beverages. Welch's Grape Juice, originally know as Dr. Welch's Unfermented Wine, was introduced for general use at the Chicago World's Fair in 1893. Instant orange juice became available in 1962. Ocean Spray introduced Cranapple juice in 1965, the same year, Gatorade, a rehydrating beverage for runners and other athletes, was formulated by a team of urologists from the University of Florida.

Seasonings and Additives

Humans have used a variety of condiments and spices to make food more palatable. Popular condiments include mustard, Worcestershire sauce, Tabasco sauce, and catsup. Mustard was first sold in paste form in London in 1720 by Mrs. Clements of Durham. Previously only seeds had been available. English mustard is made from the milder white mustard seeds while European mustards are made from stronger black mustard seeds. Worcestershire sauce was first made by John Lea and William Perrins in 1837 according to the recipe of Sir Marcus Sandys. Edmund McIlhenny began distributing Tabasco sauce made from tabasco peppers in 1868. H. J. Heinz, who first made quality horseradish, began making catsup in 1876. Catsup originated in the Orient and had been imported from Malay to England about a hundred years earlier. In the Orient it was a spicy sauce, but Americans made theirs with tomatoes sometime before 1792 when cookbooks of the period contained tomato-based catsup recipes. Heinz was also the first to market barbecue sauce nationally in 1948. In 1987, K. C. Masterpiece barbecue sauce, created by Kansas City psychiatrist Dr. Rich Davis, began national marketing.

Mayonnaise was encountered by Richlieu on Minorca in the early 17th century and introduced by him to France. French immigrants brought it to the United States in the early 1800s. Originally made at home or in restaurants, Richard Hellman of New York sold one pound containers in 1912 from his deli, and began offering bottled dressing in 1913.

Food additives may be used to preserve foods, to facilitate processing, or to enhance their appearance, taste, or smell. Additives include natural ingredients such as sugar, salt, spices, leavening agents, vinegar, or wine, as well as chemical agents that are manufactured or derived from natural materials. Chemical food additives evolved along with the evolution of the chemical sciences. Monosodium glutamate (MSG) was first prepared in a German laboratory in 1867, but its flavor enhancing properties were not discovered until 1908 by a Japanese chemist.

During the 19th century artificial sweeteners were also investigated. Saccharin, a chemically produced sweetener, was discovered by Constantine Fahlberg and Ira Remsen, Baltimore, in 1879 and was used as a commercial sweetener as early as 1910. Cyclamate was first marketed as a nonnutritive sweetener in 1949. In 1968 the U.S. Food and Drug Administration (FDA) invoked the 1958 Delaney Amendment and banned the use of cyclamate in food and beverages as a potential cancer-causing agent. By 1970 cyclamate was taken off the United States market. The FDA considered a similar ban on saccharin in 1977, but Congress acted to prevent a ban. This threat was a strong impetus to increasing research on, and to push the testing of, alternative sweeteners.

A G. D. Searle scientist discovered a new sweetener, aspartame, in 1965 while working on an ulcer drug. Aspartame is technically not an artificial sweetener since part of its constituents are natural proteins. After ten years of extensive testing, Searle began marketing Equal, a Nutrasweet sweetener, patented under the generic name aspartame, in 1981 after obtaining FDA approval for use in food and dry beverage mixes. By 1982 General Foods had introduced it into a number of their products, and it was being used in cold milk mixes and in soft drinks by early 1983 after further FDA approval for such uses.

Searle, alarmed by adverse consumer reaction to Nutrasweet-saccharin blends in soft drinks, had offered special discounts to major soft drink companies. A suit by the New York State Attorney General's Office against Royal Crown Cola for misleading advertisement in not mentioning the continued use of saccharin in its soft drinks also helped the change to full Nutraweet use.[40] By 1984 it was being used in Diet Coke and Diet Pepsi with both firms touting their 100 percent Nutrasweet formulations.

L-sugars, left-handed sugars, were discovered by George Lewis in 1980. They taste like sugar but are not metabolized by the body. Production methods are still too costly to compete with other artificial sweeteners. Another sweetener that has already been approved in Great Britain is Acesulfame K.

Debate continues over what types of food additives should be allowed. Even at the turn of the century, concern about food additives led Dr. Harvey W. Wiley to conduct a number of experiments and to condemn the use of saccharin. His work, along with Upton Sinclair's The Jungle, led to the Pure Food and Drug Act of 1906. More recently, Dr. John Olney in 1969 urged a ban on MSG in baby food when his experiments with mice showed brain damage. Baby food manufacturers voluntarily removed MSG from their products. We have been urged not to use meat products such as bacon and ham that contain nitrates and nitrites.

Many substances, natural or manufactured, may be safe in small doses or when properly processed or cooked, but can be dangerous in large amounts or when improperly used. The variety of conflicting research results and the wording of various legislation reflects basic uncertainties about where to draw the line. For example, while cyclamates have been banned in the United States, they are used in forty other countries. The scientific evidence appears to be mixed, and cyclamate was proposed in 1984 by Abbott Laboratories for reconsideration by the FDA for use in the United States. Similarly, the FDA proposed a ban on saccharin in 1977, but Congress acted to prevent such a ban. Some scientists have suggested aspartame should also be banned despite extensive tests that so far have not shown harmful effects. In 1985 the FDA proposed banning the use of sulfite to preserve fresh fruits and vegetables used in salad bars.

Health and well-being concern most people. In the publicity over potentially harmful effects of chemical substances, people often overlook that almost any substance in excess can be harmful, even water. Controversies over caffeine, vitamin C, and vitamin E have been well publicized. Present regulations ensure testing of new substances before use, and are more stringent than those applied in the past; however, long-term effects may be difficult to measure and assess based on tests on laboratory animals. No company wants, or can afford, the product liability suits that have faced drug and manufacturing firms; consequently company and government officials are conscious of the need for stringent testing. Additives are important to the quality and healthfulness of the foods we eat. Without the use of additives many foods would be unpalatable, costlier to process and purchase, or subject to rapid spoilage.

Candies and Selected Desserts

The remains of a candy-maker's tools found in the ruins of Herculaneum, 79 A.D., show that early people developed a liking for sweets. Until fairly modern times, sugar was rare and costly, so most sweets were made with honey.

Chocolate candies had to await the discovery of the New World and of chocolate. Initially used for a beverage, chocolate was not readily available for eating until produced on a factory scale by François-Louis Cailler, Vevey, Switzerland, in 1819. Fry and Sons, England, made a semisweet, dark eating chocolate in 1847.

Many of today's popular candy bars use milk chocolate. It was not until 1875 that Daniel Peter, Switzerland, Cailler's son-in-law, invented milk chocolate. Milton Hershey used milk chocolate to create the Hershey Bar in 1894, which was followed in 1896 by Leonard Heischfield's Tootsie Roll, named for his daughter. That traditional Southern favorite, Moon Pies, appeared in Chattanooga in 1919.

The 1920s and 1930s saw a number of now familiar candies introduced, including Otto Schnering's Baby Ruth and Peter Paul's Mounds in 1921, Schnering's Butterfingers in 1922, Franklin Mars' Milky Way in 1928, Snickers in 1930, and Three Musketeers in 1932. M&M's were developed in 1940 and Peanut M&M's appeared in 1954. Almond Joy was marketed in 1942.

Two popular desserts developed in the 19th century were Jell-O and Minute Tapioca. Peter Cooper devised a sweetened, flavored gelatin around 1840, but it required the labor of making gelatin first and was not popular.

Charles Knox invented packaged gelatin in 1890, eliminating the time-consuming task of making gelatin from calves hooves. In 1891 May Wait further modified gelatin by adding sugar and flavorings to produce Jell-O, which, making use of Knox's packaged gelatin, was more favorably received than Cooper's earlier version. By 1900 Jell-O was being marketed nationally.

Minute Tapioca was created in 1894 by Susan Stavors after complaints by a sailor boarding with her about her lumpy pudding. She used a coffee grinder to process the tapioca.

Snacks

Snack foods vary with geographic location and culture. In Western society they include potato chips, Cracker Jack, popcorn, peanut butter, packaged individual cakes, and pizza, to name only a few. Potato chips evolved from Saratoga Chips, devised by chef George Crumb of the Moon Lake House Hotel, Saratoga Springs, New York, in 1853 and were commercially manufactured in Albany, New York in 1925. Cornuts, toasted corn snack food, were first made in 1934. That perennial favorite of many children, Cracker Jack, caramel coated popcorn mixed with peanuts, was devised in 1871 by F. W. Rueckhein, Chicago. Prizes were added to boxes of the snack in 1912 and Jack the Sailor and his dog Bingo were added to packages in 1919. Borden purchased Cracker Jack in 1964.

Peanut butter has been used for only a little over a hundred years and has been in general use for only fifty years. Peanut butter was first devised by a St. Louis doctor in 1880 for a patient who needed protein but could not tolerate meat. Ambrose Straub of St. Louis patented a peanut butter making machine in 1903 and the same year Dr. John Kellogg of Battle Creek, Michigan, began making it in his sanatorium. By 1937 peanut butter had moved from the health food store into more general use, including in cookies and candies.

Few people outside Australia understand the Australian love for Vegemite, that salty, dark brown substance that could easily be mistaken by the unwary for axle grease. It was developed by Fred Walker in 1908. He made food supplies that would not spoil in the hot, arid outback settlements of Australia. Vegemite is made from a yeast slurry, a brewery waste product, combined with water, salt, onion, and celery flavorings and concentrated into a stiff dark brown spread. Kraft purchased an interest in Walker's company in 1926 and later hired J. Walter Thompson to promote it. It has since become an essential on almost every Australian table. Most non-Australians are unfamiliar with the proper method for eating this spread. Ideally it should be spread in a thin film on hot buttered toast or buttered saltines. A little goes a long way.

Mass-produced chewing gums are a product of the late 19th century, but gum chewing is much older. The Greeks chewed a resin, the Mayans chewed chicle a thousand years ago, and the American Indians chewed spruce gum. The first commercially produced chewing gum was State of Maine Pure Spruce Gum manufactured by John Curtis in 1848. Curtis made it in his kitchen on his Franklin stove. Chicle-based gums were developed in 1870 by Thomas Adams, a Staten Island photographer, whose Black Jack gum was the first mass-produced flavored gum. Gums were developed for purposes other than idle chewing, including health and to hide alcohol breath. A pharmacist created Beeman's for treating heartburn and it was manufactured in Merton, England, in 1894; Clove gum was developed and marketed just at the end of Prohibition to disguise alcohol indulgence. These three gums were discontinued in the mid-seventies, but reintroduced in 1986.

Snack cakes have been around for some time. Twinkies were devised by James Dewar in 1930 and named after a brand of shoes. Little Debbie Snack Cakes were introduced by the McKee Baking Company in 1960.

Pizza may have originated from *pieca*, which appeared in Naples in 1000 A.D. In 1889 Raphelle Exposito devised the modern pizza as Pizza de Margherita in honor of Queen Margherita. Pizza was brought to the United States by Italian immigrants, and the first United States pizzeria opened on Spring Street in New York City in 1895. By 1958 Frank and Dan Carrey had established the first Pizza Hut restaurant in Witchita. Frozen pizza was first marketed in 1974.

Frozen dessert and juice bars have proliferated since their inception. General Foods introduced Jell-O Pudding Pops in 1979. Dole began marketing its frozen, premium priced, Fruit 'n' Juice bars in 1984 and Coca-Cola followed with the Minute Maid Fruit Juicee in 1986. The Minute Maid bar was formerly marketed by Nutri-Foods as Guido's Ice Juicee, but was purchased by Coca-Cola in 1985. The Juicee bars are sold in a unique wrapper, a paper tetrahedron similar to the smaller ones used for individual servings of coffee creamer. The top is ripped off and the bar pushed up from the bottom, thereby eliminating the dripping of the melting bar. The packaging reflects Coke's focus on children for the flavors and parents for the no-mess packaging.

Packaging and Legislation

Although packaging is important in many respects, this review will focus on the effects of packaging on the availability and form of products. The developments listed below are representative, not exhaustive.

Paper was developed by the 8th century, but was too valuable for extended use as a packaging material until manufacturing developments in the 18th and 19th centuries made inexpensive paper available. Paper was used for bags and wrapping by the 1860s. Smith Brothers Cough Drops were the first United States product sold in factory-sealed paper packets.

Although glass bottles were made in Roman times, mass-production techniques were not applied until the late 19th century. M. J. Owens patented his automatic bottle making machine in 1895 and began bottle production with a fully automatic machine in 1903, lowering production costs and cutting wages. In 1925 a new glass bottle manufacturing machine, more efficient than the earlier Owens machine, was introduced.

A number of packaging advances were applied to milk and beverage containers. Milk bottles were introduced by Echo Farms Dairy Company, New York, in 1879, and milk cartons appeared in San Francisco in 1906. The pry-off bottle cap was patented in 1892, and aluminum foil bottle caps for milk were used in 1914. Waxed cardboard cartons and plastic jugs for milk have replaced glass bottles in most United States cities. Aluminum cans for soft drinks were introduced by Alcoa in 1960. They added tab-top opening in 1963 and the push through tab in 1973. Such cans, particularly those with the pull-off tops, have been used for a variety of products, including deodorant and even clothing.

The collapsible tube, originally made of metal, was patented by John Rand, an American artist, in 1841. Its initial use was for artists' pigments, but it was later adapted for toothpaste, cosmetics, and foodstuffs and is used in plastic form for packaging cake frostings. Dr. Washington Sheffield, a dentist of New London, Connecticut, used it for toothpaste in 1892 as did Beecham's Tooth Paste of Britain. Polyethylene, which is used for some collapsible tubes, was produced by the Imperial Chemical Industries (ICI) in 1933; manufacture began in 1939. The first collapsible polyethylene tube was made in 1953, originally for Sea & Ski Suntan Lotion. Plastic and metal tubes are used for a variety of products.

While loose tea in tins or packets is favored by the British, Americans seem to prefer tea bags. Tea bags were not developed until the 20th century; they were introduced in 1904 by Thomas Sullivan, New York, when he sent his customers samples of tea in silk bags. By 1920 Joseph Krieger, San Francisco, was selling tea in bags to commercial establishments. Lipton's bought the rights to a tea bag manufacturing machine to produce its "flo-thru" bag in 1945.

Most famous products have adopted new packaging as the techniques became economically attractive or offered some perceived value to consumers. Coke has used glass bottles, steel cans, aluminum cans, plastic bottles, and most recently is trying out plastic cans. Alexander Samuelson of the Root Glass Company, Terra Haute, Indiana, designed the famous wasp-waisted Coke bottle and patented the design in 1915. Coke paid $500 for Samuelson's design. In 1985 Coca-Cola began test marketing Coke in see-through plastic cans in Columbus, Georgia. The cans have plastic tops and were developed and used in Europe.

Cellophane and plastic film are used extensively for packaging food products. Celluloid was patented by James Hyatt, Albany, New York, in 1869; cellophane was patented in 1908 and manufacture began in 1913. Moisture-proof cellophane was invented in 1926.

Aerosol spray containers were first developed in the 1920s, but not commercially applied until 1941. Erik Ruthein of Norway invented an aerosol container in 1926, which was followed in 1939 by a disposable aerosol spray can invented by Julian Kahn of New York. Aerosol cans were used by Lyle David Goodhue in 1941 for spraying insecticides. These containers were later adapted to a variety of products. The development of a low-cost plastic aerosol valve in 1953 significantly extended their application. Debates have surfaced over the impact of the release of Freon into the atmosphere and its potential destruction of the protective ozone layer. These concerns have led to changes in aerosol propellants and to the introduction and broader use of pump dispensers.

Microwave ovens have imposed new constraints and provided new opportunities for packaging foods. The Micro Match Packaging System developed by Alcan Research Laboratories in 1986 enables food processors to engineer different heat levels for various portions of complete meals. The packaging ensure even heating and browning of crusts. Reynolds Metals Company created Container-Mate aluminum foil closures for use with retortable plastic containers developed by Continental Can Company in 1986 (retortable packages can withstand the high temperature processing used in retorts to sterilize the container and its contents). Campbell Soup Company is using this combination to package its Cookbook Classics: single serving, table-ready, microwave premium soups. The Container-Mate foil lid is peeled off prior to placing the soup in the microwave. The soups do not require refrigeration or freezing. Campbell also announced plans to use plastic containers for its Swanson frozen dinners.

Another new packaging innovation is Tug-n-Tie, a plastic bag with its own reclosure tie as part of the package seal. Removing the tie opens the package, which can then be resealed by twisting the tie around the bag top. The removable tie also eliminated nicks and tears in opening the bag and makes the package easier to open, a growing consideration for those companies that sell products for use by the elderly or handicapped. Tug-n-Tie was developed by the Boston Core Group and has been introduced for Dutch Maid Noodles by the Prince Company of Lowell, Massachusetts, in 1985.

Packaging food for consumption in space posed many new problems. Food had to be able to withstand heat, vibrations, radiation, zero gravity, and a pure oxygen atmosphere, while still satisfying human tastes. Early space flights used pureed foods, but later foods of the consistency to be eaten with a spoon were found satisfactory. Because of concern over the disposal of waste products low residue foods were used initially, but as waste disposal methods improved this was modified. Foods included roast beef, sausage, gravy, fruit, cheese, puddings, cereals, fruit juices, coffee, and tea. Freeze-dried food in plastic bags for rehydration with a water gun were also used. By using hot water, warm foods could be served.

Some foods required special treatment or packaging. Cookies had to be treated to prevent the cookie crumbs from dispersing and causing problems with instruments and filtration systems. Coating them with methyl cellulose reduced the crumb problem. Coke introduced a specially designed soft drink can for use by astronauts in 1985. The new design retains carbonation until the beverage is consumed.

Skylab included a freezer, but the space shuttle does not carry a refrigerator. Foods for the shuttle are dehydrated, irradiated, or thermostabilized so refrigeration is unnecessary. The shuttle carries an oven so foods can be heated easily. While dehydrated foods are ideal because of their light weight and small volume, taste and variety considerations have encouraged the inclusion of other food forms.

Just as cookies were coated to prevent them from crumbling in space, other foods were also processed into forms that were easier to handle and transport. Sugar was processed into cubes for more convenient handling. Sugar cubes were first produced in Great Britain by Henry Tate, Silvertown Refinery, London, in 1878. The cubes were so successful that Tate was able to found the Tate Gallery with the fortune he made from sugar cubes.

The Tylenol poisonings of 1982 and the most recent rash of packaging tamperings in 1986 have caused drug companies and others to review their packaging, and have raised consumer awareness of packaging safety. The flurry over alleged glass fragments in Gerber baby food, and the threatened tampering with Sugar-free Jell-O, and Slice soft drinks are recent examples. While there was no evidence of tampering in any of these cases, the potential for harm still remains. Companies have not found any methods to completely avoid potential tampering, despite added plastic wraps and extra closures, and have alerted consumers to watch for product tampering and contamination.[41]

One packaging advance related not to the preservation of food or to attracting buyers, but to inventory control and potentially quicker check-out processing. Bar coding was first used by United States railroads in 1967 for identifying railroad cars, while the Universal Product Code (UPC), which used a bar-coded representation for product identification, was introduced in 1973 for grocery products. With the introduction of supermarket computerized scanners in 1974 that read the product's price by identifying the bar code, UPC coding has become essential for most supermarket check-out systems. Supermarkets with appropriate systems can track product purchases and reorder or restock needed items.

Food has been subject to laws regulating its content and taxing those handling or selling it for much of human history. Fines for selling adulterated grains or oils were imposed by a 300 B.C. Sanskrit law, and Chinese records dating from the 2d century B.C. describe government officials responsible for ensuring food purity. The Chinese also imposed a tax on tea by 780 A.D.

During the Middle Ages, bread, meat, and spices, as well as other foodstuffs, were subject to adulteration or additives to increase the apparent quantity at the expense of quality, and sometimes the health of the consumer. King John of England proclaimed the Assize of Bread in 1202 prohibiting the addition of ground peas and beans to bread, while a 1266 law prohibited short weights in bread or the sale of bad meat. Spices were in great demand during the Middle Ages and often subject to extension by the addition of cheaper substances.

By the latter part of the 19th century adulteration of foods was perceived as a serious problem by many legislators and citizens.[42] In 1820 Frederick C. Accum, a German chemist and pharmacist of London, published *A Treatise on Adulteration of Food and Culinary Poisons*. After further investigation by Dr. Arthur H. Hassell in the mid-19th century, Parliament finally enacted weak legislation in 1875. Most European countries had also enacted legislation by the end of the century.[43]

American food legislation was first enacted at the state level near the end of the 18th century and at the federal level almost one hundred years later. The first American food law was passed in Massachusetts in 1784, and a California statute was enacted in 1850. The first federal food law--the Tea Act of 1883--prohibited the importation of spurious and adulterated teas. It was replaced by another Tea Act in 1897.

Dairy products and dairy substitutes were subject to a number of laws. The Oleomargarine Act was passed in 1886, taxing and regulating the manufacture, sale, and importation of margarine. The tax remained in effect until 1950. The Filled Cheese Act (1896) also taxed cheese and regulated the manufacture, sale, import, and export of "filled" cheeses, those made with milk, skim milk, added butter, animal fats, vegetable oil, or any combination of these. The Renovated or Process Butter Act of 1902 imposed a tax and prohibited unsanitary materials. The Filled Milk Act (1923) prohibited interstate shipment of milk, skim milk, or cream to which fat or oil other than milk fat was introduced. Certain infants' and children's foods were exempted.

The Meat Inspection Act (1890) provided for inspection of pork, bacon, and exported live animals. It was followed by the 1891 Cattle Inspection Act.

There was also the 1913 Imported Meat Act, the 1919 Horse Meat Act, and the 1967 Wholesome Meat Act.

In 1906, owing to the efforts of many individuals and groups--including Upton Sinclair and Harvey W. Wiley--the 1906 Food and Drug Act was passed as was a Meat Inspection Act. The McNary-Mapes Amendment to the 1906 Food and Drug Act was enacted in 1930 authorizing the establishment of standards for canned foods and labeling requirements. The first such standard was issued in 1939 for tomato products. Another amendment, passed in 1934, dealt with the inspection of seafood. In 1938 the Food, Drug, and Cosmetics Act replaced the 1906 act, incorporating the original legislation and its amendments. There were four amendments to the new 1938 act. They include: the 1954 Miller Pesticide Chemical Act, the 1958 Food Additives Amendment, the 1960 Color Additives Amendment, and the 1974 National Heart and Lung Authorization Bill. The 1958 amendment, also known as the Delaney Amendment, prohibits the use of carcinogenics. It was invoked for banning cyclamates in 1968 and would have been invoked for saccharin in 1977 had not congressional action overruled the FDA. The 1960 amendment resulted in the banning of several colored dyes: Violet No. 1 (used mainly for meat marking) in April 1973, Red No. 2 (previously used in some candies) in 1976, and later in 1976 Red No. 4 and Carbon Black. No. 4 had been used in Maraschino cherries and Carbon Black in a variety of cosmetics.

Poultry products have been subject to separate legislation not included in the 1938 act. The Poultry Inspection Act of 1957 covered foreign and interstate commerce; the Wholesale Poultry Products Act of 1968 dealt with state inspection; the 1970 Egg Products Inspection Law was further strengthened by the regulation of egg products in 1971 and packaged shell eggs in 1972.

The Fair Packaging and Labeling Act (1966) required the FDA to publish regulations related to packaging. Regulations include such requirements as portion size and number, calories per serving, grams of carbohydrates and proteins, and the percentage of United States Recommended Daily Allowances (RDA) of proteins, selected vitamins, and minerals.

These laws are intended to ensure the quality and freshness of foods, the standards used in grading, processing, and labeling as well as their consistent application. They protect the consumer, the distributor, the manufacturer, and the retailer.

Table 3.2. Food Processing, Manufacture, and Packaging

Dates	*Event*
B.C.	
4000	Beer brewed in the Near East (BBB).
	Wine made in Near East (BBB).
	Sumerians made cheese in Mesopotamia (RM).
3500	Cakes made from coarse ground grain used in Swiss lake dwellings (EB).
2600	Leavened bread made in Egypt (BBB).
2500	Egyptian records mentioned wine making (BBB).
2100	Sumerian physicians prescribed beer for illness (BBB).
2000	Egyptian and Chaldean cheese molds used (RM).
<1200	Hindu Vedas mention use of butter as a food (EB).
1000	Hops used by the Egyptians to flavor beer (BBB).
800	Chinese developed arrack from the distillation of beer (BBB).
7th c	Assyrians, Mesopotamia, imported spices from India and Persia (RM).
300	Sanskrit law imposed fines for sale of adulterated grains (MMGT).
2d c	Chinese officials responsible for the purity of foods described in Chinese classics (MMGT).
170	Bread first made by bakers of ancient Rome for public sale (DID).
A.D.	
79	Candymaker's tools left in ruins of Herculaneum (BBB).
750	Hops used as beerwort for first time in Bavaria (TH).
9th c	Brie cheese developed in France and served at court of Charlemagne (SKY).
960	Mariolles cheese produced at Abbey of Thierache, France (RM).
1000	Italians produced brandy from wine (EB).
	Naples, Italy, was the birthplace of pieca, forerunner of the pizza (RD).
1202	Assize of Bread, which prohibited peas or beans in bread, was proclaimed (MMGT).
>1300	Bernardo Buontalenti invented ice cream (EDE).
1359	Herring first cured in England (DID).
1400	Beer was imported to Winchester from Flanders (1ST).
1425	Hops first mentioned in England (DID).
1530	Bottle corks mentioned in Palsgrave's French-English dictionary (1ST).

1568   Bottled beer invented by Alexander Nowell, dean of St. Paul's, London (TH).

1573   First German cane sugar refinery began operation at Augsburg (TH).

1587   Marmalade recorded in accounts of Wollston Hall, England (1ST).

First beer brewed by Europeans in the New World made in the Roanoke Colony, Virginia (HS).

1630   English colonist sampled popcorn at first Thanksgiving (WG).

1638   Kopke, a German family, began shipping port from Oporto, Portugal (WJ).

1641   William Pynchon, Springfield, Massachusetts, founded the first U.S. meatpacking plant (WB).

1647   Captain Brocas produced the first commercially successful U.S. wine (HS).

1666   First Cheddar cheese made (TH).

1670   Parisian Café Procope introduced ice cream to the general populace (RM).

Hudson Bay Company granted a charter (EB).

1677   Ice cream was a popular dessert in Paris (TH).

1686   Ice cream bought for James II was recorded in accounts of the Lord Steward (1ST).

1730   Stilton cheese first served at the Bell Hotel, Stilton, England (RM).

1741   Artificial mineral water prepared by Dr. W. Brownrigg using carbonic acid gas (1ST).

1748   Brandy first distilled from potatoes (DID).

1767   Soda water prepared by Richard Bewley (1ST).

>1780   Camembert cheese first produced near Vimoutiers, Normandy (RM).

1784   Massachusetts law penalized seller of diseased, corrupted, or unwholesome provisions (MMGT).

1785   Soda water manufactured by H. D. Rawlings, London (1ST).

1786   Hall, New York, opened the first ice cream company in the United States (RM).

1787   Thomas Jefferson introduced spaghetti to the United States (HS).

1792   Richard Bragg's *The New Art of Cookery* contained a recipe for tomato catsup (EOOET).

>1800   Trappist monks of Notre Dame de Port du Salut manufactured Port du Salut cheese (RM).

1801	Edouard Adam, France, invented an apparatus for distilling brandy from wine (DID).
	Biscuits were called crackers after Josiah Bent opened his bakery, Milton, Massachusetts (HS).
1807	Fruit flavored carbonated soft drinks were manufactured by Townsend Speakman, Philadelphia (1ST).
	Benjamin Silliman, Yale College, began producing bottled soda water commercially (EB).
1808	Pineapple cheese first produced in Troy, Pennsylvania (RM).
1809	Joseph Hawkins patented a method for preparing imitation mineral waters (EB).
1810	Lewis Mills Norton, Troy, Pennsylvania, patented his process for making pineapple cheese (FFF).
1814	John Linebach developed a cotton seed hulling machine (WB).
1815	R. V. Effinger established a cheese factory at Bern, Switzerland (1ST).
1819	First factory-scale manufacture of eating chocolate in Vevey, Switzerland, by François-Louis Cailler (1ST).
1820	Frederick C. Accum, German chemist of London, published *A Treatise on Adulteration of Foods . . .* (MMGT).
1826	Tea sold in packets by John Horniman, Ryde, Isle of Wight (1ST).
1828	Chili con carne devised by Mexicans living in the area that later became Texas (HS).
1829	D. G. Yuengling & Son of Pottsville, Pennsylvania, oldest U.S. brewing company, was established (HS).
>1830	Reverend Sylvester W. Graham devised graham crackers as a health food (HS).
1835	Elias Durand, Philadelphia, Pennsylvania, commercially bottled soda water (FFF).
1837	John Lea and William Perrins, pharmacists, Worcester, England, made Sir Marcus Sandys' Worcester Sauce (1ST).
>1840	Peter Cooper devised first sweetened, flavored gelatin, but it was not successful (see also 1891) (HS).
1841	Collapsible tube patented by John Rand, American artist (1ST).
1845	Self-raising flour marketed by Henry Jones, Bristol, England (1ST).
1847	Ring doughnuts introduced by Captain Hanson Gregory, Camden, Maine (1ST).

	Fry and Sons, England, produced semisweet, solid eating chocolate (BBB).

Fry and Sons, England, produced semisweet, solid eating
    chocolate (BBB).

Evaporated milk made for the first time (TH).

1848   John Curtis, Bangor, Maine, developed State of Maine Pure
    Spruce Gum for chewing (1ST).

1850   California food and drug act enacted (MMGT).

>1850   Dr. Arthur H. Hanell studied food adulteration (MMGT).

Baking powder first became available (HS).

1851   James Russell, Baltimore, introduced mass-produced ice
    cream (1ST).

First large U.S. cheese factory founded by Jesse William,
    Rome, New York, to produce cheddar cheese (RM).

1852   German butchers' guild devised hot dog in Frankfurt (BW).

1853   First potato chips prepared at Moon Lake House Hotel,
    Saratoga Springs, New York, by George Crumb (1ST).

1856   Aerated bread invented by Dr. Dauglish (DID).

Butter factory was established by R. S. Woodhull, Campbell
    Hall, Orange County, New York (1ST).

1863   Viennese devised the croissant to celebrate their defeat of
    the Ottoman Turks (BBB).

1865   Union Stockyards opened in Chicago (TH).

1866   Parkesine, nitrocellulose-based plastic (like celluloid),
    produced by Alexander Parkes, Birmingham (1ST).

1867   Armour meatpacking factory opened in Chicago (TH).

Monosodium glutamate prepared in German laboratory
    (MB).

Arm and Hammer Baking Soda introduced (HS).

1868   McIlhenny Co., Avery Island, Louisana, introduced Tabasco
    Sauce (EP).

1869   Celluloid patented by James W. Hyatt, Albany, New York
    (1ST).

Margarine patented by Hyppolyte Mege-Mouries, Paris
    (1ST).

1870   Thomas Adams, Staten Island, developed Black Jack, the
    first flavored, chicle-based chewing gum (1ST).

Charles Hires, Philadelphia pharmacist, tasted herb tea and
    acquired the recipe for Hires Herb Tea (UNITD).

1871   First margarine factory began operation at Oss, Holland
    (1ST).

F. W. Rueckheim, Chicago, developed Cracker Jack snack
    food (CDIS).

First hot dog, called a dachshund sausage, was sold at
    Coney Island, but without a bun (HS).

1872   William and Andrew Smith packaged Smith Brothers's
        Cough Drops in the first factory sealed packages (HS).
1874   Robert Green, Philadelphia, invented the ice cream soda
        (RM).
1875   Daniel Peter, Switzerland, developed milk chocolate (1ST).
        British Parliament enacted food legislation (MMGT).
1876   Hires Herb Tea (later Root Beer) was sold nationally in dry
        concentrate form (UNITD).
        H. J. Heinz sold catsup nationally (RM).
        Gustave de Laval, Sweden, patented the cream separator
        (WABI).
1877   John Jossi, Wisconsin, invented American Brick cheese
        (RM).
1878   Sugar cubes were produced in Great Britain by Henry Tate,
        Silvertown Refinery, London (1ST).
1879   Saccharin discovery reported by Constantine Fahlberg and
        Ira Remsen, Johns Hopkins University, Baltimore
        (1ST).
        Milk bottles introduced by Echo Farms Dairy Company,
        New York, from bottle devised by Louis Whiteman
        (1ST).
1880   Hires Herb Tea sold as a liquid concentrate (later Hires
        Root Beer) (UNITD).
>1880   St. Louis physician developed peanut butter for patient
        intolerant of other forms of protein (RM).
        Jewish immigrants ate Kosher dried beef shaved and mixed
        with onions to form hamburg patties (HS).
1883   First U.S. Tea Act passed (MMGT).
1884   Evaporated milk patented by John Meyenberg, St. Louis,
        Missouri (1ST).
1885   Evaporated milk manufactured by John Meyenberg's
        Helvetia Milk Condensing Company, Highland, Illinois
        (1ST).
        First cream crackers manufactured by William Jacob,
        Dublin (1ST).
        Exchange Buffet, first self-service cafeteria, opened in New
        York (1ST).
        Moxie soft drink was introduced initially as a nerve tonic in
        the northwestern U.S. (WJ).
1886   Coca-Cola invented by Dr. James Pemberton, Atlanta,
        Georgia (1ST).
        Dr. Pepper (soft drink) invented (BBB).
        U.S. Oleomargarine Act passed (MMGT).

1887   Malted milk introduced by Horlick's, Racine, Wisconsin (1ST).

U.S. Tea Act replaced 1883 Tea Act (MMGT).

John Meyenberg granted a second patent on his sterilization process for evaporated milk (HH).

P. J. Towle, Minnesota grocer, devised Log Cabin pancake syrup (HS).

1889   Raffaele Esposito, Naples, Italy, devised Pizza alla Magherita in honor of Queen Magherita (RD).

1890   Packaged gelatin invented by Charles Knox (RM).

U.S. Meat Inspection Act passed (MMGT).

1891   U.S. Cattle Inspection Act passed (MMGT).

George A. Hormel, Austin, Minnesota, founded the Hormel Meat Packing Company (USAIR).

>1891   May Wait invented Jell-O, by 1900 it was sold nationally (RM).

1892   William Painter patented the pry-off bottle cap (RMRB).

Emil Frey produced Liederkranz cheese (RM).

Asa G. Candler bought the rights to Coca-Cola syrup (WJ).

James Mitchell devised a machine for extruding cookie dough, later used for Fig Newtons (EOOET).

1893   First ready-to-eat breakfast cereal, Shredded Wheat, produced by Henry D. Perky, Denver, Colorado (1ST).

Coca-Cola trademark registered with the U.S. Patent Office (WJ).

Charles Hire renamed Hires Herb Tea; Hires Root Beer marketed (UNITD).

Welch's Grape Juice, formerly Dr. Welch's Unfermented Wine, introduced at the Chicago World Fair (HS).

Waldorf Salad devised by Oscar Tschirky (HS).

1894   Charles Post developed a coffee substitute, Postum (PA78).

Coca-Cola bottled by Joseph Biedenham, Vicksburg, Mississippi (1ST).

Milton Hershey, Lancaster, Pennsylvania, introduced Hershey Bars (BBB).

Susan Stavors created Minute Tapioca by using a coffee grinder to process tapioca (WM).

Word *frankfurter* devised for the hot dog, assumed to have come from Frankfurt, Germany (CDIS).

1895   Charles Post developed Grape-Nuts, an early breakfast/health food and offered the first cents-off coupon (LB).

First flaked breakfast cereal, Granose Flakes, prepared from wheat by Dr. John Kellogg, Battle Creek, Michigan (1ST).

Michael J. Owens, Toledo, Ohio, patented a glass-blowing machine for automatic bottle production (WABI).

First U.S. pizzeria opened in New York on Spring Street (HS).

Fig Newton cookies sold by the Kennedy Biscuit Works, Cambridge, Massachusetts (EOOET).

1896 Ice cream cones first produced by Italo Marcioni, New Jersey (1ST).

Chinese ambassador Li-Hung-Chang's chef devised chop suey to appeal to Chinese and American tastes (1ST).

Leonard Heischfield introduced Tootsie Rolls candy (BBB).

U.S. Filled Cheese Act passed (MMGT).

Wahlin Company, New York, manufactured the New American Cream Separator derived from Laval's 1876 machine (WABI).

1897 Sundaes sold because soda water prohibited from sale on Sunday: ice cream with syrup only sold, Red Cross Pharmacy, Ithaca, New York (RM).

Ice cream sundae eaten at the Red Cross Pharmacy, Ithaca, New York (FFF).

1898 Kellogg's Corn Flakes developed by William Kellog, manufactured by Sanitas Food Company, Battle Creek, Michigan (1ST).

Pepsi Cola invented (BBB).

Pharmacist created Beeman's Gum for treating heartburn (WJ).

1899 Motorized milk van introduced by Eccles Co-op, Lancashire, England (1ST).

1900 Eaterie in New Haven, Connecticut, pioneered the hamburger (TT).

Milton Hershey devised the wrapped candy bar (UNITD).

>1900 Harvard beets devised, named for Harvard color, crimson (CDIS).

1901 Quaker Oats Company established to market rolled oats cereal (BUSWK).

Hearst sport cartoonist T. A. Dorgan coined the term *hot dog* for dachshund sausages (HS).

1902 U.S. Renovated or Process Butter Act passed (MMGT).

Barnum's Animal Crackers first marketed (HS).

1903   Flavor and texture of margarine improved by hardening vegetable oils (TT).

Italo Marcioni, New Jersey, patented a mold for ice cream cones (TT).

M. J. Owens, U.S., began bottle production with fully automatic machine that lowered cost and wages (TT).

Anonymous Frenchman invented enrober machine for coating candies with chocolate (RM).

Ambrose W. Straub, St. Louis, patented a peanut butter making machine (USAIR).

Dr. John Harvey Kellogg made peanut butter for his patients, Battle Creek, Michigan (USAIR).

1904   Canada Dry Ginger Ale was introduced (BBB).

Processed cheese manufactured by J. L. Kraft, Chicago, Illinois (1ST).

Ice cream cones were reintroduced in St. Louis by E. A. Hamwi, who rolled waffles in the shape of cones (1ST).

Thomas Sullivan, New York, sent customers samples of tea in silk bags, first tea bags (WB).

Iced tea served by Richard Blechynden of England at Louisiana Purchase Exposition, St. Louis, Missouri (WB).

Hamburger sandwich popularized at the Louisiana Purchase Exposition, St. Louis, Missouri (RM).

C. W. Post introduced Post Toasties (LB).

Hot dogs first sold with buns at the Louisiana Purchase Exposition, St. Louis (HS).

1905   Wafer ice cream sandwich was devised by Lewis using a rectangular brass slide to form the ice cream (1ST).

1906   U.S. Pure Food and Drug Act triggered by Upton Sinclair's *The Jungle*, based on Chicago Stockyard (1ST).

Cornflakes moved from health food market to general use (TT).

Milk cartons introduced by G. W. Maxwell, San Francisco (1ST).

U.S. Meat Inspection Act superseded 1890 Meat Inspection Act (MMGT).

1908   Cellophane patented by Dr. Jacques Brandenberger, Zurich (1ST).

Kikunae Iheda, Japanese chemist, discovered the flavor-enhancing properties of monosodium glutamate (first discovered in 1867) (MB).

Hydrox cookie first sold in U.S. by Sunshine Biscuit
Company (CDIS).

Fred Walker & Company began the manufacture and sale
of Vegemite (WJ).

1910 Saccharin used commercially as a sweetener (ACTB).

1911 General Baking Co. was formed to offer white bread; the
company later became General Host (FORT).

Nabisco introduced Oreo cookies (CDIS).

Procter & Gamble introduced Crisco shortening (WJ).

1912 Prizes first added to Cracker Jack boxes (CDIS).

Paul Sabatier, French chemist, discovered nickel made a
good hydrogenation catalyst (WB).

Richard Hellman, New York, sold one-pound boxes of
Hellman's mayonnaise (EOOET).

1913 Cellophane manufactured by La Cellophane, Paris (1ST).

Richard Hellman, New York, sold glass jars of mayonnaise
(EOOET).

1914 Aluminum foil bottle caps (for milk) produced by Josef
Jonsson, Linkoping, Sweden (1ST).

1915 Alexander Samuelson, Root Glass Company, Terre Haute,
Indiana, designed the wasp-waisted Coca-Cola bottle
(CDIS).

1916 Clarence Saunders launched a small chain of self-service
grocery stores, Piggly Wiggly (WG).

1919 Moon Pies, southern favorite, invented in Chattanooga
(CDIS).

Sailor Jack and his dog Bingo first appeared on Cracker
Jack boxes (CDIS).

Coney Island vendor devised frozen custard dessert (WJ).

1920 London firm sold energen rolls, bread substitute for
diabetics (TT).

Tea bags produced by Joseph Krieger, San Francisco (1ST).

1921 First chocolate-covered ice cream bars, called Eskimo Pie,
were marketed by Christian K. Nelson, Onawa, Iowa
(1ST).

Peter Paul Halijian, New Haven, Connecticut, introduced
Peter Paul Mounds candy bar (BBB).

Otto Schnering introduced Baby Ruth candy bar, named for
Grover Cleveland's daughter (BBB).

1922 First wrapped ice cream brickettes manufactured by
Thomas Wall (1ST).

Butterfingers candy bar introduced by Schnering (BBB).

1923 U.S. Filled Milk Act (MMGT).

1924 CO_2 process first used in manufacture of ice cream, New York (DID).

Wheaties breakfast cereal marketed (UNITD).

1925   More efficient, glass bottle manufacturing machine introduced, better than the 1903 Owens machine (TT).

Potato chips manufactured commercially in Albany, New York (WG).

1926   Moisture-proof cellophane invented by William Hale Church and Karl Edwin Pringle, DuPont (DID).

Erik Rotheim, Norway, invented the aerosol (WABI).

First White Tower restaurant opened in Milwaukee, Wisconsin (WT).

Kraft bought an interest in Fred Walker & Company, manufacturer of Vegemite (WJ).

1928   Franklin Mars developed Milky Way candy bar (BBB).

Progresso Quality Foods Company founded (BW).

1930   Wonder Bread, the first packaged, sliced bread, introduced (TT).

Mars introduced Snickers candy bar (BBB).

McNary-Mapes Amendment to the 1906 U.S. Food and Drug Act (MMGT).

Twinkies snack cake introduced by James Dewar (HS).

>1930   Bar coding system devised for identifying items, later adopted for supermarket use (WJ).

1932   Mars introduced Three Musketeers candy bar (BBB).

1933   Lithiated Lemon, English soft drink, promoted in the U.S. as 7-Up, "mixer" drink (TT).

1934   Amendment on seafood inspection added to the 1906 U.S. Food and Drug Act (MMGT).

Nabisco introduced Ritz Crackers (CDIS).

Cornuts, toasted corn snack food, first made (CORN).

1935   Kreuger Beer, Newton, New Jersey, sold beer in easy-carry cans (TT).

1936   Borden adopted Elsie the Cow as symbol (CDIS).

1937   After thirty-eight years as a health food, peanut butter was improved and marketed nationwide (TT).

Rice Krispies breakfast cereal launched (TT).

Margaret Rudkin founded Pepperidge Farms Bakery (WJ).

Harry H. Holly, U.S., developed a hamburger patty machine (CDIS).

Nabisco baked more than twenty-nine milllion Ritz Crackers per day (CDIS).

Kraft marketed Kraft Macaroni and Cheese (CDIS).

	Hormel marketed Spam, canned spiced ham and pork shoulder (CDIS).
1938	U.S. Food, Drug, and Cosmetics Act replaced the 1906 Food and Drug Act and amendments (MMGT).
1939	Julian S. Kahn, New York, invented a disposable spray can (WABI).
1940	Mars introduced M&M's, no-mess chocolate candy (BBB).
	Howard Johnson opened his first turnpike restaurant in Pennsylvania (CJ).
1942	Peter Halijian introduced Almond Joy candy bar (BBB).
	Dannon yogurt introduced to the U.S. (WJ).
1944	British extracted oil from sunflower seeds for use in margarine and other foods (TT).
1945	Bertha West Nealey with Ruth Bigelow established Constant Comment Tea Company (WM).
>1945	Lipton bought rights to use improved tea-bag producing machine to manufacture Flo-Thru tea-bags (TIME).
1948	Burton Baskins and Irving Robbins merged their ice cream chains to form Baskin-Robbins (HS).
	Heinz introduced a barbecue sauce (FF).
1949	Prepared cake mixes introduced by General Mills and Pillsbury, U.S. (TT).
	Cyclamate first marketed as a nonnutritive sweetener (ACTB).
	Charles Lubin introduced Sara Lee baked goods (HS).
1952	Institute of Margarine, Moscow, invented powdered butter (WABI).
	Herman Kirsch of Kirsch Beverages Incorporated, College Point, New York, introduced NoCal, a sugar-free soft drink (FFF).
1953	Low-cost plastic aerosol valve mechanism developed (TT).
	First collapsible polyethylene tube made by Bradley Container Corporation, Delaware, for Sea and Ski (1ST).
1954	Peanut M&M's, chocolate candy, introduced (EP).
	Miller Pesticides Chemical Act amended 1938 U.S. Food, Drug, and Cosmetics Act (MMGT).
	Procter & Gamble began marketing Fluffo yellow shortening (FORT).
1957	Ultrasonic generator, U.S., foamed bottled and canned beer to remove air bubbles that cause cloudy look (TT).
	Fast food manufacturers interested in dehydrated potato flakes for mashed potatoes (TT).

1958	Food Additives Amendment (Delaney Amendment) to 1938 U.S. Food, Drug, and Cosmetics Act enacted (MMGT).
1960	Aluminum cans used for soft drinks (BBB).
	Little Debbie snack foods introduced by McKee Baking Company, Collegedale, Tennessee (CDIS).
	Color Amendment to the 1938 U.S. Food, Drug, and Cosmetics Act was enacted (MMGT).
1961	Coke introduced Sprite, lemon-lime drink (MR).
	Campbell Soup Company bought Pepperidge Farms Bakery (WJ).
1962	Diet Rite Cola introduced (BBB).
1963	Alcoa introduced the tab-top can for beverages (BBB).
	Pepsi followed Coca-Cola's lead in introducing Tab and introduced Diet Pepsi (BBB).
	Tab diet soft drink introduced by Coca Cola Company (BBB).
1964	Borden's bought Cracker Jack Company (CDIS).
1965	G. D. Searle scientist discovered aspartame (ACTB).
	Ocean Spray introduced Cranapple juice (HS).
	Gatorade, a rehydrating beverage, was developed by a team of University of Florida urologists (MKTGN).
1966-72	Plastic bags containing dried food for one way rehydration using a water gun were used in space (EB).
1967	Imitation milk marketed in Arizona (HS).
1968-72	Cookies coated with methyl cellulose to prevent crumbs developed by NASA for use in space (EB).
1969	Dr. John Olney urged a ban on monosodium glutinate in baby food (MB).
1970	Cyclamate withdrawn from the U.S. market (ACTB).
1971	Manny Wesber invented Canfield's Diet Chocolate Soda (CJ).
1972	"Natural" breakfast cereals introduced by major food companies: Post, Kellogg, General Mills (HS).
1973	Push-through tabs on cans reduced litter (TT).
	Hershey added nutritional information to Hershey candy bar wrapper (CDIS).
	Violet No. 1 food dye banned in the U.S. (MMGT).
1974	U.S. National Heart and Lung Authorization Bill enacted (MMGT).
1975	General Mills introduced granola bars (FORT).
	Wally Amos, Hollywood, California, opened the first Famous Amos Cookie Store (WJ).
	Nabisco introduced Double Stuf Oreo cookies (CDIS).

1976   Plastic aluminum was developed (TT).

Red No. 2 food dye banned in the U.S. (MMGT).

Red No. 4 and carbon black food dye banned in the U.S. (MMGT).

1977   U.S. Food and Drug Administration proposed banning saccharin use, but U.S. Congress imposed a moratorium (ACTB).

1978   Sophia Collier founded the American Natural Beverage Corporation (BUSWK).

1979   Jell-O Pudding Pops introduced (GF).

1980   Hershey began using foil wrapper on Hershey candy bar (CDIS).

Pillsbury began work on Toaster Strudel (FORT).

1981   Searle introduced Equal sweetener based on aspartame (Nutrasweet) (CDIS).

Gilbert Levin, U.S., patented L-sugar, same taste as sugar, but not metabolized by the body (FORT).

Foods using Yellow No. 5 food dye began carrying a warning label (MMGT).

Aspartame approved by U.S. Food and Drug Administration for use in food and dry beverage mixes (ACTB).

Low calorie Aunt Jemima Pancake Syrup introduced (BUSWK).

California Wine Cooler introduced (MKTGN).

1982   General Foods used Nutrasweet in its products (CDIS).

Pepsi Free, caffeine-free Pepsi, introduced (MR).

Diet Coke debuted (MR).

Quaker Oats Company introduced Chewy Granola Bars (FORT).

Campbell Soup Co. launched Prego Spaghetti Sauce (WJ).

Milk Marketing Board, U.K., developed a cheese substitute (WABI).

Tylenol poisonings case led to tamper-proof packaging for nonprescription drugs (NYT).

1983   Nutrasweet used in cold milk product mixes (CDIS).

Procter & Gamble introduced Duncan Hines cookies, soft and chewy inside, crunchy outside (WJ).

Caffeine-free Coke debuted (MR).

Coke adopted Nutrasweet as sweetener in Diet Coke (MR).

Dannon introduced a French-style yogurt (Custard) (WJ).

Aspartame approved by the U.S. Food and Drug Administration for use in soft drinks (ACTB).

Crisco introduced a yellow, butter-flavored shortening, resulting in the demise of Fluffo shortening (FORT).

1984 Pepsi and Coke switched to 100 percent Nutrasweet formulations in Diet Pepsi and Diet Coke (CDIS).

Coke available on the space shuttle (MR).

Pepsi introduced Slice while Coke test marketed Minute Maid Orange Soda in Canada (BW).

Nabisco introduced mint flavored Oreo cookies (CDIS).

Dole introduced Fruit'n'Juice bars (FORB).

1985 La Vie de France, Virginia, introduced a low cholesterol croissant made with margarine (WJ).

Bob Greene, columnist, popularized Canfield's Diet Chocolate Soda (CJ).

Source Perrier announced plans to add citrus flavor to Perrier (ECON).

U.S. Food and Drug Administration proposed ban on the use of sulfite to preserve fresh fruits and vegetables in salad bars (CJ).

Coke introduced a special can for use by astronauts on space shuttle (CDIS).

Coke introduced new formula Coke nationally (WJ).

Coke reintroduced original formula Coke as Coke Classic (WJ).

Coke introduced Cherry Coke nationally (WJ).

Campbell Soup Company launched Prego Plus Spaghetti Sauce (WJ).

U.S. Department of Agriculture developed electronic device to measure fruit and vegetable ripeness (BUSWK).

Prince Company, Lowell, Massachusetts, introduced Tug-n-Tie package for Dutch Maid Noodles, easy open, tie reseal (MKTGN).

Harry H. Holly developed a new hamburger patty machine, Universal System, that interwove meat strands to retain moisture during cooking (CDIS).

McDonald's introduced McDLT, devised by Will May at Texas franchise, primarily a packaging change for hamburger separating the lettuce and tomato (FORT).

Pillsbury introduced Toaster Strudel (FORT).

Brassiere du Pecheur developed a syrup for making beer by adding water (BUSWK).

T. J. Cinnamons Bakery, Kansas City, opened in January selling large, fresh-baked cinnamon rolls (TIME).

Coca-Cola test marketed in clear plastic cans (WJ).

Peachtree Schnapps introduced (FF).
1986   Warner-Lambert reintroduced Beeman's, Black Jack, and
    Clove chewing gums (WJ).
    Alcan Research Labs announced the MicroMatch
    packaging system (MKTGN).
    Reynolds Metals Company introduced the Container-Mate
    foil closure for packaging use (MKTGN).
    Campbell's Soup Company test marketed Cookbook Classic
    Soups, single-serve, table-ready, microwave (MKTGN).
    Coca Cola introduced Minute Maid Fruit Juicee (formerly
    Guido's Ice Juicee) (FORB).
    Second Tylenol tampering led Johnson and Johnson to
    withdraw capsules and market caplets, flattened, oval
    shaped tablets (WJ).
1987   Nabisco baked more than sixty million Ritz Crackers per
    day (CDIS).
    K. C. Masterpiece Barbecue Sauce, created by Rich Davis,
    Kansas City, was marketed nationally (FF).

Food Stores and Restaurants

People first gathered food for themselves, later trading for foods and goods
they did not have or could not grow, make, or easily find themselves. Early
markets provided a means for buyer and seller to exchange goods, just as
stores do. Families also prepared their own food for much of history.
During the Middle Ages monasteries occasionally served as inns providing
travelers with shelter and food. Food vendors have hawked ready-to-eat
items in markets, at fairs, and along the roads for much of our history, and
"cook-shops" selling cooked foods existed in ancient Rome. Proper
restaurants, however, like stores, are a recent institution in Western
countries. Food has often continued to be sold or bartered in the market
even after other goods have been removed to stores.[44] This was true in
England and is still true in many of the lesser-developed nations. In the
United States, we still have roadside stalls, and even some farmers' markets.
People like to see and test fresh foods, which may be part of the reason
roadside vegetable stands are so popular in the summer.

Stores

Early stores appeared in China, Germany, and Japan. The Chinese had
developed retail store chains ca. 300 B.C.; the Fugger family of Germany
operated stores in the 15th century; and a chain of pharmacies was founded
in Japan in 1643.

Figure 3.4. Stores and Restaurants

YEAR

YEAR		
1987	Self-service check-out	
		Fast food breakfasts
1975	Scanners	
	Combination stores	
		Pizza Hut
1950		McDonald's (Chain)
		McDonald Brothers
	Supermarkets, shopping carts	White Castle, White Tower
1900	Self-serve stores, vending	Hamburger stand
	Paper bags	Cafeteria
	Cooperative stores	
1800		
	Hudson Bay Co.	Restaurants (Paris)
	Japanese pharmacies	Covent Garden Market
1600		
		Coffeehouses (Europe)
	Fugger family	
1200		
400		
		Cook shops (Rome)
	Chinese stores	
-1200		

The records of the founding of guilds, societies, and companies provide clues about developments in retailing. The Society of Apothecaries and Grocers and the Fruiterer's Company were both founded in 1606 in London, followed by the establishment of Covent Garden Market in 1634. The Hudson Bay Company operated a chain of trading outposts in Canada following the issuance of its charter in 1670. Today, the Hudson Bay stores are large, sophisticated department stores, although some of the more remote outposts still operate.

Cooperative stores, owned by and operated for the benefit of their customers, have existed for over 150 years. The first cooperative stores in America were established in Philadelphia and New York in 1829, and the Cooperative Wholesale Society was established in Liverpool in 1831.

A number of familiar grocery chains were established in the 19th century. The Great Atlantic and Pacific Tea Company, later A&P, was established in 1869. Kroger was founded in 1882, and the Gristede Brothers went into the grocery business in 1891.

The supermarket is a 20th-century phenomenon that evolved over a number of years as retailing practices changed from outdoor markets to grocery stores to the ever larger self-service food stores. The first self-service grocery store was opened by Alpha-Beta in California in 1912. Piggly Wiggly stores, following similar principles, were established by Clarence Saunders in 1916; Woodward's, of Vancouver, Canada, introduced a self-service food department in 1919. The cash register, introduced by John Patterson, founder of NCR, in 1884, assisted the check-out function of these self-service stores by recording sales and controlling cash. The first true supermarket appeared in 1930: the King Kullen food stores, on Long Island, operated by a former Kroger store manager and offering self-service, a wide selection of foods, and lower prices than the traditional grocery stores.

The development of the Universal Product Code and its use with supermarket scanners was discussed above. The first scanners were installed in a Marsh Supermarket in Troy, Ohio, in 1974, and by 1987 Kroger's was testing self-service scanning units. Currently over 15,000 stores are using scanners, including 50 percent of U.S. supermarkets.[45] Some stores are experimenting with self-service check-out. Kroger has installed Service Plus scanners that enable customers to scan their purchases and produce a printed record, which is then paid at a separate cashier's station.

Neighborhood grocery stores have almost disappeared except in the larger cities, but have been replaced by the new convenience stores.[47] These stores began to proliferate in the 1960s and generally carry a limited variety of foods; they concentrate on certain staples and high profit items. Many are combined with former service stations and sell gasoline as well as food.

The first United States hypermarche, Biggs, was opened near Cincinnati, Ohio, in 1984. Based on similar stores operating in France as early as 1963, these stores contain some 200,000 square feet, 65,000 of which is devoted to foods; compare the area of the typical supermarket, which ranges from 30,000 to 50,000 square feet. The hypermarche stores combine a supermarket and a department store for one-stop shopping. Safeways in California opened similar stores in the 1960s. Supermarkets are currently experimenting with the "boutique concept" of specialty areas or even leased space for a variety of services. This development in retailing represents a return to the concept of the general store so dear to early settlers.

Among the devices that make shopping easier are shopping carts for in-store use and grocery bags. Shopping carts were first devised by Sylvan N. Goldman of Humpty Dumpty stores in Oklahoma City in 1937. The carts consisted of a folding chair on wheels with a basket fastened to the seat to hold groceries. Recently, some stores have added straps to hold children into the seats that are now part of many shopping carts.

Paper bags for holding purchases have been used since the 1860s. Before then clerks would often place small purchases for those customers without baskets, bags, or other containers in cones made of rolled paper and twisted at the bottom to hold their shape. Some merchants began to create these containers in advance, and in the 1860s a number of machines were devised to make paper containers, including William Goodale's 1859 patent for a paper bag machine, an automatic machine by Charles Hill Morgan in 1860, and a more popular version of it by S. E. Pierre. In 1865 Pierre licensed his design to a number of printers. Inspired by his success, others developed competing machines. The Union Company of Pennsylvania made a superior machine in 1869 that combined features of several earlier designs. By 1875 the Union Company had changed its name to the Union Bag and Paper Company and had begun to manufacture the bags. Over 606 million bags were made in its first year of production. The square-bottomed brown paper bag with pleated sides was devised by Luther G. Crowell in 1872, and Charles Stilwell patented a machine for making a similar bag he called S.O.S. (self-opening sack) in 1883. The paper bag continued to be the main container used for groceries and other merchandise until the late 1970s, when a number of grocery stores began to experiment with plastic grocery bags. Some discounters and other general merchandisers had used plastic bags somewhat earlier.

Restaurants

As noted above, after the collapse of the Roman Empire access to food and lodging depended on the homes of the wealthy, the monasteries, and the few inns and taverns that existed, usually in towns or cities. With the rise in trade and travel more taverns, inns, and hotels were established. By the 16th century, taverns and hotels supplied lodging, food, and drink to travelers, and coffee shops also served some foods. The first coffeehouse opened in 1554 in Constantinople, followed by similar institutions in Oxford, England in 1650, in Hamburg in 1678, and in Vienna in 1683. The first tea shop was established in 1884, but tea had been sold, since its introduction to the West, in many coffeehouses.

The first restaurant, operated primarily to sell cooked food, was established by A. Boulanger in Paris in 1765.[47] Since *boulanger* is the French word for baker, A. Boulanger may not really be a man's name, but rather the name of a bakery shop. In any event, the first luxury restaurant, La Grande Taverne de Londres, opened in Paris in 1782. The French Revolution led to a decline in the homes of the wealthy and to an accompanying decrease in private employment for chefs, who, seeking a means of earning a living, established public restaurants. By 1804 there were over 500 in Paris. Early American restaurants evolved from taverns and inns, which appeared about 1650; among these was the Blue Anchor Tavern, Philadelphia.

Self-service restaurants and food vending machines began in the 19th century and were developed in the early 20th century. Cafeteria-style restaurants started with food service during the Gold Rush in California, but the first true self-service cafeteria was the Exchange Buffet in New York, operating in 1885. Automatic coffee vending machines were installed in 1898, peanut vending machines were installed in 1901, and an early automat, using German machinery, opened in 1902 in Philadelphia. Automats provided unattended access to food: the customer could lift a small door or flap that was released by depositing the required number of coins in a slot.

Although we think of fast-food restaurants as a recent phenomenon, White Castle offered fast counter service in 1921 in Wichita, and White Tower was established in 1926 in Milwaukee. Both were fixtures in many Midwestern towns in the forties. They focused on low-cost hamburgers and limited menus.

Pizza Hut restaurants were also established in Wichita by Frank and Dan Carrey in 1958. After the purchase of Pizza Hut by Pepsi-Cola, the competing Wendy's chain announced it would no longer serve Pepsi fountain syrups.

McDonald's first restaurant opened in Des Plaines, Illinois, in 1955. Their early stores were drive-in restaurants with little or no seating. They gradually evolved into the internationally operated, family-oriented restaurants found along every U.S. interstate highway. Breakfast was added with the introduction of Egg McMuffin in 1973, first in Santa Barbara and nationally in 1977.

Other fast food restaurants also introduced their own breakfast menus to compete with McDonald's. Wendy's tested breakfast service in 1981 and began offering it nationally in 1983. By 1985 they were joined by Burger King's Croissan'wich in competing with McDonald's for the breakfast customers. With fierce competition, Wendy's announced a revision in 1986 allowing each franchise to decide whether to offer breakfast service.

More people are eating out and doing it more frequently. Single heads of households and working wives have contributed to this trend. The proliferation of fast-food restaurants offering low cost, consistent food and service have not only responded to increasing consumer demand, but have also contributed to the trend encouraging families to eat out.[48]

Table 3.3. Stores and Restaurants

Dates	Event
A.D. 15th c	Fugger family, Germany, founded stores (EB).
1554	First recorded coffeehouse opened in Constantinople (1ST).

1600	East India Company obtained charter (USAIR).
1606	Society of Apothecaries and Grocers and Fruiterer's Company, London, were founded (TH).
1634	Covent Garden Market opened in London (TH).
>1637	Conopios, disciple of Cyrill, patriarch of Constantinople, fled to Oxford and introduced coffee there (BCT).
1643	Coffee drinking popular in Paris (TH). Chain of Japanese pharmacies founded (EB).
1650	First coffeehouse in England opened in Oxford, England (TH).
1652	First London coffeehouse opened by Pasqua Rosee (BCT).
1671	First Marseilles coffeehouse opened (BCT).
1678	First German coffeehouse opened in Hamburg (TH).
1679	London merchant opened Hamburg coffeehouse (BCT).
1680	Blue Anchor Tavern opened in Philadelphia, later became one of earliest U.S. restaurants (RM).
1683	First coffeehouse in Vienna opened and Viennese coffee served (TH).
1765	First restaurant opened in Paris by A. Boulanger (EB).
1782	First luxury restaurant, Le Grande Taverne de Londres, Paris, opened (EB).
1804	More than 500 restaurants were operating in Paris as a consequence of the French Revolution (see text for fuller explanation) (EB).
1829	First cooperative stores in America founded in Philadelphia and New York (TH).
1831	Cooperative wholesale society, North of England Co-operative Company, Liverpool, established (1ST).
1859	William Goodale, Clinton, Massachusetts, obtained U.S. patent 24734 on a paper-bag-manufacturing machine (FFF).
1860	Charles Hill Morgan invented the first commercially feasible automatic machine to make paper bags (HS).
1865	S. F. Pierre licensed his paper-bag-making process to printers (WB).
1869	George H. Hartford formed Great Atlantic and Pacific Tea Co., A&P, and opened a chain of stores (RM). Union Co., Pennsylvania, devised a successful paper-bag-making machine combining the best features of others (WB).
1871	Thomas Lipton opened Lipton's Market, Glasgow, Scotland, and used coupons, "Lipton's Pounds" (TIME).

1872	Luther C. Crowell devised the square bottomed brown paper bag with longitudinal folds (U.S. patents 123811 and 123812) (HS).
1875	Union Company changed its name to Union Bag and Paper Co. and began manufacture of paper bags (WB).
1879	James Ritty, Dayton, Ohio, devised his "incorruptible cashier," the first cash register (MITU).
1880	James Ritty improved his cash register, making it more accurate (see 1879 and 1884) (FFF).
1882	H. Kroger opened his first grocery store (BW).
1883	Charles Stilwell developed a machine for making a self-opening, pleated, flat-bottomed grocery bag (U.S. patent 279505) (EOOET).
1884	First tea shop opened in England, earlier tea was sold in coffeehouses (RM).
	John Patterson, founder of National Cash Register Company, began to sell cash registers after having bought the rights from James Ritty (MITU).
1898	Coffee vending machines installed, Leicester Square, London, by Pluto Hot Water Syndicate (1ST).
1901	Peanut vending machine installed at the Pan-American Exposition, Buffalo, New York, by Mills Novelty Company (1ST).
1902	Early automat offered food for a "nickel in the slot," Horn & Hardart Baking Co., Philadelphia (TT).
1912	Alpha-Beta established self-service grocery stores in California (1ST).
1915	U.S. grocery store installed a turnstile and checkout (TT).
1919	Woodward's, Vancouver, Canada, introduced self-service food department (1ST).
1921	White Castle restaurant chain established in Wichita, Kansas, and began selling its hamburgers (FORB).
1925	Howard Johnson opened his first ice cream shop, Wollaston, Massachusetts (RM).
1929	Howard Johnson opened his second restaurant (CJ).
1930	Michael Cullen operated King Kullen food stores, Long Island, New York, as first supermarkets (1ST).
1937	Humpty Dumpty Stores, Oklahoma City, provided shopping carts--a basket on a folding chair on wheels (WABI).
1940	Richard and Maurice McDonald opened their first restaurant in Pasadena, California (HS).
1948	McDonalds, operated by the McDonald brothers, became a self-service restaurant (HS).

1955 McDonald's restaurant chain opened in Des Plaines, Illinois (RM).
1958 Frank and Dan Carney, Wichita, Kansas, opened the first Pizza Hut restaurant (RD).
1963 Carefour, France, hypermarche, opened (WABI).
1967 Railroads adopted bar codes for tracking railroad cars (WJ).
1968 Hooleys' opened first Cub store, Fridley, Minnesota (WJ).
1973 Universal Product Code introduced for identifying grocery products (WJ).
      McDonald's began serving breakfast with Egg McMuffin (WJ).
1974 Computer controlled scanners were installed in Marsh Supermarkets in Troy, Ohio (BUSWK).
1977 McDonald's launched Egg McMuffin nationally (CDIS).
1980 West Edmonton Mall, Alberta, Canada, shopping entertainment center with amusement park addition opened (WJ).
1981 Wendy's experimented with a breakfast menu (CDIS).
1983 Wendy's introduced a new breakfast menu nationally (CDIS).
1984 First U.S. hypermarche, Biggs, opened near Cinncinati (CDIS).
1985 Burger King began national introduction of its new breakfast menu with Croissan'wich (WJ).
1987 Kroger Supermarkets experimented with Service Plus, customer-operated scanner units for grocery check-out (BUSWK).

We have progressed from gathering and preparing our own food to buying it raw, precooked, or ready-to-eat. We can buy food from many places and in many different shapes, sizes, or forms. We can buy it, at least in some localities, at any hour of the day or night and any day of the week. The proportion of our time devoted to earning money to buy food has decreased and so has the time and energy needed to procure, fix, and serve it. Chapter 4 will consider some of these changes.

4
Cooking

Having solved the problem of acquiring and preserving food, people were free to use their ingenuity to make their food more interesting, edible, and palatable by using a variety of techniques including cooking. Many family and societal rituals surround eating and meal preparation. Breaking bread and sharing salt are two examples of such rituals and in some societies were regarded as almost sacred; guests participating in such rituals were protected.[1] In advanced societies, cooking became an art, not merely a necessity or convenience, and gourmets (and gourmands) appeared early in human history.

How did the human race learn to cook food? Where did stoves originate? Where and when did particular cooking utensils and tools appear? When were cookbooks first produced and used? Although we may not have complete answers to these questions, we can trace where and when many utensils became generally available or were subject to patents. These developments are discussed in three sections: Cooking Facilities, Cooking Utensils and Tools, and Cookbooks. Good sources include Harrison's *Kitchen in History*,[2] Lifshey's *The Housewares Story*,[3] and Franklin's *From Hearth to Cookstove*.[4]

Cooking Facilities

Cooking facilities are essential to the application and use of both utensils and cookbooks and range from outdoor fires to sophisticated electronic kitchens. The first cooking and heating source, the open fire, remained the primary means of cooking in Western society until the 19th century. Although braziers were used in Assyria and Babylonia by 750 B.C. and in fireplaces and campfires in other areas, the source of heat was still an open fire, usually at ground level.[5]

In some instances the fire was used to heat stones, which cooked food either directly by food placed on them, or indirectly if stones were dropped into containers with the food. In the latter case, the hot stones heated water or liquid in a container that did the actual cooking. In remote parts of Scotland and in parts of Ireland, women continued to use this method until the late 19th century.[6]

The development of cooking stoves is closely tied to the development of heating stoves, which are treated in greater detail in chapter 8. Because of fuel availability, wood and coal burning stoves developed first, followed by gas, and last by electric stoves. All four were in use by the early 20th century. Gas stoves use an adjustable flame, making it easier to control the temperature than with wood or coal and may, with an automatic pilot light, be restarted easily. There are no residual products such as ashes from either gas or electric stoves.

Wood and Coal Stoves

The Chinese were using stoves for heating by 600 B.C. The Russians and the Germans also used stoves. German title makers were making stove tiles by 1350 A.D.

As cast iron manufacture developed, it was applied to many things. By 1490 cast iron stoves were used in Alsace, but widespread use for stoves did not occur until the late 18th century. Cast iron ovens began to appear in England by 1770 and were followed by cast iron boilers for water that began appearing in 1780. That same year, Thomas Robinson patented his "open range," providing more convenience for the cook, but still using an open fire. In some ranges the fire was elevated to place the cooking surface nearer the height of a modern stove. It usually incorporated an oven and, in later models, a boiler for heating water as well.[7]

The first practical cooking stove was designed by Count Rumford around 1790 when he improved both the stove and the double boiler. Rumford's stove was made of brick and was designed to make maximum use of the heat generated by a wood or coal fire. Rumford had recommended the use of firebrick because of its insulating properties, but most of the imitation Rumford ranges were made of cast iron which radiated heat into the kitchen, often making summer cooking unpleasant. The only element of Rumford's design these manufacturers included was the oven.[8]

Figure 4.1. Cooking Stoves

YEAR

1987		
	Induction hot plate	
		Convection oven
1975		
	Ceramic cooking surface	Self-cleaning oven
		Domestic microwave
1950		
		Microwave oven
	Aga stove	Oven thermostat
1900	Nickel chromium wire	Electric range
	Oberlin stove	Electric oven
	Mott's cast iron stove	Sharp's gas stove
1800	Bodley's closed range	
	Cast iron ovens (U.K.)	Rumford's range
1600		
	Cast iron stove (Alsace)	
1200		
400		
	Stove (China)	
-1200	Braziers (Middle East)	
-4400		
-10800		
-23600		
-59200		
-110400		
-212800		
-407600		
	Fire	

One factor in Britain, and to a degree elsewhere, that inhibited the adoption and use of stoves was that in addition to the cost and maintenance of the stove, the kitchen in most lower-class homes was also the main living room of the house, and the fireplace was the main source of heating. Although change was slow for some, it did occur. By 1814 a patent steam kitchen and range, which supplied one to fourteen gallons of boiling water, was being advertised in the *Morning Post* (U.K.).

In the early 19th century, a cheap source of fuel and lower cost production methods for stove plate made the production and manufacture of cooking stoves practical in the United States. In 1819 Jordan Mott, New York, developed the 'nut coal' burning stove that used the until then useless anthracite coal, and which significantly reduced the cost of fuel for cooking. In 1830 Mott completed the development of stove plate from pig iron, lowering the cost of stove manufacture. Mott patented his stove in 1833. Philo Penfield Stewart patented his Oberlin cast iron stove, named for Oberlin College, in 1834 and sold over 90,000 stoves in a thirty-year period.

In Britain, George Bodley, an Exeter iron-founder, patented the closed top cooking range in 1802. The fire was still open in front. By 1810 the closed range, with the fire completely enclosed, was being manufactured commercially in England, but it took a few years before it was adopted widely. Brown's Patent Universal Cooking Apparatus, a closed range, appeared in 1840, and in 1864 self-filling water boilers were available. The Aga stove invented by Gustav Galen, Swedish physicist, appeared in 1924. Initially designed for solid fuel such as wood or coal, it was later adapted for gas or oil and was widely used in Europe. The introduction of the cleaner and more easily controlled forms of fuel, gas, and electricity eventually led to the replacement of wood and coal stoves for cooking.

Gas Stoves

Gas use developed along three fronts: heating, lighting, and cooking. Basic to all three applications was the development of suitable gas fuel sources. The early sources of gas were developed from coal gas.

The precursor of the gas cooking stove dates to the early 19th century. The first record of gas use for cooking was by Albert Winsor, Germany, who gave a dinner party in 1802 with food cooked by gas. Winsor did not really use or develop a gas stove. Samuel Clegg described a device designed specifically for cooking that was made at the Aetna Iron Works near Liverpool in 1824. It was gun-barrel shaped with numerous small holes and could be used vertically for roasting and horizontally for frying. This was followed near the end of the 1820s by John Robinson's gas burner for heating water and other liquids, described in 1831 in the *Mechanic's Magazine*. It was never patented or manufactured.

The first proper gas cooking stove was designed by James Sharp, Northampton, England, and installed in his home in 1826. He also installed the first commercially produced gas stove in the Bath Hotel, Lemington, England, in 1834. Earl Spenser's interest in gas cooking influenced Sharp to continue his development of the stove and to establish a gas stove factory in 1836.[9]

Sharp's stove was upright; meat was suspended on hooks in the ceiling above a circle of gas burners in the base. The English were fond of roasts and continued to demand cooking facilities producing fire roasted meats. Gas stoves remained primitive until mid-century.

By 1852 Bower's Registered Gas Cooking stove with an insulated oven and a cooking top was being sold. In 1855 R. Bunsen invented his laboratory burner which was later adapted for both heating and cooking.

Other fuels used for cooking included kerosene, oil, and alcohol, and a methylated spirit gas pressure stove was invented by B. Whangton in 1865.

With older ovens, including cast iron ovens, it was difficult to monitor foods. The test of good cooks was their ability to judge accurately the temperature and condition of the oven and the progress of the food being cooked. If the oven was opened too soon, the bread or cake would not rise properly. If left too long, or at too high a temperature, food became dry, overcooked, or even burned. Glass oven doors, patented in 1855, enabled the cook to monitor the progress of food without interfering with its cooking.

Cleaning the stove was a time consuming, dirty, and difficult task. The advent of vitreous panels in 1869 reduced such work significantly. The smooth, glassy surface was easier to wipe clean because dirt did not adhere as readily. The change to gas and electric cooking also reduced the more onerous cleaning chores associated with the smoke residue or ashes of wood and coal stoves.

By 1869 in England commercial gas stoves were being rented for 6d per annum. This comparatively low rent, coupled with the introduction of the English coin-in-the-slot gas meter, put gas use within the reach of most people. Previously, the deposits required for access to the gas and the equipment were prohibitive to those with small incomes. With the new meter, people paid in advance for use so the risk to the gas companies was nonexistent. The declining cost of the equipment, together with the looming threat of the electric industry, contributed to the declining rates.[10]

Conditions in the United States were somewhat different. The new country had started with isolated homesteads, each of which had had to be self-sufficient, and did not begin the 19th century with a largely urbanized population, as in England. Settlers had to have the means of producing most of what they needed to survive. Fireplaces with built-in brick ovens or portable Dutch ovens were used for cooking, and early settlers relied more heavily on pan breads such as Johnny cake than on leavened breads cooked in ovens. Stoves appeared after 1819 and became popular when Mott's "nut coal" stove was marketed.

Gas ranges began to appear in urban homes as soon as piped gas supplies became available starting in the 1860s, but rural homes had to wait for the introduction of gas in pressurized containers. As a result, rural use did not become common until the 1920s.

Although thermostats had been applied to heating in the 19th century, neither they nor time controls were adapted for use on stoves until the 20th century. In 1915 the first thermostatic oven regulators for gas ovens were developed, and by 1923 Regulo, the first commercially successful thermostat for domestic ovens, was used by the Davis Gas Company. By 1930 a gas range with automatic oven and top burner lighting was introduced, followed in 1931 by one with oven time control. During the 1920s and early 1930s, developments in gas and electric stoves proceeded side by side.

Electric Stoves

Widespread use of electric stoves required both proper heating elements and adequate supplies of electricity. The first electric cooking device was an oven installed in the Hotel Bernina, Samaden, Switzerland, in 1889; the first commercially produced electric oven was sold by the Carpenter Electric Heating Manufacturing Company of St. Paul in 1891. U.S. patents for electric cooking elements and stoves were issued to George Simpson in 1859 and to William S. Hadaway in 1869. Hadaway's stove used a ring shaped, spiral coiled burner for better heat distribution. The General Electric Company was marketing an "electric rapid cooking apparatus" that boiled a pint of water in twelve minutes in 1890. The first electric kitchen demonstrating a number of electric appliances in addition to the electric stove was exhibited at the Colombian Exposition, Chicago, in 1893.

Early stoves required a great deal of electricity and burned out their heating elements quickly. A. L. Marsh patented nickel-chromium resistance wire in 1906. This provided the basis for modern cooking elements in the electric stove, and in 1926 Hotpoint developed Calrod, a superior heating element for electric stoves.

Among early stoves, the General Electric range appeared in 1905, and by 1910 the Carron electric range was being sold. After World War I, Belling introduced the Modernette, an electric stove, based on the design of a submarine unit. This was one of the first popular electric stoves in Great Britain.[11]

Development of thermostats for electric stoves followed and then paralleled those for gas stoves. By 1919 an oven thermostat was in use in an electric range, and by 1930 a range with an automatic timing control had been introduced.

More recent developments provide new methods of cooking by microwave, induction, and convection. The microwave oven was patented in 1945 by Perry Spencer and first manufactured by Raytheon for commercial use in restaurants and hospitals in 1947. Marketing of domestic use units began in 1965. Other developments were ceramic cooking surfaces, introduced in 1966; induction cooking, demonstrated by Westinghouse in 1970; the self-cleaning oven, introduced in 1970; the convection oven, offered in 1977; and in 1984 the induction hot-plate developed by Anton Seelig. Induction cooking heats food without the cooking surface itself becoming hot. The convection oven reduced cooking time by ensuring an even distribution of heat throughout the oven.

The microwave was not an immediate success because it required different cooking methods and utensils and was not well suited to producing baked products such as rolls, cakes, and bread. Few foods packaged for microwave use were available and metal containers block and reflect microwaves, interfering with cooking. But by 1982 a number of foods packaged for the microwave were being marketed, and the rapid cooking, as well as the reduced need for extra utensils, have encouraged its use. Many cooks have learned to work with the idiosyncrasies and to appreciate its convenience for cooking fresh vegetables and reheating foods. The microwave is also being more heavily used and may yet fulfill the wonders predicted for it in revolutionizing the kitchen.

Developments in cooking units have been extensive. Many cooks now use the electric frypan or the crockpot in preference to the multipurpose stove.

Table 4.1. Cooking Facilities

Dates	Event
B.C.	
500,000	Fire used (HEP).
6000	Pottery used (WB).
750	Assyrian and Babylonian carvings depicted cooking over brazier (WB).
600	Chinese used heating stoves (EB).
A.D.	
1350	German tile makers made stove tiles (EB).
1490	Cast-iron stoves used in Alsace (WB).
>1770	Cast iron ovens began to appear in England (CD).
1780	Thomas Robinson patented the open range in England (CD).
1780-82	Cast iron boilers for heating water made in England (CD).

1782   Josiah Wedgewood developed a pyrometer to check temperatures in pottery furnaces (TH).

>1790   Count Rumford developed an improved cooking stove and double boiler (HFB).

1797   Meat roaster developed by William Strutt, Derby (CD).

1802   Albert Winsor, Germany, gave a dinner party with food cooked by gas (DID).

George Bodley, Exeter iron founder, patented a closed top cooking stove (EH).

1810   Closed range manufactured in Britain (CD).

1814   Patent steam kitchen and range advertised in the *Morning Post*, supplied 1-14 gallons of boiling water (CAD).

1819   Jordan Mott's (New York) "nut coal" burning stove, used then useless anthracite coal, reducing the fuel cost (RC).

1824   Samuel Clegg described a gas cooking device made at Aetna Iron Works near Liverpool (CD).

1826   James Sharp, Northampton, designed and installed the first gas cooking stove in his home (1ST).

1830   Mott completed the development of stove plate from pig iron, thereby reducing the cost of stove manufacture (RC).

Cooking stoves using wood and coal were common (RC).

1831   *Mechanics' Magazine* described John Robinson's gas burner for cooking (CD).

1834   First commercially produced gas stove installed in Bath hotel, Lemington, England (1ST).

Philo Penfield Stewart patented the Oberlin cast iron stove, manufacturing 90,000 stoves over thirty years (SG).

1836   Gas stove factory was established by James Sharp in Northampton, England (1ST).

1840   Brown's Patent Universal Cooking Apparatus, a closed range, was sold in England (HFB).

1850   Thomas Masters devised the first oven thermometer (FHTC).

1852   Bower's Registered Gas Cooking stove with insulated oven and cooking top sold (1ST).

1855   R. W. Bunsen produced a gas burner that was later adapted for heating, cooking, and lighting (TH).

Glass oven doors used (CD).

1859   George B. Simpson, Washington, D.C., patented an electric stove (patent 25532) (FFF).

1864   Self-filling water boiler available (HFB).

Cooking

1865	Methylated spirit gas pressure stove invented by B. Whangton (DID).
1869	Vitreous enamel panels for stoves patented (CD).
	Commercial gas stove rented for six shillings/annum in England (DID).
	William S. Hadaway, Jr., New York City, patented an electric stove with a ring-shaped, spiral-coiled burner for uniform heat conduction (FFF).
1880	Scotch and Irish still used hot stones in certain locales for heating and cooking (CD).
1889	First electric oven installed at Hotel Bernina, Samaden, Switzerland (1ST).
1891	First commercially produced electric oven sold by Carpenter Electic Heating Manfacturing Company, St. Paul, Minnesota (1ST).
1892	Coin operated gas meters installed by South Metropolitan Gas Co. led to general adoption of gas (1ST).
1893	First electric kitchen shown at the Colombian Exposition, Chicago (WB).
1905	General Electric range marketed (HFB).
1906	A. L. Marsh patented nickel-chromium resistance wire, used in modern cooking element of electric stove (TAC).
1910	Carron electric range sold (TT).
1912	Belling introduced the electric geyser, a rapid immersion heater, but with heavy power consumption (TT).
1915	Thermostatic oven regulators for gas ovens developed (1ST).
1919	Oven thermostat used in an electric range (TAC).
	Belling introduced the first popular U.K. domestic electric stove, Modernette, based on submarine unit (TT).
1920	First practical immersion water heater with thermostatic control introduced (TT).
1923	Regulo, the first commercially successful thermostat for domestic ovens, was used by Davis Gas Co. (TT).
1924	Aga Stove invented by Gustav Galen, Swedish physicist, using solid fuel, later models use gas or oil (TT).
1926	Hotpoint developed Calrod, a superior heating element for electric stoves (TAC).
1930	Electric range with automatic timing control introduced (TAC).
	Gas range with automatic oven and top burner lighting introduced (TAC).

1931 Gas range with time control introduced (TAC).
1945 Microwave oven patented in the U.S. by Perry LeBaron
      Spencer (1ST).
1947 Microwave oven manufactured by Raytheon Co., Waltham,
      Massachusetts (1ST).
1959 Kew Gardens Hotel installed a microwave oven (1ST).
1965 Microwave oven introduced for domestic use (TT).
1966 Ceramic cooking surface for domestic use available (TT).
1970 Self-cleaning ovens available (CREPT).
      Westinghouse demonstrated the induction cooking surface
      (IEEE).
1977 Convection oven first available (MI).
1984 Induction hotplate invented by Anton Seelig, Germany; it
      heated food without getting hot itself (WABI).

Cooking Utensils and Tools

A plethora of cooking utensils is available; we are often accused of having more gadgets than we need. This is not a 20th-century phenomenon, however; the abundance of 19th-century gadgets is equally striking. Many 19th-century utensils are no longer used--some proved too complex to use, others were used so infrequently they were not worth the storage space; in some cases the function they were used for is no longer performed. Furthermore, we have tools, like the food processor, that do the tasks of many formerly separate tools.

The various items discussed in this section have been divided into broad categories: dishes, pots and pans, tableware, gadgets and tools, and small appliances. Developments in some areas overlap: glassware is treated under dishes, as is Corningware, although it is used primarily for cooking. And with the advent of the microwave, almost any nonmetallic item can be used for cooking--including paper towels and paper plates.

Figure 4.2. Selected Cooking Utensils and Tools

YEAR

YEAR			
1987			
	Fiestaware reintroduced		
1975			Silverstone
	Fiestaware withdrawn	Mr. Coffee	Rival Crockpot
	Corningware	Electric can opener	Teflon pans
1950			Electric fry pan
	Electric knife	Bunker can opener	Teflon
	Fiestaware	Temperature controlled coffee maker	Pressure cooker
1900	Stainless steel	Pyrex glass	Electric fry pan
	Boron silica glass	Lower-cost aluminum process	Aluminum pans
	Pressed glass	Enameled cast iron	Vacuum coffee maker
1800	Coffee pot	Enameled saucepans	
	Fine porcelain (Europe)		
1600	Cut glass	U.S. glass factory	Pressure cooker
	Venetian glass	Enamelware	
	Fine porcelain (China)		
1200			
	Crystal glass (U.K.)	Fork (Italy)	
400			
	Glass blowpipe		
			Iron pots (China)
-1200	Glassware		
	Stoneware (China)		
	Potter's wheel (Sumeria)		Stone pots (Egypt)
-4400			
	Pottery		

Dishes

Dishes can be, and have been, made of wood, pottery, glass, plastic, paper, or metal. All of these substances continue to be used. The earliest dishes may have been of wood. Wood was cheap and, in many areas, readily available and easily replaced. It was used by pioneers and in colonial homes in the U.S. Metal dishes were generally more expensive and, while durable, sometimes imparted a metallic taste to food.

Pottery has remained a popular choice throughout history. It is durable if fired at a high enough temperature, inexpensive, and easy to clean. Unlike wood, it does not pick up stains or flavors. The spread of pottery owes much to techniques adopted in the late 17th and early 18th centuries.[12]

Pottery was developed by 6000 B.C. and the potter's wheel was in use by 3250 B.C. in Sumeria. The Egyptians were using glaze by 3000 B.C.; it sealed surfaces and provided a more durable finish. The Chinese were using the potter's wheel sometime between 1500 and 1027 B.C. and were producing fine white stoneware by 1400 B.C. The Chinese experimented with porcelains in 618 to 907 A.D. and were producing fine porcelains between 1279 and 1368. Inspired by examples of Chinese porcelain that had reached the West,

Italians manufactured soft porcelain by 1575, and a faience (earthenware) pottery was opened in Nevers in 1578. After considerable experimentation, hard porcelain was developed in Europe in 1707 by E. Tschirnhaus and J. Bottger. A porcelain factory was established in Meissen in 1710 and another at Sevres in 1756.

Technical advances and more pottery designs followed rapidly. Josiah Wedgewood founded his pottery works in 1760, and in 1782 he developed a pyrometer for checking the temperature in pottery furnaces, providing better control of the firing process. In 1780 Thomas Turner introduced the popular Willow pattern china. The expansion of the pottery industry made pottery dishes inexpensive and consequently available to a broader community. Gradually, pottery replaced wooden dishes for all but the very poor.[13]

In the U.S. a number of inexpensive lines of dishes have been produced including Fiestaware, originally introduced by the Homer Laughlin China Company in 1936 and discontinued in 1972. The dishes were designed by Englishman Frederick Hurten Rhead, who joined the firm in 1927 as its head designer. Prices for the dishes increased as collectors sought rare pieces such as red Fiestaware, made with uranium oxide and discontinued in 1942 during World War II. In response to this surge of interest, Homer Laughlin reintroduced the line in 1986.

The first man-made glass, in the form of pottery glaze, was developed by the Egyptians about 3000 B.C. The first glass vessels appeared about 1500 B.C. and the blowpipe, which greatly increased the production and variety of glass pieces, was in use by 30 B.C.

Glassmaking continued steadily in the Middle East and Byzantium, but after the decline of the Roman Empire, glass in Europe was made by isolated craftsmen carrying on the old traditions. Venice was well established in glass making by about 700 A.D. and by 1300 held a dominant position in Europe. By 1100 English glassmakers had made a crystal glass, the secret of which they guarded carefully, but the process became more widespread with the process developed by French glassmaker Saint-Louis in 1782. Glass bottles and windows were manufactured at Chiddingfold, Surrey, by Laurence Vitracius in 1226. Venetian glass, known for its high transparency, was developed in 1463 by Beroverio of Murano, Italy.

Various individuals experimented with glass and developed new processes or types of glass. Caspar Lehman, jewel cutter to Emperor Rudolf, began the development of the cut glass process in 1600. By 1609 he had been granted a patent on the process. In 1674 the English glassmaker George Ravenscroft developed lead glass, and in 1719 John Akerman began making cut glass in England.

The first glassmaking factory in the U.S. was established at Jamestown, Virginia, in 1608. Through a series of misfortunes, including the failure of the colony itself, glassmaking was abandoned and not reestablished until 1739 when Caspar Wistar built a glassmaking plant in Salem, New Jersey.

Pressed glass was developed as a cheaper alternative to cut glass and was most often used in decorative pieces. The first U.S. firm to produce pressed glass was Bakewell, Page, and Bakewell of Pittsburgh in 1825. They were followed the same year by Deming Jones's Boston and Sandwich Glass Company. The latter also made the first pressed glass tumbler in 1827.

Glass was not suitable for cooking until the development of more durable boron-silicon or Pyrex-type glass capable of withstanding the heat of cooking and the changes in temperature. Carl Zeiss, Germany, developed boron-silicon glass, the forerunner of Pyrex glass, in 1884. Pyrex glass, originally used for battery jars, was perfected for cooking by Corning Glass Labs in 1915. Mrs. Ann Bridges acquired in 1909 the rights to a vacuum coffee maker, later to become Silex, and used the new Pyrex glass to make the first glass vacuum coffee pot.

Corning also developed other glass and ceramic cooking materials. They offered oven-proof pyroceramic cookingware, called Corningware, in 1957. Pyroceram was developed by accident when some material left to cure was processed at a much higher temperature than normally used. Corning also applied glass and pyroceram to dinnerware and introduced Corelle Livingware in 1970. Corningware is used for cooking, serving, and storage.

Today the householder has many choices for temporary food storage. Among these are plastic containers, aluminum foil, and plastic bags. Temporary food storage was improved with the development of polyethylene by R. O. Gibson of Imperial Chemical Industries, England, in 1933. Tupperware Corporation, an early producer and marketer of plastic storage containers, was established in 1945. Reynolds Aluminum Company demonstrated aluminum foil for home use in New York in 1944 and began manufacturing aluminum foil for home use in 1947. Baggies, plastic bags, were first marketed in 1962.

Most paper household products were developed after the advances in wood pulp technology described more fully in chapter 10. Paper plates, napkins, cups, and even clothing were all manufactured by 1868. Charles Francis Jenkins invented the conical paper cup after 1900, and such cups were being manufactured and made available in 1908 by the Public Cup Vendor Company of New York. The first recorded use of the paper napkin in Great Britain was at a dinner given by John Dickinson & Company at the Castle Hotel at Hastings in 1887. Martin Chester Stone of Washington, D.C., devised the paper drinking straw in 1886, patented it in 1888, and invented a machine for their manufacture in 1905. Paper towels were popularized in the United States by the Scott Paper Company in 1907 when seeking a use for paper too wrinkled to be converted into toilet paper.

Pots and Pans

Stones were used early as cooking surfaces and with hollow depressions even as pots; the Egyptians were using soapstone pots by 1600 B.C. Among the early iron artifacts were iron cooking pots.[14] Their development, like many other cooking utensils made of metal, including cutlery, followed the development of metals technology.

Aluminum was a costly metal until the discovery in 1886 of the electrolytic process significantly reduced manufacturing costs. At one time it was more highly prized than silver or gold; Napoleon III used a set of aluminum forks and spoons for his most honored guests.[15] With cheaper sources of the metal available Henry W. Avery, Cleveland, Ohio, made the first aluminum saucepan in 1890, and stamped and cast aluminum pans were being made in 1892 by Pittsburgh Reduction, later Alcoa. Sears was selling aluminum cooking utensils in its catalog by 1897, but aluminum utensils were not used much until after 1910.[16]

The earliest record of the use of enameling is 1200 B.C., but the technique was not applied to utensils until the 16th century. In 1545 A.D. Bernard Palissy made his first enamelware, and by 1609 enameled tin ware was being made in Delft. By 1799 Dr. Hinkling invented an enameling process for saucepans, and in 1839 Clarke patented a new method for enameling kitchenware. Jacob Vollrath, Sheboygan, Wisconsin, was making enameled cast iron ware in 1874.

Nonstick cookware was based on the discovery of a means of bonding a nonstick coating, Teflon, to a metal surface. Teflon, discovered in 1938, was commercially produced in 1948, and the Tefal Company of Paris used it to make the first nonstick saucepans in 1955. Harbenware introduced Teflon pans to Britain in 1956, and Tefal introduced the pans to the U.S. market in 1960. An improved Teflon, Teflon II, was introduced in 1968 and the nonscratch Teflon, Silverstone, was introduced by Dupont in 1976.

The pressure cooker was first developed in the 17th century, but did not become practical until the 20th century. Denis Papin demonstrated the pressure cooker in 1679, and in 1938 Alfred Vischer, Chicago, designed a pressure cooker with an interlocking lid eliminating the bolts formerly required. Vischer's work actually started near the end of World War I, but materials and manufacturing problems plagued his efforts. A low pressure cooker was patented by Hautier, France, in 1927. Although the crockpot, electric pans, and microwave have, in many households, supplanted the pressure cooker, it still fills a niche for rapid cooking, particularly at higher altitudes, and is still energy efficient.

Tableware

Tableware or eating implements for daily use are a fairly modern innovation. Cutting tools and knives have been used since the Stone Age, but spoons and forks are much more recent innovations.

There are conflicting stories about the origin of the fork. According to the *Encyclopaedia Britannica* the Romans devised a two-pronged fork that was used mainly for serving and continued to be used for that purpose during the Middle Ages. The *People's Almanac* lists the first eating fork as a small gold one used in 1057 A.D. by the Venetian daughter of Byzantine emperor Constantine X Ducas. According to the *World Book Encyclopedia,* the honor of using the first fork belongs to the fastidious wife of Doge Domenico Silvio of Venice in 1100.

In any event, it took some time before the fork came into common use. While a number of forks were recorded in death inventories in the 14th century, most Europeans did not use them until the Renaissance. A silver fork was also mentioned in the will of John Baret of Bury St. Edmunds in 1463, but the honor of introducing the fork into use in England is usually accorded to Thomas Coryate in 1608. Matching sets of spoons and forks were common by the mid-18th century.

Knives and spoons were used by the Egyptians, the Celts, and the Romans. The development of cutting instruments paralleled the development of metal working and weapons. Early knives were of flint, followed by knives of copper and bronze and later of iron and steel. Until the late Middle Ages, iron and steel were highly prized because of their durability and relative scarcity. As metal technology improved, particularly in the 17th and 18th centuries, iron and steel tools became more common.[17] In 1904 the first stainless steel was made by Leon Guillet, but he failed to promote it. Stainless steel was first cast by Harry Brearley, Sheffield, England, in 1913 and applied in 1915 to cutlery. The International Steel Company, Meriden, Connecticut, was manufacturing knives with stainless steel blades by 1921. The hand-held electric knife was first marketed in the United States in 1939, but, like many other appliances, not really exploited until after World War II.

Gadgets and Tools

The 19th-century kitchen was filled with a variety of gadgets, some useful for general purposes, others specialized. A host of inventors turned their attention to kitchen and household chores. The earliest were simple mechanical aids to enable a housewife to do specific tasks faster or more easily. Urban centers provided the necessary manufacturing and distribution facilities.[18]

Two examples of 19th-century gadgets are the Universal Tool, patented by William Henry Thayer, Cleveland, in 1881, and another patented by Mr. Youngs, in 1876. The Universal Tool could function as a trivet to hold hot pressing irons, a meat tenderizer, pot lifter and carrier, hot pie plate lifter, stove lid lifter, and candle holder. It could also be used as a weapon against burglars, intruders, or mashers. Youngs's tool combined a nail puller, meat tenderizer, pie crust fluter, pastry sealer, stove lid lifter, punch, can lid lifter, and tack hammer. An inventory of the many interesting 19th-century household devices is provided by Loris Russell in *Handy Things to Have around the House*.[19] I have chosen only a few representative examples.

One of the most striking and varied tools is the apple parer in its many forms. Several writers attribute the origin of the apple parer to Eli Whitney in 1778, to Joseph Sterling in 1781, or to Daniel Cox and sons in 1785. The earliest patent of an apple parer is a mechanized version devised by Moses Coates in 1803. S. Crittenden patented a parer and corer in 1809, and a combined parer, corer, and slicer was patented in 1838. A screw parer was patented by Charles Carter in 1849, and a popular hand apple parer and corer was patented by W. Brock in 1877. This sampling does not include the many homemade devices that were used during this period.

Apparently the irregular shape of the potato was more daunting and only a few potato parers were developed. A potato paring knife was patented in 1860.

Another popular device for inventors was the egg beater. One of the earliest patented was the Monroe egg beater in 1853. A wire one followed in 1859 and was fairly successful, a metal strip model appeared in 1863, and in 1874 Monroe improved his earlier model and assigned the patent to the Dover Stamping Company. In 1891 Whitney and Kirby patented an egg beater that incorporated principles still in use today.

Similar principles were used in developing the ice cream maker. Nancy Johnson invented an ice cream maker in 1846, and it was patented in 1848 by William Young. The electric mixer and ice cream maker had to wait for the development of electric power supplies and did not appear until 1918. The first portable mixer appeared in 1923, and Sunbeam offered the Mixmaster in 1930. Kenneth Wood, England, devised the Kenwood Chef in 1947, the forerunner of the modern food processor.

The flour sifter was also subject to several patents offering different features: some were hand-held with cranks, some were designed to sit on counter- or tabletops, some were round, others rectangular, some had rollers.[20] Tilden's Universal Sifter, a table model made of wood with a crank, was patented in 1863. Earnshaw's Patent Flour Scoop and Sifter, made of tinned sheet iron and hand-held, was patented in 1865.

Exactly when measuring cups and spoons specifically for cooking were first made is uncertain. It is likely they evolved from the measures used to measure beer and ale at the public houses. F. A. Walker was making tin and copper measures in 1870 and Matthai-Ingram were offering a graduated measure in 1890. A set of measuring spoons, quarter, half, and one teaspoon, dates from 1900. Fannie Farmer's cookbook, published in 1896, was the first to use standard measures. By 1925 the American Home Economics Association, concerned with the variations in supposedly standard measures, asked the U.S. Bureau of Standards and the Bureau of Home Economists to investigate. As a result a code of specification for standard measures was developed and adopted.

The first commercial canned goods developed by Appert, the inventor of the canning process, were actually bottled. Cans were first used by Bryan Donkin and John Hall in 1812. Canned goods did not come into common use until late in the 19th century, and can be attributed in part to the Civil War and to the westward movement of people, both of which hastened the spread of canned goods in the United States. These events increased the need for foods that could be transported and stored for extended periods without spoilage.

Although bottles could be opened by removing the cap, cans required a special opening device. The can opener became practical when lighter weight tin plate was used for cans. The first U.S. patent for a can opener was issued in 1858 to Ezra J. Warner, Waterbury, Connecticut. The key opening can was devised by J. Osterhoudt in 1866, but was only used for a limited number of canned goods. A more general-use can opener was still needed for most canned foods and several other can opener patents followed: one using a cutting wheel was patented in 1870 by William W. Lyman; Brock patented his can opener in 1880; and by 1885 the Army and Navy Stores catalog was offering can openers. The wall-mounted can opener appeared in 1927.

Modern can openers encompass both hand-operated models and portable electric can openers. Familiar to many modern users is the Bunker, introduced in 1931 and still in use today, using pivoted handles to hold the can and a key-type handle to turn the cutting wheel. The first portable electric can opener was patented the same year, but it was apparently never produced. Ronson produced the first practical portable electric can opener, the Can-Do, in 1964. By 1948 a magnetic lid lifter had been added, and a removable cutting wheel to ease cleaning was introduced in 1949.

The electric can opener was introduced by two California firms in 1956 with patents granted in 1957 and 1959 respectively. During the 1950s and 1960s a variety of opener combinations were offered including ones with knife sharpeners, food grinders, ice crushers, and juicers.

Pastry crimpers have a long history. One dating from 200 A.D. was found in Roman Gaul, indicating that the French love of pastries is long standing. Other devices have a more uncertain history. Several potato/vegetable mashers including a rotary model date from the 1870s. A bread slicing machine was marketed in 1879. Wooden cake turners for pancakes and other fried foods were used in Colonial times. A number with varying features were offered in the 19th century. An egg lifter (turner) was patented by Ann E. Smith in 1890.

Storing liquids at their desired temperature for extended periods of time is now possible because of the thermos. James Dewar invented the thermos flask for laboratory work in 1892, and it was applied to domestic use in 1904. A boon to travelers and to the energy conscious, it enables storage of hot or cold liquids for a period of several hours. The steel thermos, used by many working men, will keep liquids hot overnight. It makes preparing an early morning breakfast while camping bearable. Hot coffee or tea can be made without having to wait for water to heat.

The thermos will keep water hot, but the water still must be heated first. The tea kettle has remained a popular device for heating water, but the whistling tea kettle is fairly recent. The idea for the whistling tea kettle is claimed by Joseph Block, New York, who while visiting Gebruder Hansel, Westphalia manufacturer, suggested the addition of a whistle to alert the homemaker when the water boiled. Block began importing the new tea kettles in 1922 and they were an immediate success. Also for tea, the first identifiable silver tea pot dates from 1690.

Coffee was initially prepared by boiling hulls and later whole beans. By the 1400s coffee was prepared by mixing a ground form with spices and sugar in hot water and allowing the mixture to seep to bring out the flavor and for the grounds to settle. The first Western coffee grinder appeared by 1687, but the coffee mill was not patented until 1829, and the electric coffee grinder did not appear until 1937 when Kitchen-Aid introduced their first model.

Several sources indicate the French *biggin,* a drip coffee pot, was invented in 1800.[21] Joel Schapira credits Jean Baptise de Belloy, archbishop of Paris, with inventing the first percolator that was a drip percolator, not the pump percolator so familiar in the United States.[22] Count Rumford further improved De Belloy's coffeepot by introducing a device to keep the grounds compressed. Valerie-Anne D'Estaing stated that Descroisilles, a Rouen pharmacist, invented the cafeolette, a form of drip coffee maker, in 1802; a porcelain model was developed shortly after by Antoine Cadet, a French chemist.[23]

Improvements in coffee making continued with the first practical pump percolator, which used a tube in the handle to spray hot water over the ground coffee and was manufactured by Jacques-Augustin Gandais, Paris jeweler, in 1827. In the same year Nicolas Felix Durant invented a percolator with the tube inside the pot. James Nasor devised the familiar pump percolator in 1865.

Vacuum coffee makers began in 1840 when Scottish Marine Engineer Robert Napier developed the Napier vacuum machine. The present Silex coffee maker is a modification of Napier's original design. As mentioned earlier, the U.S. rights to the Silex-type design were acquired in 1909 by Mrs. Ann Bridges and Mrs. Sutton of Connecticut, who manufactured the coffee maker with the then just developed Pyrex glass. Coffee filters were devised by Mellita Bents, Germany, in 1908 and the Chemex coffee maker was developed by Peter Schlumbohm in the period 1938 to 1941.

Electric coffee makers experienced a similar evolution and adaptation process as had manual coffee makers. The cold water pump electric percolator was introduced by Landers, Frary & Clark as the Universal Percolator in 1908. Knapp-Monarch Company introduced the Therm-A-Magic percolator with an automatic shut-off based on a preset time and temperature in 1931, but it did not keep the coffee warm after making it. In 1936 Farberware began marketing the Coffee Robot, a vacuum-type brewer that was thermostatically controlled and kept the freshly brewed coffee hot. By 1972 popular drip coffee makers such as Mr. Coffee had been introduced and were in use.

The end of the 19th century and the early part of the 20th century saw a wave of small appliances. In part, the inadequacies of the electric range may have contributed to the number of small, special purpose appliances developed. The depression and World War II put a damper on further developments until the 1950s and 1960s.

After the end of the war, the rising level of disposable income, the spread of electrical power sources, and the expanded, now excess, factory capacity led to a surge of new products. The proliferation of gadgets resulted in cluttered kitchen counters and impelled the development of, and advertisements for, "appliance garages" and for units that fastened to the wall, the side, or bottom of wall cabinets. Products seldom wore out before they were replaced. Constantly improving new features, coupled with the scarcity and high cost of repair services, made the United States a throw-away society.

Small Appliances

The development and expansion of small appliances began at the end of the 19th century, driven in part by the development and expansion of electrical energy supplies. The Crystal Palace and Colombian Exposition introduced a variety of such devices, and a rapid succession of new small appliances appeared in the 1890s. The first electric kettle appeared in 1891 and was followed by a series of others. The automatic electric tea maker was first patented in 1902, but the first all electric one was not available until 1933. By then an electric light and clock had also been incorporated into the design, even though the first automatic tea kettle with thermostatic control was not marketed until 1955.

Toasters were popular early electric devices. The first electric toaster was marketed by Crompton Company, Chelmsford, in 1893. General Electric's toaster appeared in 1909, and the pop-up toaster was introduced in 1926 by the McGraw Electric Company, of Minneapolis, Minnesota. A pop-down toaster was introduced by Beardsley and Wolcott in 1930 as was the Proctor toaster with a bimetallic heat sensor to test the surface temperature of the bread instead of relying on a timer to regulate toasting time.

The toaster oven was introduced in 1922 by Best Stove, Detroit, Michigan. The first toaster oven to use horizontal toasting was introduced by General Electric in 1961 followed in 1971 by a king-size version. General Electric's Toaster Oven was advertised for heating frozen foods as well as making toast.

The electric food mixer was introduced in 1918, and the blender was developed in 1932 by Stephen Poplawski. Fred Waring, the famous orchestra leader of the 1930s, patented Fred Osius's blender in 1938. The successor to the blender, the food processor, had it roots in the Kenwood Chef (1947), and a new commercial unit, the French Robot Coupe patented in 1963. A home model of the Coupe, the Magimix, was introduced in France in 1971, but was not successful. Its successor, the Cuisinart, imported to the U.S. in 1975, combined the features of a blender, shredder, and slicer, providing a popular multipurpose tool.

The electric fry pan began early in the century, but was not popular until the 1950s. Westinghouse developed the first electric frying pan in 1911, but it was not until Sunbeam introduced theirs in 1953 that the electric fry pan became popular. The immersible electric fry pan was introduced the following year by Farber.

The Naxon Beanery, the forerunner of the Crock-pot, appeared in 1969. The Crock-pot was an electric kettle, usually with a ceramic inner container, for slow cooking of foods. The Rival Crock-pot was marketed in 1971.

Modern time-saving devices like those listed above indicate the cultural gap between modern and early societies, who did not perceive time as a quantity to be saved. Their first concern was to obtain sufficient food. It was only later that people, having secured their sources of food, became concerned with its preparation and sharing successful methods with others. How early this occurred is not known. Our first written record of cooking methods come from the 2d century B.C.

Table 4.2. Cooking Utensils and Tools

Dates	*Event*
B.C.	
6000	Pottery used (WB).
3500-3000	Potter's wheel was used in Mesopotamia (TH).
3250	Evidence in Sumeria of vases made on potter's wheel (DI).
3000	Egyptians glazed pottery (WB).
2500-2000	Potter's wheel and kiln used in Mesopotamia (TH).
1600	Soapstone pots used in Egypt (WB).
1500	Glass vessels made in Egypt and Mesopotamia (EB).
1500-1027	Chinese used the potter's wheel (WB).
1400	Fine white stoneware made in China (EB).
1200	Earliest record of fused enameled ware (DID).
800-200	First iron utensils used (TH).
50	Potter's wheel used in southern England (DID).
30	Blowpipe for glassware used in Sicily (EB).
A.D.	
200	Pastry crimper was used in Roman Gaul (EL).
618-907	Early form of porcelain made in China (EB).
>1057	Venetian daughter of Byzantine emperor, Constantine X Ducas, used a small gold fork (PA78).
1100	Fork introduced by the wife of Italian nobleman, Doge Domenico Silvio of Venice (WB).
12th c	English glassmakers first made crystal glass (WABI).
1226	Glass bottles and windows manufactured at Chiddingfold, Surrey, by Laurence Vitracius (1ST).
1279-1368	Chinese developed hard porcelain (EB).
1307	Inventory of Edward I, England, mentioned seven forks (WABI).
1380	Inventory of Charles V, France, mentioned twelve forks (WABI).
1384	Ordinance of Pewters' Guild, London, founded (EDE).
1389	Inventory of the duchess of Touraine mentioned two forks (WABI).
1463	Silver fork mentioned in the will of John Baret of Bury (1ST).
	Beroverio, Murano, Italy, developed Venetian glass (WABI).
1535	First glass made in the western hemispere (WB).
1545	Bernard Palissy made his first enameled ware (DID).
1575	Soft paste porcelain manufactured by Bernado Buontalenti, Florence (1ST).
1578	Conrade brothers opened a faience pottery in Nevers (1ST).

1600	Caspar Lehman, jewel cutter to Emperor Rudolf, began cut-glass process (TH).
1608	Forks introduced to Great Britain from Italy by Thomas Coryate (1ST).
1609	Caspar Lehmann of Prague granted a patent on his cut-glass process (1ST).
	Tin enameled ware made at Delft (TH).
1614	Glass industry established in England (TH).
1642	First cooking utensils made in the United States at Saugus, Massachusetts (EL).
1674	Lead glass made by George Ravenscroft of England (WB).
1679	Denis Papin, France, demonstrated the steam digestor with safety valve, first steam pressure cooker (DID).
1687	Coffee grinder invented (WABI).
1690	First identifiable English silver teapots (500).
1702	Corkscrew mentioned in an analogy in a scientific article on the tadpole (RM).
1707	E. W. von Tschirnhaus and J. F. Bottger discovered the secret of making "hard" porcelain like Chinese (TH).
1710	Porcelain factory established in Meissen, Saxony (TH).
1719	John Akerman made cut glass in Great Britain (1ST).
1720	First tinware made in United States at Berlin, Connecticut (EL).
1750	Johann H. G. von Justy, Germany, suggested use of enameling for cooking utensils (EL).
1756	Porcelain factory established at Sevres (TH).
1760	Josiah Wedgewood founded his pottery works at Eturia, Staffordshire, England (TH).
	The modern tablespoon with a rounded back adopted (EB).
1778	Eli Whitney at age thirteen is supposed to have devised an apple parer (FHTC).
1780	Thomas Turner introduced Willow pattern china (1ST).
1781	Joseph Sterling, South Woodstock, Vermont, developed an apple parer (FHTC).
1782	Saint-Louis, France, successfully fabricated crystal glass (WABI).
1784	George Washington's expense ledger recorded the purchase of a "cream machine" that used ice (RM).
1785	Bench-type apple parer made by Daniel Cox and Sons, West Woodstock, Vermont (FHTC).
1799	Dr. Hinkling invented an enameling process for saucepans (DID).

1800	French "biggin," drip pot for making coffee and named for the inventor, introduced to U.S. (EL).
	Jean Baptise de Belloy, archbishop of Paris, devised the first percolator (dripped through to pot below) (BCT).
>1800	Count Rumford improved de Belloy's coffee pot by adding a device to compress grounds (BCT).
1802	Descroiselles, Rouen pharmacist, invented the cafeollete, a form of drip coffee maker (WABI).
>1802	Antoine Cadet created a porcelain coffee pot (WABI).
1803	Moses Coates, Chester County, Pennsylvania, patented the first mechanized apple parer (LR).
1809	S. Crittenden, Guilford, Maine, patented a parer and corer (LR).
1825	Bakewell, Page, & Bakewell Company, Pittsburgh, produced the first pressed glass (WB).
	Boston & Sandwich Glass Company founded by Deming Jarves, early maker of pressed glass (WB).
1827	Boston & Sandwich Glass Company made the first pressed glass tumbler (WB).
	Jacques-Augustin Gandais, Paris jeweler, devised the first practical pump percolator (tube in handle) (BCT).
	Nicolas Felix Durant invented a pump percolator with an internal tube (BCT).
1829	Coffee mill patented by James Carrington, Wallingford, Connecticut (FFF).
1834	Melamine first developed (PM).
1838	R. W. Mitchell, Springfield, Ohio, patented a combined corer, parer, and slicer (LR).
1839	Clarke patented a new method for enameling kitchenware (DID).
1840	Napier vacuum machine for coffee making developed by Scottish marine engineer, Robert Napier (BCT).
1844	George Kent invented a knife-cleaning machine (DID).
1846	Nancy Johnson invented an ice cream maker (LR).
1848	William Young patented Nancy Johnson's ice cream maker (LR).
1849	Charles P. Carter, Ware, Massachusetts, patented a screw apple parer (LR).
1853	James F. and Edwin P. Monroe patented the egg beater (KC).
1858	First U.S. patent for a can opener was issued to Ezra J. Warner, Waterbury, Connecticut (EL).

1859	Wire egg beater patented by J. F. and E. P. Monroe was one of the more successful models (LR).
1860	C. Degiton, Philadelphia, patented a potato paring knife (LR).
	First U. S. patent for the corkscrew issued to M. L. Byrn, New York (RM).
1863	Timothy Earle, Smithfield, Rhode Island, patented a metal strip egg beater (LR).
	Tilden's universal sifter was patented (flour sifter) (KC).
1865	Earnshaw's Patent flour scoop and sifter patented (KC).
	James H. Nason, Franklin, Massachusetts, patented a coffee percolator (PA78).
1866	J. Osterhoudt, New York, invented the key opening can (WABI).
1868	Many paper products available including paper plates, cups, collars, vests, napkins, and towels (DH).
1869	Cornelius Swartout patented the first U.S. waffle iron (HS).
1870	William W. Lyman, U.S., patented the first can opener to use a cutting wheel, not a spike or knife blade (EL).
	F. A. Walker offered tin and copper measures for home use including cooking (FHTC).
	F. A. Walker sold a potato masher (FHTC).
1872	U.S. patent issued to Silas Noble and James P. Cooley, Granville, Massachusetts, for toothpick-making machine (WJ).
1874	Monroe improved his egg beater and assigned the patent to Dover Stamping Co. (LR).
	Jacob Vollrath, Sheybogan, Wisconsin, made enameled cast iron ware (SS).
1877	W.E. Brock, New York, patented a popular hand apple parer and corer (LR).
1879	Bread slicing machine manufactured by Summerscales, Keighley, Yorkshire (MH).
	Rotary masher for vegetables manufactured by George Kent Co. (MH).
1880	Can opener patented by William E. Brock, New York (KC).
1884	Carl Zeiss, Jena, Germany, developed boron-silicon glass later developed by Corning into Pryex glass (WABI).
1885	First appearance of the can opener in Army and Navy Stores catalog (DBE).
1886	Marvin Chester Stone devised the paper drinking straw (HS).

1887   Paper napkins used at John Dickinson & Company annual
       dinner, Castle Hotel, Hastings (1ST).
1888   Marvin Chester Stone, Washington, D.C., patented the
       paper drinking straw (WABI).
1890   Egg lifter for poached eggs patented by Ann E. Smith (KC).
       First aluminum saucepan made by Henry W. Avery,
       Cleveland, Ohio (1ST).
       Matthai-Ingram offered a graduated measuring cup (FHTC).
1891   E. H. Whitney and J. L. Kirby, Cambridge, Massachusetts,
       patented an egg beater incorporating principles still
       used (LR).
       Electric kettle marketed by Carpenter Electric
       Manfacturing Company, St. Paul, Minnesota (1ST).
1892   Thermos flask invented by James Dewar, Cambridge,
       England (1ST).
       Pittsburgh Reduction, later Alcoa, made first stamped and
       cast aluminum cookware (EL).
1893   Electric toaster marketed by Crompton Company,
       Chelmsford, England (1ST).
1894   First electric kettle shown in Chicago (TT).
       House Furnishing Review (January 1894) described machine
       similar to Silex coffee maker (EL).
1897   Sears offered aluminum cooking utensils in its catalog (SS).
1900   Boker Coffee offered quarter, half, and one teaspoon
       measuring spoons (FHTC).
>1900 Charles Francis Jenkins invented the conical paper cup
       (HS).
1902   Frank Clarke, Birmingham gunsmith, patented the first
       automatic tea making machine (TT).
       Electric kettle introduced into Britain (TT).
1904   Vacuum flask, originally designed for lab work, first used in
       the home (TT).
1905   Marvin Chester Stone invented a machine for making paper
       drinking straws (WABI).
1907   Bakelite invented by Dr. Leo Baekland, Belgium (1ST).
1908   Paper cups introduced by the Public Cup Vendor Company,
       New York (1ST).
       Melitta Bentz, Germany, devised the Melitta coffee filter
       (WABI).
       Landers, Frary & Clark introduced an electric percolator,
       Universal Percolator, with cold water pump (EL).
1909   First commercial manufacture of Bakelite undertaken (TH).
       General Electric made the first electric toaster (TT).

	Mrs. Ann Bridges and Mrs. Sutton acquired rights to a coffee maker and used new Pyrex glass (EL).
1911	Westinghouse introduced the first electric frying pan (EL).
1915	Pyrex glass produced by Corning Glass Laboratory, U.S., resistant to chemicals and heat, used for cooking (TT).
	Long-life stainless steel designed specifically for cutlery patented (TT).
1918	First electric food mixer available in U.S. (TT).
	Landers, Frary, & Clark, New Britain, Connecticut, introduced an electric waffle iron (EL).
1920	Armstrong Electric, Huntington, West Virginia, introduced a waffle iron with a heat indicating light (EL).
>1920	Cartier introduced personalized silverplated swizzle sticks (CDIS).
1921	Swan electric kettle, with double the heating power of others, introduced by Bulpitt and Sons, England (TT).
	International Silver Co., Meriden, Connecticut, produced knives with stainless steel blades (EL).
1922	Best Stove, Detroit, introduced "oven type" toaster in which bread was held vertically (EL).
	Joseph Block, U.S., suggested use of whistle on tea kettle to German manufacturer, Gebruder Hansel, Westphalia (EL).
1923	Air-o-Mix, Delaware, marketed Whip-All, a portable mixer (WABI).
1924	D. A. Rogers, South Minneapolis, announced the first automatic toaster with a dial to regulate timing (EL).
1925	Star Can introduced the first can opener with a toothed can wheel to rotate can against cutting wheel (EL).
	American Home Economics Association requested the U.S. National Bureau of Standards to investigate standard measures for home (JHC).
1926	Pop-up electric toaster was marketed by McGraw Electric Company, Minneapolis, Minnesota (1ST).
	Waters Genter Co., Minneapolis, introduced Toastmaster, the first successful automatic ejector toaster (EL).
1927	Wall-mounted can opener marketed by Central States Manufacturing Company, St. Louis, Missouri (1ST).
	Polar Ware Company, Sheboygan, Wisconsin, made the first stainless steel cookware (EL).
	Hautier, France, patented the first controlled low pressure cooker (WABI).

Englishman Frederick Hurten Rhead was named art
director for Homer Laughlin China Company (he later
designed Fiestaware dinnerware) (USAIR).

1930   First electric kettle with safety shut off available from
General Electric, England (TT).

Proctor Electic introduced the first automatic toaster with
heat sensor to monitor toast cooking (EL).

Beardsley and Wolcott, Waterbury, Connecticut, offered
"drop down" toaster, bread inserted at top, drops down
(EL).

Sunbeam, Chicago, marketed Mixmaster, the first stationary
mixer (WABI).

Waters Genter Co. introduced their Waffle Master, an
automatic waffle iron (EL).

1931   Preston C. West, Chicago, patented a portable electric can
opener, but it was apparently never made (EL).

Bunker, the first can opener to use pivoted handles to hold
can with key-type handle for cutting wheel, was
introduced (EL).

Knapp-Monarch Company introduced Therm-a-Magic
percolator that shut off after making coffee (EL).

1932   Stephen J. Poplawski developed a blender from his mixer
(EL).

1933   Polyethylene produced at Northwich Cheshire Laboratories
of Imperial Chemical (ICI) by R. O. Gibson (1ST).

First all-electric tea maker introduced by Goblin; it included
an electric light and clock (TT).

Samson United, Rochester, New York, designed a toaster
to accommodate rolls and bread (EL).

1936   Faberware Coffee Robot, a vacuum-type brewer with
thermostatic control that kept coffee warm indefinitely,
was introduced (EL).

Fiestaware dinnerware introduced by Homer Laughlin
China Company (BUSWK).

1937   Kitchen Aid introduced the first electric coffee grinder (EL).

James Kennedy developed the process used to make the
first copper-clad bottom on stainless steel cookware
(EL).

Toast-o-Lator, New York, introduced a toaster using a
conveyer device (EL).

Alfred Vischer, Chicago, designed a pressure cooker as
interlocking pan and lid eliminating bolts (TT).

	Roy J. Plunkett, Dupont, discovered Teflon (WJ).

Roy J. Plunkett, Dupont, discovered Teflon (WJ).

Fred Waring patented the blender developed by Fred Osius (EL).

1938-41 Peter Schlumbohm invented the Chemex coffee maker (EL).

1939 Polyethylene manufacture begun by Imperial Chemical Industries (ICI) (1ST).

Hand-held electric knife marketed in the U.S. (TT).

First hand-cranked can opener with "positive single-action," no separate piercing lever, made by Vaughan Manufacturing (EL).

1940 Swing-a-Way, the first wall-mounted can opener with mounting permitting it to swivel right or left (EL).

1942 Red Fiestaware dinnerware made with uranium oxide discontinued because of World War II (BUSWK).

1944 Reynolds Metal Company demonstrated aluminum foil for home use in New York City (full production began in 1947) (MITU).

1945 Tupperware Corporation founded (TT).

1946 Gaggia, Italy, invented the expresso machine for coffee (WABI).

1947 Kenneth Wood, England, designed Kenwood Chef, the forerunner of the food processor (WABI).

1948 Teflon first commercially produced by DuPont (TT).

Swingmaster, the first can opener with magnetic lid-lifter, introduced by Steel Products Manufacturing (EL).

1949 Sunbeam introduced a toaster with a bimetallic sensor control; bread lowered itself automatically (EL).

Rival Manufacturing offered the first can opener with removable cutting wheel for easier cleaning (EL).

1953 Karl Ziegler invented a cheaper process using atmospheric pressure for producing polyeythlene (TT).

Sunbeam marketed automatic electric fry pan (KC).

1954 S. W. Farber, England, introduced an immersible electric frying pan developed by H. K. Foster (EL).

1955 Tefal Co., Paris, manufactured nonstick saucepans (1ST).

General Electric introduced a toaster with lower drawer for heating buns (EL).

Automatic kettle with thermostat, automatically switched kettle off after boiling, introduced (TT).

1956 First electric can openers produced by Klassen Enterprises and Udico Corp, California (EL).

Nonstick saucepans introduced to Great Britain by Habenware (1ST).

Proctor introduced a four-slice toaster (EL).

1957 Karl Zyssert marketed a vegetable shredder (TT).

Ovenproof pyro-ceram cookingware introduced, accidentially discovered when oven curing temperature was too high (TT).

1959 General Electric introduced a toaster with improved safety using a douple-pole main switch (EL).

1960 Teflon pans introduced into the U.S. by T-Fal (EL).

Robot Coupe, developer of the commercial food processor, established in France (later Cuisinart) (NYT).

1961 General Electric introduced the first toaster oven using horizontal toasting, also used for frozen foods (EL).

1962 Colgate-Palmolive introduced Baggies plastic sandwich and storage bags (CP).

1963 Robot-coupe, a restaurant food processor, patented by Pierre Verdun, France (WABI).

1964 Ronson Corporation, Woodbridge, New Jersey, introduced the Can Do, the first portable electric can opener with mixer (EL).

1965 Salton-Hotray, New York, marketed an electric bun warmer (EL).

1968 Dupont introduced Teflon II, improved Teflon (WJ).

1969 Naxon Beanery, forerunner of the Crockpot introduced (KC).

Samsonite Corporation, Denver, introduced a can opener with removable can-piercing lever for easier cleaning (EL).

Home yogurt maker marketed (TT).

1970 Corelle Livingware dinnerware introduced by Corning (EL).

1971 Rival Crockpot marketed (KC).

Pierre Verdun marketed Magimix, French food processor, for home use (WABI).

1972 Homer Laughlin China Company discontinued Fiestaware dinnerware (BUSWK).

Automatic filter drip coffee makers marketed (CT).

1975 Cuisinart, food processor developed by Robot Coupe, introduced to the U.S. (NYT).

1976 Silverstone, nonscratch Teflon, introduced by Dupont (WJ).

1986 Fiestaware dinnerware reintroduced by Homer Laughlin China Company (BUSWK).

Cookbooks

In many primitive societies the cooking falls to the women, but in advanced societies most professional chefs have been men, although women cook in the home. The earliest cookbooks of which we have knowledge were written by men. One of the earliest cookbooks, Athenaeus's *Deipnosophistai*, in Greek, dates from the 2d century B.C. and mentions several earlier texts no longer extant, including one from 350 B.C. by Archestratus entitled *Hedypatheia*. The first cookbook by a woman, Hannah Woolsey's *The Queen-like Closet or Rich Cabinet*, appeared in 1670.

Other notable cookbooks among those surviving include one written by Apicius, Rome, around 35 A.D. Huou, a royal Chinese chef, wrote his cookbook sometime between 1215 and 1294. The first French cookbook of importance, *Le Viander*, appeared in 1374. A chef of the court of Richard I wrote The Form of Cury in 1390, and another early French cookbook, La menagier de Paris, was published in 1394.

After the invention of the Gutenberg printing press and the development of paper technology, cookbooks were more readily available for household use. The first printed cookbook, *Honesta Volupate* by Bartholomaeus Platina, Venice, appeared in 1475. The first American cookbook, *The Compleat Housewife*, compiled by Mrs. E. Smith, Williamsburg, Virginia, was published in 1736. Two other cookbooks popular in the United States appeared in 1796, Susanna Carter's *The Frugal Housewife* and Amelia Simmons's *American Cookery*. Mrs. Carter's book was illustrated by Paul Revere.

An increasing number of cookbooks appeared during the 19th century. Catherine Beecher published *Miss Beecher's Domestic Receipt Book* in 1846. It was the first to describe how to perform basic cooking techniques. Mrs. Beeton's *Book of Household Management* was printed in 1861. *The Boston Cooking-School Cook Book*, by Mary Lincoln, appeared in 1883, and the Fannie Merritt Farmer edition was issued in 1896. Mrs. Farmer's book was the first to use standardized methods and measurements, and thus allowed a variety of cooks to produce consistent results.

Table 4.3. Cookbooks

Dates	Event
B.C.	
350	Archestratus's Greek cookbook, *Hedypatheia*, written (EB).
2d c	Athenaeus, Greek, wrote *Deipnosophistal*, Greek cookbook (EB).
A.D.	
14-37	Apicius, Roman epicure, wrote a cookbook (EB).

1215-1294	Huou, China, wrote *The Important Things to Know about Eating and Drinking,* an early Chinese cookbook (EB).
1374	First French cookbook of importance, *Le Viander,* published (EB).
1390	Chef of Richard I, wrote *The Form of Cury,* which also contained some apple recipes (EB).
1394	One of early French cookbooks published, *La Menagier de Paris,* also contained apple recipes (EB).
1475	Cookery book, *Honesta Volupate,* Bartholomaeus Platina, Venice, published (OCLC).
1670	Hannah Wooley wrote first cookbook by a woman, *The Queen-like Closet or Rich Cabinet* (EB).
1736	First American cookbook, *The Compleat Housewife,* complied by Mrs. E. Smith, Williamsburg, Virginia (WG).
1740	Cookery school established by E. Kidner, London (1ST).
1796	Susanna Carter, Philadelphia, published *The Frugal Housewife; or The Complete Woman Cook* (FHTC). Amelia Simmons, Hartford, Connecticut, published *American Cookery* (FHTC).
1846	Catherine Beecher published *Miss Beecher's Domestic Receipt Book* describing basic cookery techniques (WL).
1861	*Mrs. Beeton's Book of Household Management* published (TH).
1883	Mary Lincoln published the first *Boston Cooking-School Cookbook* (WL).
1896	Fannie Merritt Farmer's *The Boston Cooking-School Cookbook* published; it used standardized measures (EB).
1984	Recipe Writer, software package for home use, with master index, menus, and shopping list, introduced (AMERW).

Cooking techniques have evolved over the centuries from the slow, tedious process of holding meat on sticks over an open fire to the modern use of a microwave oven that cooks food in a few minutes. We now use plastics and even paper for cooking utensils that are often disposable. Meal preparation can be compressed to a much smaller proportion of our time. We can prepare and cook our food or buy it prepared for cooking or even precooked. We have the option of doing things by choice rather than by necessity. We may preserve fruits or vegetables by canning, drying, or freezing, bake fresh bread, or make pasta if we choose. But it is a matter of choice; we are not forced to do it. Our choices are limited only by our income and by the time available to us.

5
Clothing

We can only speculate as to when and why people began to wear clothing. Tantalizing clues tug at our imagination: remnants of fibers found at Stone Age sites; cave drawings; bone needles and buttons. Clothing and fashion seem to be a condition of civilized living. Clothing is necessary in cold climates and is useful in warm climates to protect the skin from sun, insects, and brush, and to preserve social mores associated with nudity. It has long been used to differentiate, to decorate, and to indicate status; clothing and fashion are natural companions.

Cave drawings depict early people wearing clothing, most likely skins laced together. Skin and leather have been, and continue to be, used for clothing, but unless leather is well cured, it can be difficult to shape and work. It can be stiff and may irritate the skin, although leather garments prepared by North American Plains Indians were soft and pliable. At some point early man discovered various plant and animal fibers could be twisted to form threads and these threads could be woven to form fabrics. Cloth is softer, more absorbent, lighter weight, and generally more comfortable to wear; it drapes and shapes itself to the wearer. Little or no remains have been found of early clothing, but traces of fabric have been identified at some Stone Age sites.[1] Remains of felt, a nonwoven fabric made of wool or animal hair, have also been found at early settlements.[2]

Clothing was worn by peoples of early civilizations in the Mediterranean and Near East: various surviving pictures, carvings, sculptures, monuments, and documents depict a variety of styles and fabrics. At first fabrics were woven in relatively narrow widths and short lengths and early garments were probably constructed of uncut pieces of fabric stitched together and fastened with pins or brooches.[3] While sarongs and saris are wrapped and draped, most clothing construction requires cloth or other flexible material, tools to shape or cut the material to the desired size and form, and a means to fasten the individual pieces together. As people mastered the technology of creating cloth and clothing, they also became interested in its potential for enhancing appearance and in fashioning cloth into different shapes. Paralleling this process, this overview is divided into five categories: fibers, the process and production of fabrics, clothing construction, items of apparel or adornment, and a short section on stores, the purveyors of fashion.

Fibers

The term *fiber* can refer to either natural or manufactured fibers. Natural fibers used for clothing include wool, silk, cotton, ramie, jute, and flax (the fiber used for linen), as well as animal hair fibers such as cashmere, mohair, camel, alpaca, and angora. Wool and silk are animal fibers whereas cotton and flax are plant fibers. Manufactured fibers are derived from carbon-based compounds, some mineral fibers, and other chemicals, and include rayon, polyamide (nylon), acrylic (orlon, Acrilan), polyester (Dacron, Kodel), triacetate (Arnel), and others. While each fiber may be treated differently in the early manufacturing of fiber preparation or creating threads or yarns, later stages of processing, such as the making and weaving of fabric, becomes more and more similar.

Natural Fibers

Natural fibers are organic matter and deteriorate readily so most remains have been found in drier regions or on corpses in peat bogs where conditions preserved such fibers. No one knows which fibers were first used by humans, but both wool and flax are among the earliest found in archaeological excavations.[4] Traces of wool have been found in Anatolia dated at about 6500 B.C., and ceramic seals related to the wool trade have also been found in Iraq, 4200 B.C., indicating trade in wool from Central Asia. Flax was cultivated in India and China around 5000 B.C. and traces of wool and flax have been found at Swiss lake dwellings dated at 4000 B.C. Linen, made from flax, appeared in Egypt about 3500 B.C. and a very fine (540 threads/inch) linen was found there dated at 2500 B.C.

The evolution and changes in sheep breeding were discussed in chapter 2, and it was noted that by 3000 B.C., white woolly sheep were being raised in Sumer. A major advance for increasing wool production came with the development of the shearing machine in Australia. While James Highen, a

Melbourne printer, patented a shearing machine in 1868, the first successful machine was developed by Frank Wolseley and John Howard in 1883 after Wolseley had patented three earlier machines. The machine was first used in 1888 at the Dunlop Station.

Figure 5.1. Fibers and Cloth Production

YEAR		
1987		
1975		Air jet loom
	Arnel, triacetates	
1950	Orlon, Dacron (polyester)	Sulzer loom
	Nylon, Merinova	
	Cotton picking machine	Draper loom
1900		
	Rayon, Wool shearing machine	Northrop loom
	Rubber cloth	Fancy goods loom
1800		Jacquard loom
	Cotton gin, casein	Power loom, steam factory
		Spinning jenny, water frame
1600		Cotton mill (U.K.)
	Silk (France)	Saxony wheel
		Spinning flyer
1200		Spinning wheel (Paris)
	Silk (Italy)	
	Cotton (Greece), silk (Europe)	
400	Silk (Korea)	
	Silk Road	
-1200		
	Cotton (Peru)	Spinning wheel (India)
	Silk (China), cotton (India)	
	linen (Egypt, Europe)	
-4400	Wool (Europe)	Horizontal loom (Egypt)
	Linen (China, India)	
-10800		

The raw materials for wool and linen were readily available in Europe. While fine woolens and linens were prized and handed down from generation to generation, they were not as rare or as costly as cotton and silk.

Traces of cotton use were found in Mohenjo-daro, India, dated about 3000 B.C. along with spindle whorls too light for wool that were most likely used for cotton. Cotton grave cloths have been found in pre-Inca Peru, 2000 B.C., showing that cotton also was cultivated in the New World. By 700 B.C. King Sennacherib, Assyria, was planting cotton "trees." It was imported to Europe by 300 B.C. but was very costly. Until the 7th century A.D., India and Egypt were the main sources of cotton, which was a luxury fabric in European countries until it became available from the Americas.

Cotton growing and manufacture spread slowly westward. By 630 A.D. Arab countries were producing cotton. During the 8th century it was grown on the Greek mainland, and by the 12th century Mediterranean cotton was processed into cloth on the Italian peninsula. Cotton was manufactured in Spain in 1225, and by 1641 it was also being manufactured in Manchester, England, which later became a major center for the English textile industry. A cotton mill was established by English settlers in the United States in 1638.

Of major importance to the spread and use of cotton was the cotton gin invented by Eli Whitney in 1793. It significantly reduced the labor of removing seeds and waste during the processing of raw cotton. The idea for the gin was suggested to Whitney by his landlady, Catherine Littlefield Greene, who had spent part of her life on a cotton plantation and was familiar with cotton harvesting and processing. She also suggested the use of metal teeth to clutch the seeds but slide through the cotton fiber.[5]

Another labor intensive operation was the hand picking of cotton until the development of mechanical pickers. Angus Campbell, a Scot, invented the first spindle-type picking machine in 1899. Hiram M. Berry invented a barbed-spindle machine in 1925, but the most successful was the picker developed by John and Mack Rust in 1927.

Silk was, and remained, a luxury fabric until modern times. Silkworm cultivation, sericulture, and silk manufacture were carefully guarded secrets. Silk appeared in China around 2640 B.C. By 1766 B.C. the Chinese were producing silk twill damask. Traces of silk were found in Swiss Hallstatt sites of 699-601 B.C.; whether it originated in the East or came from Greece is uncertain. Since Alexander's conquests and Roman trade occurred later, if the silk was produced in the East, the circuitous route would provide some interesting historical data. The Greeks did have a form of raw silk, called *tusseh*, which came from wild silkworms, but it was not of the same quality as Chinese silk.

The Silk Road to China was established by 100 B.C. In the opinion of some critics, the importation of costly silk fabrics contributed to the fall of the Roman empire by draining the imperial treasury and preoccupying some citizens with luxury instead of the affairs of state. The cost of raw silk was one pound of silk for three pounds of gold.[6] However, Ernst Samhaber, in *Merchants in History*, disagrees and suggests instead the major contributor was the decline of the merchant class and of trade generally.[7] By 300 A.D. Chinese peasants had carried sericulture to Korea and from there it made its way to Japan. A Chinese princess took silkworms, hidden in her headdress, to Khotan in western China, a major trading center with the West, in 419 A.D.

While silk manufacture was slow to reach the West, the technology spread gradually throughout most of Europe starting in 552 A.D. when agents of the Byzantine Empire smuggled silk worms out of China. By the 8th century the Arabs had also learned sericulture and in 827 introduced silk manufacture to Sicily. Sometime in the 1100s silk manufacture began in Italy. By the 1400s it was also established in France. Silk is now produced mainly in the Far East, India, and the Mediterranean. Of the 56,128 metric tons produced in 1984, more than half came from China and Japan.[8]

While rubber is a natural substance, it is not a fiber. In some ways it is similar to various plastics that are not spun, but cast or rolled into sheets. Thomas Hancock of England developed a process for making rubber in 1820 which he used to produce such articles of dress as suspenders, trouser straps, and garters. He also developed sheet rubber for hospital sheets.

The development of manufactured fibers in the late 19th and throughout the 20th century led over time to a decline in the use of natural fibers partly due to the low cost and wrinkle shedding properties of the new fibers. Since many manufactured fibers relied on petroleum by-products, however, the oil embargo of the 1970s led to increased costs, which when coupled with new finishes developed for natural fibers caused a resurgence of interest and use of cotton, wool, and other natural fibers. For example, cotton blends with new finishes have become popular because of their wrinkle-free properties. Second, natural fibers breathe, allowing perspiration to evaporate; they are therefore cooler in summer heat. Wool retains warmth even when wet, making it particularly popular for outdoor clothes worn in cold, wet weather. It sheds rain, yet absorbs moisture from the air or body. Sheer wools such as "nun's veiling" are used for summer garments, and similar woolen fabrics are used in hot, dry climates. New finishes and the durability, comfort, resiliency, and warmth of wool continue to make it a popular fabric. The Wool Board's promotions have also helped. Wool, like cotton and silk, is also a renewable fiber: the sheep grow a new coat every year and, if properly pastured, do little damage to the environment. Silk, despite numerous less expensive imitations (some of the first manufactured fibers were developed as silk substitutes), has retained its position as a luxury fabric.

Table 5.1. Natural Textile Fibers

Dates	Event
B.C.	
6500	Woolen cloth used at Catal Huyuk, Anatolia (found adhering to human bones) (DBE).
5000	Flax cultivated in India and China (EB).
4200	Tall-al-Asmor, Iraq, seals indicated trade in wool from Central Asia (EB).
4000	Flax used by lake dwellers in Switzerland (WB).
3500-3000	Linen woven in Middle Egypt (TH).
3000	Cotton used in India (EB).
2640	Legend reported silk developed by Chinese Empress Hsi-ling Shih (WB).
2500	540 threads/inch linen made in Egypt (WB).
	Cotton textile traces left in Mohenjo-daro, India; spindle whorls too light for wool also left (DBE).
2400	Cotton products used in Mexico (EB).
2000	Pre-Incan Peru cotton grave cloth used (EB).
700	King Sennacherib, Assyria, planted cotton plants (DBE).
699-601	Traces of silk fabric left in Hallstatt sites, Switzerland (DBE).
300	Cotton imported into Europe, but was costly (TH).
100	Silk road to China established (EB).
75	Pompey, Roman general, returned to Rome from China carrying silk (EB).
A.D.	
552	Europe silk industry was established (TH).
630	Cotton was introduced to Arab countries (TH).
8th c	Arabs learned how to raise silk worms (EB).
	Cotton grown on Greek mainland (EB).
>768	Charlemagne introduced linen manufacture to France (WB).
>827	Arabs introduced silk manufacture to Sicily (EB).
942	Linens and woolens were manufactured in Flanders (TH).
12th c	Italians introduced silk culture into Europe (EB).
	Venice was the major cotton manufacturing city processing Mediterranean cotton (EB).
>1100	Florence was known for fine woolens (EB).
1225	Cotton manufactured in Spain (TH).
1253	Linen first manufactured in England (TH).
1315	Florence captured Lucca and took Sicilian weavers to Florence (EB).
	Lyons silk industry established by Italian immigrants (TH).

1530	Pietro Martyre d'Anhiera described rubber balls in *De Orbo Novo*, Alcala, Spain (1ST).
1638	English settlers built a cotton mill in the United States (EB).
1641	Cotton goods manufactured in Manchester (TH).
1685	French Huguenots began silk manufacture in Great Britain (TH).
1762	Rubber solvent, turpentine, discovered by François Fresneau, Marennes, France (1ST).
1811	J. N. Reithoffer, Vienna, manufactured rubber goods (1ST).
1830	Crinoline, woven horsehair fabric, invented by Oudinot, France, to stiffen soldiers' collars (WABI).
1839	Charles Goodyear developed vulcanized rubber which made possible the commercial use of rubber (TH).
1865	Pasteur cured the silk worm disease and saved the French silk industry (TH).
1868	James Highen, Melbourne (Australia) printer, patented a sheep shearing machine (EI).
1872	Wolseley, Australia, developed the first practical sheep shearing machine (EI).
1878	Fur farming began in Canada (TH).
1888	Wolseley sheep shearing machine used at Dunlop Station, Australia (EI).
1899	Angus Campbell (Scot) invented spindle-type cotton picking machine (DID).
1925	Hiram M. Berry invented the barbed-spindle cotton picking machine (DID).
1927	John and Mack Rust invented a mechanical cotton-picker capable of harvesting a bale of cotton per day (TT).
>1930	Women's briefs, underwear, introduced (GBL).
1955	Jesse E. Harmond, U.S., developed the continuous process for flax fiber, thereby reducing time to produce fiber (WB).
1957	Mothproofing of wool by dipping on-the-hoof confirmed (TT).
1964	New long-lasting, cheap, moth-proofing agent, Siven 55, developed by Fibre and Forest Institute, Israel (TT).

Manufactured Fibers

The first suggestion for the development of manufactured or artificial fibers was made by Dr. Robert Hooke in 1664 when he suggested that a process for mechanically extruding fibers similar to a silkworm spinning be used. The technology then available was not suitable, and it was not until the late 19th century that a practical method was developed. After Hooke, the next major development in the history of manufactured fibers was in 1780, when C. W. Scheele discovered casein, a protein found in milk and cheese and later used for plastics and imitation wool yarns. In the period from 1855 to 1892 a number of European patents were granted related to the production of rayon or artificial silk. Chardounet patented the spinneret process for producing rayon in 1885. The first rayon factory producing women's underwear began operation in 1892 at Besancon, France, and rayon stockings appeared in 1910. Rayon was the first manufactured fiber to be produced in the U.S. Since then a total of twenty-one generic manufactured fibers have been developed.[9]

The quality and natural feel of rayon was continually improved and new manufactured fibers were developed in the 1930s, but most did not become generally available until after World War II when postwar production and manufacturing techniques boomed. In 1930 Walter of Germany produced man-made fabrics from acetylene, and by 1935 Merinova, an inexpensive imitation wool produced from casein, was manufactured in Italy. Nylon was developed by Dupont in 1937, and the first nylon stockings appeared in 1940. Polyester fiber, later known as Dacron, was discovered by E. Whinfield, England, in 1941, but not marketed until 1951 in the United States. Orlon, another polyester fiber, was introduced by Dupont in 1948, while triacetate fibers Acrilan followed in 1952 and Arnel was introduced in 1955.

Many manufactured fibers are made with petroleum and petroleum by-products. Thus, petroleum prices are a factor in the cost of production. After sharp rises in the 1970s, prices have again declined. Other organic substances can also provide the raw materials base and chemists continue to find new sources for producing such fibers.

Manufactured fibers are not subject to disease or affected by weather conditions as are plant and animal fibers. They are often less costly than natural fibers; they may be stronger, are wrinkle-free, and many are easier to clean, although they do not breathe. Continued improvements to manufactured fibers and man-made fabrics make it unlikely their use will be abandoned, despite the ongoing appeal of natural fibers. Convenience, improved properties, and cost are powerful determinants.

Table 5.2. Manufactured Textile Fibers

Dates A.D.	Event
1664	Dr. Robert Hooke suggested the production of artificial silk (DID).
1780	Casein was discovered by C. W. Scheele (later used in making plastics and nylon yarns) (DID).
1855	Georges Audemars, Switzerland, obtained the first patent for producing artificial silk (rayon) (DID).
1865	Schutzenberger was aware of process reactions to make rayon acetate (Celanese) (DID).
1883	First man-made fiber, cellulose-based artificial silk (rayon), developed by Joseph Swan, but was not used (1ST).
1885	Comte Hilaire de Chardonnet patented a process for making artificial silk (rayon) (DID).
1890	Schutzenberger, Germany, first made rayon (DID).
1892	First artificial silk (rayon) factory established at Besançon, France (1ST). Viscose process for rayon patented by Edward Bevan and Charles Cross (DID).
1898	Weaveable, dyeable viscose rayon yarn developed by C. H. Stearn and C. S. Cross, Kew, Surrey (1ST).
1899	Spinnerets for Bevans and Cross 1892 process patented (DID).
1902	W. H. Walker, U.S., patented man-made textile yarn, acetate rayon, later used for women's underwear (TT).
1905	Rayon viscose yarn manufactured commercially by Courtauld, Coventry (1ST).
1909	Synthetic rubber developed by F. Hofman, Bayer Co., Elberfeld, Germany (1ST).
1910	Synthetic rubber manufactured by Bayer Co., Leverkusen, Germany (1ST). First stockings made from synthetic fiber of rayon, Bamburg, Germany (1ST).
1914	Protein plastic casein made from skim-milk used to make plastic knitting needles and crochet hooks (TT).
1919	Stretch-spinning of rayon invented by Edmund Thiele and Dr. Elsaessor, Germany, by cuprammonium process (DID).
1930	J. Walter Reppe, Germany, made artificial fabrics from acetylene (TH).

Polystyrene commercially developed by I. G.
Farbenindustrie, Germany, used for toilet articles, cases
(TT).

1930 Polyvinyl chloride developed by W. L. Semons, of B. F.
Goodrich, U.S. (TT).

1935 Cheap, imitation wool, Merinova, manufactured of casein in
Italy, resistant to moths, heat, mildew (TT).

1937 Nylon developed at DuPont Laboratories and patented by
William Caruthers (TH).

1939 Nylon yarn produced commercially by DuPont at Seaford
factory (1ST).

1941 Polyester fiber (Dacron) discovered by E. Whinfield, Calico
Printers Association, Accrington, England (TT).

1946 Lurex, aluminum-based yarn, produced more cheaply than
other metallic yarns such as lamé (TT).

1948 DuPont introduced Orlon (polyester) (DID).

1951 Terlyene (Dacron, polyester) products manufactured in
Great Britain (1ST).
DuPont produced the first orlon (polyester) garments (TT).
Dacron (polyester) men's suits marketed by Hart,
Schaffner, and Marx (FFF).

1952 Acrilan, acrylic fiber, announced by Chemistrand Corp.,
U.S. (TT).
Nylon stretch yarn introduced from Switzerland by
Heberlein Patent Corporation, New York, and used in
men's stretch socks (FFF).

1955 Triacetate fibers, Arnel, Celanese, with luxurious feel and
appearance at lower cost, introduced (TT).

Cloth Production and Processing

Human creativity and innovativeness is amply demonstrated by the advances
made in cloth processing and production. To early people the development
of the loom was a revolution, as was the later application of water-driven
mills in the Middle Ages and still later the application of steam and electric
power to textile processes. Innovations often have a long history of trial and
error behind them, and developments in clothing production have arisen
from a natural progression of events rather than from isolated discoveries.

Just as fibers disintegrate when exposed to moisture, time, and weather,
so too do the tools used to create them. These tools developed gradually,
were carried from place to place, and passed from generation to generation
making it difficult to identify precisely where and when they first appeared.
Although little is known about the earliest methods, much has been written

about the clothing industry since the 18th century, particularly the period of the industrial revolution.

This overview of machines and processes is highly selective, focusing on those that replaced domestic techniques or had a domestic application. Some examples of such tools are spindles, looms, loom weights, bobbins, shuttles, carding tools, knitting needles, and spinning wheels. Other machines such as the sewing machines are used, not in cloth production, but in clothing construction, and are discussed in the section Sewing Equipment.

Machines

The process for weaving cloth may have developed from or along with basketry, given the obvious similarities in technique.[10] The first looms were most likely developed during the Stone Age and were probably waist looms with one end conveniently tied to a nearby stake or vertical looms tied to tree branches with weights holding the bottom threaded bar. Waist looms were used in Peru before 2000 B.C. Evidence of vertical looms has been found in the Near East: loom weights for vertical looms have been found at archaeological sites dating from 3000 B.C. The first depiction of a horizontal loom is on pottery found at al-Badari, Egypt, dated at 4400 B.C.

Since their invention, looms have continued to evolve, but little documentation is available. Although the physical remains of advanced looms have not been located, the surviving fragments of fine fabrics disclose evidence of sophisticated weaving techniques and bear witness to the existence of more complex looms. For example, silk twill damask fragments dating from 1766 to 1122 B.C. indicate that the Chinese must have developed advanced looms and sophisticated weaving techniques. Similarly, the Celts produced and traded with the Romans fine woolens made with check and plaid patterns.[11]

The drawloom, which could produce more complex patterns by raising and lowering groups of threads, was introduced to Europe during the Middle Ages, probably from China. Since the Byzantines stole the silkworm from China, it is also likely that they observed and copied Chinese loom technology. In any event, drawlooms were used in the Italian silk industry during the Middle Ages, around the 12th century.[12]

The application of new technologies came slowly and in some places met with firm resistance. Authorities in some European communities, fearing unrest and possible revolt, preferred to avoid large gatherings of workers and encouraged cottage crafts. Spinning of thread and weaving of cloth were among these cottage crafts. Wool merchants and those controlling material sources felt they could better control the price of cloth by dealing with individual workers.[13] A loom for simultaneously weaving a number of ribbons was developed in Danzig in 1579, but its inventor was killed and the loom destroyed because of fear it would reduce employment. Nonetheless, by 1682 a weaving mill with one hundred looms had been established in Amsterdam.

The textile industry was a major source of economic wealth in the Middle Ages, which is why some governments attempted to curtail the use and export of technology, while others sought to exploit it for their own advantage. Various monarchs sought to encourage or acquire textile resources including raw materials, skilled workers, and access to the secrets of technical processes. England was a major source of high quality wool during the Middle Ages, and Edward I instituted taxes and export restrictions to promote the development of a textile industry in England. Finished goods (e.g., bolts of cloth) were more profitable than raw materials (e.g., wool). When the Florentines invaded Lucca in 1315 they took the Sicilian weavers there to Florence. Henry IV of France established a tapestry factory in Paris in 1601.

By the 18th century attitudes toward technology began to change and improved looms were developed. Managing complex patterns often required additional help to operate the loom. The "draw-boy," later a part of the drawloom, was originally the human assistant who helped manage the raising and lowering of warp threads to create such patterns. In 1725 an improved mechanical draw-boy was developed, and in 1728 perforated cards were used to control the draw-boy. John Kay patented the flying shuttle loom in 1733, reducing the number of operators required and increasing weaving speed by eight times. The first power driven loom was developed by Edmund Cartwright in 1785. Joseph-Marie Jacquard introduced an improved loom in 1801 and the jacquard loom in 1805. The loom, controlled by punched pattern cards, made practical the automatic weaving of complex patterns with less labor; it is often cited in both the history of factory automation and the development of computers and programming.

Improvements of the loom continued throughout the 19th and 20th centuries. William Crompton introduced a fancy goods loom in 1837. John H. Northrop placed a new loom in operation in 1889 that featured an automatic supply of filling and thread followed in 1895 by an automatic bobbin changing loom. The Draper high-speed loom was introduced in 1930 and improved loom speeds by twenty percent. The modern Sulzer shuttleless loom was introduced in 1950, and in 1975 a faster loom using air jets to move threads was developed.

The raw material for the loom is yarn which the loom transforms into cloth. Silk could be unwound as a single, long filament from a cocoon and did not require spinning unless a thicker yarn was desired. The process of unwinding the cocoon was called reeling. Cotton, wool, and flax are composed of shorter fibers and require twisting to create yarn suitable for weaving.[14]

The technology of spinning fibers into yarn developed more slowly than that of weaving. The spinning wheel appeared in India about 1300 B.C. The Chinese added a treadle and used it for silk reeling by 200 A.D. There is some argument as to whether and how it traveled to Europe.[15] In any event, the spinning wheel appeared in Paris in 1268, and by 1273 Francesco Borghesano of Bologna had developed a hydraulic version. Leonardo da Vinci developed the spinning flyer in 1516, and his design was used in 1533 by Johan Jurgen to develop the Saxony wheel, a heavy duty wheel for spinning coarse wool and cotton. The spinning wheel was in common use by 1530.

The 18th century contributed a number of innovations to spinning, including the application of steam power. John Hargreaves invented the spinning jenny for mechanically spinning yarn in 1764, and Richard Arkwright invented the water frame, water-powered spinning machinery, in 1769. Further developments rapidly followed with Samuel Crompton patenting the "mule," 1,000 spindles operated by one person, in 1779. Watt and Boulton installed the first steam engine in a cotton spinning factory in 1785.

Many natural fibers must be processed before they are suitable for spinning. Wool and cotton are carded; the fibers are separated and straightened. A detailed description of manual carding processes and equipment is given in *Foxfire 2*.[16] Carding technology underwent a similar evolution. The first English carding machine developed in 1553, like the Danzig ribbon loom, was considered a threat and was banned by the English Parliament. During the late 18th century, however, just as faster looms created a demand for more rapid spinning, so too was it necessary to improve carding. Consequently, carding developments paralleled those in spinning and included water- and mule-powered devices.

Heckling, a process similar to carding and combing, but used for flax, separates and straightens flax fibers. A heckling machine for flax was developed in 1832, and a continuous process for flax fiber processing was developed by Jesse E. Harmond, U.S., in 1955, which reduced to ten days the time required to process fiber. Older methods could take as long as two years since part of the process included partial rotting of flax plants.

Processes used to finish fabric were mechanized earlier than cloth production. Nicias, Roman governor of Greece (146 B.C.), was credited by Carter in the *Dictionary of Inventions and Discoveries* with developing fulling, the process of removing residual grease from woolen cloth, of shrinking or sizing the wool to retain its shape, and of felting or smoothing the surface, thereby strengthening the cloth.[17] Roman laundries, however, used fuller's earth, a claylike material that absorbs grease and oil, to clean garments during Roman times. Too, much Roman wool was purchased from the Celts so it is likely the Celts also knew about or may have used fulling, and the Egyptians used natron, a form of soda, for washing and scouring linen cloth.[18] Consequently, while the process of fulling and felting woolen cloth may owe its English name to the Latin launderers, the process was also used by other, earlier cloth makers, thus casting doubt on Nicias's contribution.

The first water-powered fulling mill is recorded in 1010 A.D. During the Middle Ages, in some regions, the peasants were required to use the fulling mill controlled by the clergy or the nobility, adding additional cost to the production of cloth for the peasant and producing added income for the abbot or lord. By the end of the 14th century fulling was done almost exclusively by mills.[20]

Water power, which has had a variety of applications, was used to drive both Borghesano's spinning wheel and Arkwright's water frame. Water power was also applied early to the silk industry. By 1700 northeastern Italy boasted 100 silk mills. John Loombe copied Italian technology when he and his brother Thomas established their silk mill at Derby in 1719.

Water mills were also used for the scutching (separating the rotted flax fibers), the washing, and the beetling (finishing the cloth) process in linen production. Water power was not used for cotton production until 1770. By then, steam technology had become practical and was quickly adapted to cotton manufacture, as was described above in spinning and loom developments. More modern developments in cloth finishing include finishes to prevent fabric from wrinkling; these were developed for cottons in 1932 by R. S. Willows and for wool in 1957 in Australia.

It is not always possible to trace the first appearance of a particular type of fabric, but the following examples provide a rough guide. Fine wool crepe was made in Bologna, Italy, in 680 A.D., wool broadcloth was made in England in 1197, Joinville mentions velvet in 1277, but cotton velvet was not made in England until 1756. Lace was made in France and in Flanders in 1320 and in England in 1626. Oil cloth was first made by Nathan Taylor, Knightsbridge, London, in 1754, while brocade was made at Lyons, France, in 1757.

Fabric Decoration

Cloth has been dyed from earliest times: traces of colored fibers have been found in Stone Age sites.[20] The Egyptians used alum as a mordant for fixing natural dyes.[21] Until the 19th century, dye processes changed little, and dyes and techniques used in ancient times are similar to natural dyes and processes used in the 20th century in Appalachia.[22]

Natural dyes are made from a variety of substances including plants, animals, and minerals. The color of the dyed cloth depends on the raw materials used including the dye matter, the fabric, the mordant or pretreatment, and the finishing or fixing process, as well as the method of dyeing and drying. Some natural dyes cannot be used without mordants; they will not penetrate fabric fibers or remain colorfast. Different mordants used with the same dyestuff and same fabric will produce different colors. Mordants used in ancient times included alum (an aluminum based compound) and salts of various metals, including iron, tin, and copper. It was not until 1850 that salt of chronium was used. The raw material for some dyestuffs were in limited supply; Tyrian purple was no longer available after the mid-15th century and one dyestuff was made from the dried bodies of a species of female insects found in Mexico.[23] As a consequence there was considerable interest in finding substances that could be manufactured from inexpensive, readily available substances.

Manufactured dyes were developed in the latter half of the 19th century, and modern dyers owe much to the methods for the extraction of aniline from indigo by Otto Unverdorben in 1826 and to the process using potassium hydroxide developed by C. J. Fritzsche in 1841. The first aniline dye, produced by William H. Perkins in 1856, changed the colors used in fabrics in the West. With aniline dyes it was possible to produce brilliant colors relatively cheaply. Some of the more garish colors produced during the late 19th century were a direct consequence of the new aniline dyes. Among the early colors produced were mauvine and fuchsine (magenta). Initially these new colors were not as long lasting as natural dyes, but as chemical processes improved, the dyes became better and eventually replaced natural dyes for most purposes. The cost of dyestuffs and dyeing also declined, enabling even the poorest to wear bright colors.

Manufactured fibers required special dyes because many were less absorbent than natural fibers. A dyeable viscose rayon yarn was developed in 1898 by C. H. Stearn and C. S. Cross, Surrey, England, while A. Clevel produced a new class of dyes suitable for rayon in 1922. Some newer processes allow insertion of dyes or coloring before the fibers are formed during the spinning process.[24]

Fabric decoration also has a long history. In addition to dyeing, embroidery, applique, and printing have also been used. Printed fabrics may have been produced in India by the 4th century B.C., while printing blocks were discovered in Egyptian burials at Akhmin dated at 300 A.D. Pre-Colombian printed textiles (<1492) were produced in Peru and Mexico.[25] Printing and painting of fabrics was also practiced in Mughal India in the 16th and 17th centuries.[26] A machine for printing fabrics was devised in 1783 and became the basis for similar modern machines. A silk screen print process was developed in Manchester by Samuel Simon in 1907.

Table 5.3. Cloth Production and Processing

Dates	Event
B.C.	
4999-4001	Spindle whorls used at Grotte du Pontal, Herault (EB).
	Plaited, woven cloth, Robenhausen, Switzerland (EB).
4400	Earliest evidence of use of the horizontal loom, depicted on pottery dish at al-Balari, Egypt (EB).
4th c	Fabric printing done in India (EB).
A.D.	
200	Chinese added treadle to spinning wheel and used it for silk reeling (EB).
300	A print block probaby used for textiles was included in a burial at Akhmin in Upper Egypt (EB).
680	Crepe (fabric) first made, Bologna, Italy (DID).
1010	First useful application in the West of a fulling mill on River Serchio in Tuscany (DID).
1138	Tanning mill belonging to Notre-Dame de Paris mentioned (MM).
1197	Broadcloth first made in England (DID).
1266	Sicilian weavers fled to Lucca, Italy, following French invasion and unrest in Sicily (EB).
1268	Spinning wheel used in Paris (RF).
1272	First spinning wheel was probably used, Bologna (UZ).
1273	Francesco Borghesano, Bologna, built a hydraulic spinning machine (UZ).
1277	Velvet first mentioned by Joinville (DID).
1320	Lace first made in France and Flanders (DID).
1400	Improved felt-making process developed (WB).
1500-1700	Mughal Indians painted and printed fabrics (EB).
1516	Dyestuff indigo imported to Europe (TH).
	Leonardo da Vinci invented the spinning-flyer, first continuous movement in textile manufacturing (EB).
1530	General use of the spinning wheel in Europe (TH).
1533	Johann Jurgen built a flyer spindle/rotary swift that included a treadle, using da Vinci design, Saxony wheel (DBE).
1553	Carding machine constructed, but English Parliament forbade its use (UZ).
1571	Linen damask first made in England (DID).
1579	Loom for weaving a number of ribbons at once developed in Danzig, but the inventor was killed and the machine destroyed (UZ).

1589	Reverend William Lee, Cambridge, invented the stocking frame, the first knitting machine (TH).
1626	Lace first made in England at Great Marlow, England (DID).
1631	Calico introduced into England by the East India Co. (DID).
1667	Feltmakers Company, London, incorporated (TH).
1682	Weaving mill with 100 looms established in Amsterdam (TH).
1712	Calico printery established by George Leason and Thomas Webber, Boston, Massachusetts (FFF).
1717	Silk throwing machine invented by John Lombe, copied from an Italian machine (DID).
1719	Lombe Brothers built the first British textile factory, the Derby Silk Mill (EB).
1725	Basile Boucher improved the "draw-boy" on a loom, which was important for pattern weaving (EB).
1728	Perforated card used to control the loom "draw-boy" (EB). Lewis Paul invented and patented two carding machines (DID).
1733	John Kay patented the flying shuttle loom, which increased loom speed eight times (TH).
1742	Cotton factory established in Birmingham and Northampton (TH).
1743	East Indian yarns imported to England for finer cotton goods (TH).
1754	Oil cloth first made by Nathan Taylor at Knightsbridge, London (DID).
1756	Cotton velvet first made at Bolton, Lancashire, England (TH).
1757	Brocade first made at Lyons, France (DID).
1758	Ribbing machinery for the manufacture of hose invented by Jedediah Strutt (TH). Lace making machine invented by Strutt (DID).
1764	Spinning jenny invented by John Hargreaves (AO).
1769	Richard Arkwright patented the water frame, water powered spinning machinery (EB).
1770	Sixteen-spindle spinning jenny patented (AO).
1771	Sir Richard Arkwright operated the first spinning mill in England (TH). First synthetic dye for cotton invented by Dr. R. William (DID).
1773	Richard Arkwright produced all-cotton calico (EB).
1775	Silk winding machine invented by Leture, Lille (DID).

1779	Samuel Crompton's mule with 1,000 spindles operated by one person (EB).
	Steam-driven factory equipment installed at James Pickard's button factory, Birmingham (1ST).
1782	Antonine Germondy of Lyons patented a hand- or water-operated carding machine (DID).
1783	Copper-cylinder calico printing developed by Henry Bell (TH).
1784	Eighty spindle spinning jenny patented (AO).
	Fournier de Granges, France, improved Germondy's carding machine (DID).
	Simon Pla, France, invented a mule-powered carding machine (DID).
1785	James Watt and Matthew Boulton installed a rotary motion steam engine in a cotton spinning factory (1ST).
	E. Cartwright patented the first power driven loom (EB).
1789	First steam driven cotton factory operated in Manchester, England (TH).
1790	Edmund Card invented a comber carding machine (DID).
1791	Serrazin, France, invented a carding machine for wool and felting (DID).
1792	Thomonde, Paris physician, invented a cotton carding machine (DID).
1793	Eli Whitney invented the cotton gin (TH).
1795	Silk winding machine was invented by Biard, Rouen (DID).
1796	Louis Martin patented a carding machine for wool (DID).
1797	Tellie, France, improved Martin's carding machine (DID).
1801	Joseph-Marie Jacquard demonstrated an improved loom (EB).
1803	Improved methods of cloth take-up from loom devised (EB).
1805	Jacquard introduced the jacquard loom (EB).
1812	Philippe Gerard invented a machine for spinning flax (TH).
1813	Felt first made from shoddy at Botley, Yorkshire (DID).
1815	James Syme, Scotland, invented a process for making waterproof cloth (DID).
1820	Thomas Hancock, London, patented a process for the manufacture of rubber (1ST).
1822	Differential carding machine was invented by Asa Arnold, U.S. (DID).
1823	Macintosh patented Syme's method of making waterproof cloth (1ST).
	Mansfield tinsmith introduced differential roving winding gears into carding machine (DID).

1826   Otto Unverdorben obtained aniline from indigo (TH).

1832   Heckling machine invented by Phillipe de Girard, France (DID).

1836   Dubus-Bonnel, Paris, patented spinning and weaving of glass for fiberglass (WABI).

1837   William Crompton introduced a fancy goods loom (EB).

1840   William Rider, Springfield, Massachusetts, used Goodyear process of vulcanization for shirring cloth (1ST).

1841   C. J. Fritzsche produced aniline oil by treating indigo with potassium hydroxide (TH).

1845   Joshua Heilman, French inventor, patented a machine for combing wool and cotton (TH).

1847   Latch hook knitting machine developed (EB).

1850   Le Chevalier P. Clausser, France, invented method of making short-staple flax and mixed cotton/wool goods (DID).

1856   William H. Perkin prepared the first aniline dye (TH).

1889   James H. Northrop's new automatic loom in operation automatically suppling filling and thread (ET).

1895   Northrop added the first bobbin changing to his loom (ET).

1901   First synthetic vat dye, Indanthrene Blue, successfully made, a fast dye that did not fade or washout (TT).

1907   Silk-screen print process used by Samuel Simon, Manchester (1ST).

1915   Manchester Yellow, common dye ingredient found to moth-proof wool, by E. Meckbach, German Dye Trust (TT).

1922   A. Clevel discovered a new class of dyes for coloring rayon fibers (TT).

1930   The Draper high speed loom introduced, improving loom speed by 20 percent (ET).

1932   Wrinkle-free fabric announced after fourteen years of development by R. S. Willows, Manchester, England (1ST).

1934   Phthalocyanine dyes prepared (TH).

1950   Sulzer weaving machine introduced, modern automatic shuttleless loom (TT).

>1950   Fusing used in clothing manufacture (EB).

1957   Permanent-press treatment for wool via chemical treatment developed in Australia (TT).

1964   Permanent press fabrics marketed (ET).

1975   New faster-weaving loom using air jets to move threads introduced (TT).

1985   Textile designers used computers for design work (CDIS).

Clothing Construction

Most clothing was produced in the home until the early 20th century, although some ready-made clothing was sold during Roman times. A variety of different tools and methods are used in the creation of clothing. Fasteners, sewing machines, patterns, and related aids are described below.

Fasteners

Many methods have been used to fasten clothing or hold various separate pieces together. Among these are cords and ribbons, points, buttons, pins, snaps, hooks and eyes, zippers, and, more recently, Velcro.

Buttons appeared first in Mohenjo-daro, India, 3999-3001 B.C. They were also found in England and Scotland, 2999-2001 B.C. Early buttons were made from bone, horn, wood, leather, shell, stone, gems, glass, metal, or pottery. Mechanization of button manufacture, like so many other processes, began in the 18th century. Metal buttons were made from Sheffield plate developed in 1743 by Thomas Boulsover. As steam power was applied to cloth production and processing, it also was used for related needs including button manufacture. Steam-driven factory equipment was used in James Pickard's button factory in 1779. Technical problems had to be solved before mass production techniques could be fully used. Among these problems was a means of fastening the separate parts of a button together. Bertel Sanders solved this problem in 1807 when he perfected a means of locking button parts together. In 1825 his son developed the flexible canvas shank for buttons particularly useful in undergarments.

Pins as fasteners, like buttons, have been traced to the 2d millennium B.C., but they had no covering over the sharp point. Bone pins were used in England in 1483. Brass pins were imported to England from France in 1540, and pin manufacture in England began in 1543. Lemuel Wright invented a pin-making machine in 1824, and John I. Howe patented one in 1832. The safety pin was invented and patented in the United States by Walter Hunt, New York, in 1849 and patented in Great Britain six month later by Charles Rowley.

Two fasteners used both in the home and in industry include the hook and eye and the snap. The hook and eye was patented by Camus in 1808 and manufactured in the United States in Waterbury, Connecticut, by Holmes and Hotchkiss in 1836. Snaps were developed for gloves by Paul-Albert Regnault, France, in 1855, and sew-on press-studs or snaps were devised in Paris in 1901.

A new type of fastener, the zipper, was developed by W. L. Judson in 1891 and patented in 1893. It was first used in clothing by Firma Vorweck & Sohn, Wuppertal-Barmen, Germany (1912). Judson's zipper worked but occasionally came apart, sometimes to the embarrassment of the wearer! The zipper became practical only after modifications made by Gidden Sundback, Hoboken, N.J., in 1913. Judson's device was a series of hooks and eyes with a slide clasp for opening and closing, while Sundback used spring clips in place of the hooks and eyes. The U.S. Navy helped popularize the zipper by using it in flying jackets in 1917. Zippers were used on jeans in place of buttons in 1926 and finally incorporated in men's trousers in 1935.

The idea for the Velcro fastener was devised in 1947 by Georges de Mestral, patented in the U.S. in 1955, and patented worldwide in 1957. While more expensive initially than zippers, they were easy to sew and easier for the handicapped to use. They were inspired by the burr that sticks so readily to clothes but can be pulled off easily, provided you are wearing gloves or can find a way to protect your fingers from the sharp projections. The nylon loops on the Velcro strip stick to each other just like the projections on burrs and can be separated easily, but do not separate unless pulled apart.

Sewing Equipment

Sewing needles and thread appeared about the same time. Fine bone needles dating from 40,000 B.C. have been found. Pliny, 23-79 A.D., mentioned bronze needles. Iron needles dating from the 14th century had a closed hook, not an "eye." Needles with eyes were manufactured in the Low Countries in the 15th century, while "Spanish" steel needles were made in London by an Indian in 1545. Automated manufacturing was helped by a needle manufacturing machine patented by the Excelsior Needle Company of Wolcottville, Connecticut, in 1886.

Modern needles are made from coils of steel wire cut initially into pieces of sufficient length for two needles. The short pieces are heated, straightened, and then pointed on one end at a time by grinding. After creating fine, sharp points at both ends, the wire is flattened in the middle, eyes for two needles punched, and a wire threaded through both eyes. The wire is then cut into two needles, the heads or eye ends are rounded and smoothed, and the needles are tempered for strength, polished, sorted, and packed. Special purpose needles may be processed for added features, such as a split in the top for easier threading, a curved point, an eye at the bottom as for sewing machines, or a hook for crochet needles.[27]

Thimbles, the adjunct to the needle, have been found in the ruins of Pompeii and Herculaneum, 79 A.D. The modern thimble was devised by Nicolas van Benschoten, Amsterdam, in 1684, while John Lofting, Islington, manufactured thimbles in 1695. Thimbles and other hand protection devices were necessary when all or most clothing was made by hand. With the decline of hand sewing and mending, they are probably of more interest to collectors than householders. Many people prefer to buy a new pair of socks rather than darn or patch a hole. If the cost of labor is considered, the cost of replacement is often cheaper than the cost of repair.

Figure 5.2. Sewing Equipment

YEAR		
1987		
1975		
1950	Lycra (elastic)	Necchi zigzag sewing machine
	Velcro	
1900	Improved zipper	
	Zipper	Electric sewing machine
	Snap, hook and eye	Domestic sewing machine
1800		Sewing machine (France)
		Sewing machine (England)
	Modern thimble	
1600		
	Steel needles	
	Iron needles	
1200		
400		
	Thimble	
	Bronze needles	
-1200		
	Buttons (Mohenjo-daro)	
-4400		
-10800		
-23600		
	Bone needles	
-59200		

Early fabrics were either used uncut or cut with knives. Scissors, specially designed tools for cutting fabric and other materials, were developed in the Bronze Age. Most modern scissors are made of two pieces

164

fastened together or hinged below the handles, whereas the first scissors were made in one piece and the two cutting blades were joined above the handles by a c-type spring. Scissors of the spring type appeared 500 B.C.; a fine bronze Egyptian pair from the 3d century B.C. has survived. Pivoted scissors of bronze and iron were used in China, Japan, Korea, and ancient Rome, but pivoted scissors were not used domestically in Europe until the 16th century. Mass-produced, modern cast steel scissors for domestic use were developed by Robert Henchcliffe, Sheffield, in 1761.

Cords, thongs, rope, and thread were used to lace or stitch furs and leather together before fabrics were developed. With the development of cloth, thread was made by hand twisting and later using wooden spindles to help pull and twist the fibers into yarn or thread. Once the spinning wheel came into use, it mechanized the production of thread. Cotton thread in the United States was first made by Hannah Wilkinson of Pawtucket, Rhode Island, in 1793 and was superior to the linen thread previously used. Commercially produced sewing cotton thread was first made in 1812 by James and Patrick Clark, Paisley, Scotland. By 1820 reels (thread wound on a core, but not wooden spools) were introduced by J&J Clark for 1/2d deposit. In the United States, silk thread was manufactured in 1819, and by 1849 it was offered on spools instead of skeins. Nylon thread was sold in 1946 as Monocord and Nymo.

The discovery and use of rubber eventually led to its application to fabrics and its use in sewing elastic. Thomas Hancock was making rubber suspenders, trouser straps, and garters in 1820. Rattier and Guibal, Paris, manufactured the first elastic in 1830. Lycra, man-made elastic, was sold by Dupont in 1958.

Another sewing aid, the tape measure, was made more practical to produce and less costly by a machine for printing ribbon for tape measures patented by Lavigne of France in 1847.

The 18th century brought the first development of the sewing machine. Thomas Saint patented his design in 1790, but there is no evidence he actually constructed the machine. His design--intended for sewing leather--incorporated many of the features of later, successful machines.

Barthelemy Thimmonier built the prototype for the first commercially produced sewing machine in 1829, but he was unable to persuade the tailors to adopt it. French tailors, incensed by the perceived threat to their livelihood, destroyed all but one of the eighty machines in use at a Paris clothing factory for the manufacture of military uniforms. Thimmonier tried again in 1845, but three years later a mob again destroyed his work; the revolution of 1848 effectively ended the French sewing machine industry.

A number of American inventors built and patented sewing machines in the mid-19th century, and later, because of court suits, pooled the patents. Elias Howe, Spencer, Massachusetts, developed his sewing machine in 1841 and patented it in 1846. It was adapted by John Nicols in 1851 for sewing shoes and also modified to provide continuous stitch sewing by Isaac Singer. William Hunt had built a similar machine earlier (1834), but had never patented it. Isaac Singer manufactured his first machine in 1851 and by 1858 he had introduced a model for domestic use. An industrial machine to stitch buttonholes was patented in 1854 by Charles Miller of St. Louis, Missouri. For home use, either buttonhole-maker attachments are used or the later zigzag machine developed in the mid-20th century.

Other improvements included the use of electricity to power the sewing machine continuously and the development of new, more flexible features. Singer produced the first electric-powered sewing machine in 1889, while the self-threading sewing machine needle was invented by Frank G. Carter, Brighton, Sussex, in 1908. The zigzag machine, first marketed in the United States by Necchi in 1950, was well suited to knit fabrics, and useful both for making buttonholes and for decorative sewing.

The commercial production of cloth during the 18th and 19th centuries freed much time for other tasks, but sewing and mending remained major occupations. Sewing consumed significant portions of the housewife's time in the spring and fall when seasonal changed focused efforts on the production of new garments. One 19th-century housewife, together with a hired dressmaker, spent two days a week during May and June making clothes for her family of five. She undertook a similar sewing session in the fall.[28]

The sewing machine was a major time saver for women in making family clothes. *Godey's Lady's Book* estimated a man's shirt required fourteen hours twenty-six minutes by hand and only one hour sixteen minutes by machine. A woman's chemise took ten hours thirty-one minutes by hand and one hour one minute by machine.[29] McCall and Simplicity now market paper patterns for garments they claim can be constructed in one evening.

A significant contribution for home seamstresses was the paper pattern. One of the earliest books showing diagrams for cutting cloth was Juoan de Alegas's *Libro de geometrica y traca* (1589). More elaborate diagrams were provided in *Description des arts et metiers*, by M. Garrault (Paris, 1769), and *The Taylor's Complete Guide* (1796). The first full-sized paper pattern was included in the magazine *Bekleidungskunst für Damen--Allgemeine Muster Zeitung* in 1844. In the 1850s another magazine, the *World of Fashion*, also printed paper patterns in its issues. The packaged paper pattern was developed by Ebenezer Butterick in 1863, followed by James McCall in 1870, and by *Vogue* in 1882. At this time many people still made their own clothing or used a tailor or dressmaker. Just as standard measuring tools made cookbooks practical and enabled an ordinary person who could read to achieve consistent results, so too did the paper pattern make it easier for the home seamstress to produce better fitting garments. Ready-to-wear clothing had been sold in Rome during the Roman Empire, but did not come into use in the United States and Europe until the end of the 19th century, although ready-to-wear men's clothing including jeans was available earlier.

Table 5.4. Clothing Construction and Aids

Dates	Event
B.C.	
40,000	Fine bone needles and yarn using threadable portion of plants, fur, or animal hair used (WB).
3999-3001	Buttons found in Mohenjo-daro (?2500-1700) (DBE).
2999-2001	Safety pin used (but without protected end) (DBE).
	Buttons used in northern England and Scotland (DBE).
1500	Spring-type, pivoted scissors used in Europe (EB).
3d c	Bronze Egyptian scissors used (RM).
A.D.	
77	Pliny mentioned bronze needles (DID).
79	Needles were left in the ruins at Pompeii and Herculaneum (EB).
100	Pivoted steel scissors were used in Rome (RM).
5th c	Seville tailors used pivoted scissors (RM).
14th c	Closed hook iron sewing needles used (EB).
15th c	"Eyed" sewing needles made in the Low Countries (EB).
1483	Bone pins used in England (DID).
1540	Brass pins imported to England from France (DID).
1543	Metal pins made in England (DID).
1545	Fine steel "Spanish" needles made by an Indian in Cheapside, London (DID).
1589	Juoan de Alega's *Libro de Geometrica y Traca* contained small-scale illustrations for cutting cloth (DBE).
1684	Modern thimble invented by Nicolas van Benschoten, Amsterdam (1ST).
1695	Thimbles manufactured by John Lofting, Islington (1ST).
1743	Sheffield plate, originally used for buttons, developed by Thomas Boulsover (1ST).
1761	Robert Henchliffe, Sheffield, England, made cast steel scissors for domestic use (EB).
1769	*Description des Arts et Metiers*, issued by M. Garsault, Paris, had diagrams for cutting cloth (DBE).
1775	Charles F. Weisenhall patented a double-pointed needle (EDE).
1790	Thomas Saint patented the sewing machine, but there is no evidence the machine was ever constructed (1ST).

1796   *The Taylor's Complete Guide* contained instructions for cutting cloth (DBE).

1804   *Chester Chronicle* reported "drawers of light pink now the ton among our dashing belles" (WC).

1807   Bertel Sanders, a Dane, perfected a means of locking together button parts for mass production (EDE).
Mme Tallien popularized long trousers with lace as underwear (GBL).

1808   Camus patented hooks and eyes fasteners (PM).

1809   Bobbin-net machine invented (DID).

1812   James and Patrick Clark, Paisley, Scotland, manufactured cotton sewing thread (1ST).

1819   Rodney and Horatio Hanks, Mansfield, Connecticut, manufactured silk thread (FFF).

1820   J&J Clark introduced cotton reels (spools), Paisley, Scotland 1/2d deposit (1ST).

1824   Lemuel Wellman Wright invented a pin-making machine (DID).

1825   Sanders's son replaced the metal button shank with a flexible canvas one (DBE).

1827   Samuel Williston invented a machine for producing fabric covered buttons (PM).

1829   First sewing machine was built by Barthelemy Thimmonier (1ST).

1832   John I. Howe patented a pin-making machine (RMRB).

1834   William Hunt, New York, constructed one of the first sewing machines with vibrating arm and curved needle (TH).

1836   Hooks and eyes were manufactured in Waterbury, Connecticut, by Holmes and Hotchkiss (FFF).

1838   Samuel Sloan, Poughkeepsie, New York, established a pin-making factory, that made 100,000/day (RM).

1841   Elias Howe invented his sewing machine (patented in 1846) (MITU).

1844   First full-sized paper pattern printed in *Bekleidungskunst für Damen--Allgemeine Muster Zeitung,* Germany (DBE).

1846   Elias Howe patented his sewing machine (TH).

1847   Lavigne, France, patented a machine to print ribbon for making tape measures (WABI).

1848   Joseph Wiss founded his scissors-making firm (RM).

1849   Safety pin patented by Walter Hunt, New York (1ST).
Safety pin patented in Great Britain by Charles Rowley, six months after Hunt patented his (1ST).

General Merritt Hemingway produced silk thread on
    spools, U.S. (FFF).

1850   Isaac Singer devised the continuous stitch sewing machine
    for domestic use (TH).

>1850   *World of Fashion* included printed patterns (DBE).

1851   Isaac Singer patented his sewing machine, an improvement
    of the earlier (1846) design by Elias Howe (MITU).

1854   Sewing machine to stitch buttonholes was patented by
    Charles Miller, St. Louis, Missouri (FFF).

1855   Snap fastener invented by Paul-Albert Regnault, France, for
    fastening gloves (WABI).

1863   Ebenezer Butterick developed the first packaged paper
    dressmaking pattern (TH).

1867   Ebenezer Butterick founded the E. Butterick Company,
    New York City, to manufacture his paper patterns
    (FFF).

1870   James McCall began manufacture of packaged paper dress
    patterns (ET).

1882   *Vogue* first issued paper patterns (EDB).

1886   Excelsior Needle Company, Wolcottville, Connecticut,
    manufactured uniform sewing needles by machine at a
    lower cost (FFF).

1889   First electric sewing machine was produced by Singer in
    Elizabethport, New Jersey (1ST).

1891   W. L. Judson, U.S., invented the clothing zipper (TH).

1901   Sew-on press-studs (snaps) invented in Paris (TT).

1908   Self-threading sewing machine needle introduced, invented
    by Frank Griffin Carter, Brighton, Sussex (TT).

1912   Zippers used in clothing by Firma Vorweck & Sohn,
    Wuppertal-Barmen, Germany (1ST).

1913   Gidden Sundback, Hoboken, New Jersey, patented
    "separable fasteners," improved and successful zipper
    (1ST).

1926   Zippers replaced buttons on jeans (TT).

1935   Zippers introduced on men's trousers (1ST).

1946   Belding Hemingway Corticelli, Putnam, Connecticut, sold
    commercially nylon threads, Monocord and Nymo
    (FFF).

1948   Georges de Mestral developed idea for Velcro fastener
    (WABI).

1950   Necchi zigzag sewing machine marketed in the U.S. (TIME).

1955   Velcro fastener patented (TT).

1957   Velcro fastener patented worldwide (WABI).

1958   Lycra, man-made elastic, sold by DuPont; it was stronger,
        longer lasting than rubber (TT).

Apparel and Adornment

Once cloth production and the tools to make clothing were developed, more effort could be devoted to attractively shaping the clothes and to fashion. Excellent summaries on dress and the history of clothing are contained in the *World Book*, the *Encyclopaedia Britannica*, and Kohler's *A History of Costume*.[30] Almost every library has a variety of books on costume. This section will touch on only a few items of apparel: underwear, outerwear, and adornment.

Underwear

Linen was often chosen for undergarments since it was gentler to the skin and could be cleaned more easily than   leather or wool.   Early undergarments consisted of a shirt for men and a dresslike garment for women. By 1150 men wore drawers as an undergarment, and these evolved over years in varying lengths and fabrics. Woolen garments were highly touted in the 19th century and the x-shaped overlapping frontal fly was introduced in 1910.[31] Cooper & Sons of Wisconsin introduced the Jockey Brief in 1934 based on a popular French men's swimsuit, and in 1938 they were marketed in the U.K. by Lyle and Scott.

Women adopted drawers much later than men. M. Leloir described *les calecons*, a drawerlike undergarment worn by French women around 1560, while Fynes Morrison in his *Itinerary* (1605-17) noted that Italian women wore drawers as an undergarment. The Italian princess Catherine de Médicis, who became the French queen in 1533, popularized women's drawers in France. They were patterned after men's britches, but in more feminine fabrics and trimmed with lace. Later, padding was added to enhance the figure and make horseback riding more comfortable.[32] By the 17th century in France women's drawers had gone out of fashion. Drawers for dancers were in use by 1779, however, and the late 18th and 19th century saw a resurgence in their popularity.

The Germans had adopted pantaloons during the 1770s, and by 1804 the *Chester Chronicle* reported "drawers of light pink now the ton among our dashing belles." Mme. Tallien of France in 1807 popularized long lace-trimmed trousers for French women. These garments remained in use for the next hundred years with some variations in length and fullness. Amelia Bloomer adapted "Turkish" trousers or pantaloons in 1851, but by 1859 she abandoned them. Many women were repelled by her strong feminist viewpoint. Her name, however, was later applied to knickerbockers or trousers worn under a skirt for cycling and later to many variations of female undergarments.[33]

Figure 5.3. Underwear

YEAR

1987

     Leg warmers
1975
     Control top panty hose
     Seam-free stockings, panty hose
1950 Bikini panty
     Nylons
     Briefs, panty girdle, padded bra,
       bra cup sizes
1900 Rayon stockings and undergarments
     Bust improver, breast supporter

1800

1600
     Women's drawers, corsets (Spain),
      knit stockings

1200
     Men's drawers

 400

-1200
     Corsets (Minos)

-4400

Changing dress fashions, including shorter hems, affected the design of undergarments. The 1912 Army and Navy Store catalog advertised "golf-shape knickers." By 1914 short "panties"--or more accurately, short divided petticoats--were introduced and remained popular throughout the 1920s. This loose style of panty or "tap pants" was still in vogue in 1929 when a version of the panty girdle was introduced. With the absence of stockings to hold them down, due to the "bare legged look" then in vogue, girdles took on more of a panty shape and style.[34] Women's briefs were developed in the 1930s, and by 1934 the term "briefs" was in use.[35] About this time, shorts had become popular, and so shorter, closer fitting underwear was essential. The bikini panty, patterned after the bikini swimsuit of 1946, was being worn around 1950.

The corset, whose function was to reshape the body, to display or flatten the breasts, to cinch the waist, or to flatten the stomach, was used as early as the second millennium B.C., by the Minoans. Corsets were not used in Greece or Rome, but did appear in 16th-century Spain. The corset went in and out of fashion and became a controversial garment in the 19th century when feminists and doctors criticized the more extreme forms of corsets, citing broken ribs, collapsed lungs, prolapsed uterus, and weakened abdominal walls as potential problems.[36] The corset faded from general use with the advent of the girdle and the Roaring Twenties, but is still used by men and women for medical purposes.

The girdle, using improved elastic fabrics, appeared in 1912 and led to the eventual demise of the whalebone corset. A panty-type girdle was in use in 1929, and the girdle in its turn was replaced by control top panty hose around 1965.

The other function of the corset, to shape the breasts, was replaced by the brassiere. By 1884 bust improvers were advertised, and by 1895 Montgomery Ward's catalog depicted a "breast supporter" particularly useful for women engaged in sports. The first brassiere for dressy occasions was mentioned in *Vogue* in 1909. Otto Titzling developed a brassiere, in 1912, but failed to obtain a patent. In 1914, Mary Phelps Jacob patented her brassiere, created from two halves of a handkerchief and a ribbon. Jacob claimed to be the originator of this brief form of upper chest wear, despite Titzling's earlier design. D. J. Kennedy patented a padded bra, originally developed to protect female sports participants, in 1929. The same year Phillipe de Brassiere combined Titzling's functional design with fashionable fabrics, and gave his name to the brassiere. By 1939 Warner had introduced cup sizes for their bras, thus improving comfort and fit. Wacoal Corporation, Japan, introduced a machine-washable wired bra using a shape-memory wire in 1985. Even if twisted during machine laundering, the wire resumes its original shape.

Outerwear and Accessories

This section treats a diverse group of outerwear and accessories including rainwear, swimwear, footwear, gloves, handkerchiefs, and hats. Protection from rain was initially provided by oiled and tightly woven fabrics. It was not until the discovery and use of rubber that modern waterproof fabrics were developed.

The first waterproof fabric was postulated by François Fresneau in 1748; however, the first manufactured raincoat, Fox's Aquatic, sold by G. Fox of London in 1821, was made of Gambroon, a type of twill material made with mohair. In 1823 Charles Macintosh, who had experimented with various processes for waterproofing, patented the waterproofing process developed in 1815 by John Syme, Scotland. Charles Goodyear developed the rubber vulcanization process in 1839, and it was used in 1843 by Thomas Hancock to improve Macintosh's waterproof cloth. The new waterproof garment took its name from Macintosh, but with the addition of a *k* to be called the mackintosh. In 1851 another firm, Bax & Co., developed Aquascutum, a chemically treated woolen fabric that shed water. It was popularized by the escape during the Crimean War of General Goodlake, who credited his good fortune to his gray Aquascutum coat, which was similar in appearance to those worn by the Russian soldiers. Modern waterproof garments may be made of rubber, plastic, or plastic-coated fibers, such as the Teflon coating used for Gore-Tex garments.

Umbrellas have a long and prominent history. Egyptian pharaohs and Assyrian kings used them; a frieze from 1350 B.C. illustrates an Assyrian king shaded by a long-handled parasol carried by a servant. A collapsible umbrella was found in a Korean tomb dated 25 B.C. The first record of waterproof umbrellas is given by an inventory of King Louis XIII, which listed several oiled silk umbrellas. Jonathan Swift mentions an umbrella in *A Tale of a Tub* (1696). In 1715 Marius, Paris, was offering collapsible shaft, folding umbrellas.

Some recent fashion trends include suits for women, especially at work, and ever-changing dress lengths. The lady's suit was created by English designer Redfern in 1885. Hemlines rose after World War I and then went down to more modest length with the depression in the thirties but rose again during World War II. Christian Dior introduced the New Look with longer skirts in 1947 only to be reversed by Mary Quant, an English designer, who introduced the mini-skirt in 1965. While hemlines went down again in the 1970s, some minis reappeared for the younger set in the 1980s.

As sea bathing became popular in the mid-19th century, swimsuits appeared. Some of the first swimsuits for women were based on Amelia Bloomer's bloomers, first devised in 1851 and later used for women's sportswear. Women wore black stockings, bloomers, and a short-sleeved dress. In 1907 Annette Kellerman wore a one-piece, skirtless bathing suit at Revere Beach, Boston, and was arrested for indecent exposure. After World War I, a clinging one-piece knit suit for women was introduced in France. During the 1930s women's briefs and men's Jockey Briefs were modeled on a close-fitting, popular French men's swimsuit and women's swimsuits became shorter, eventually evolving into a two-piece garment, brief shorts, and a halter top. Louis Reard, French couturier, introduced an even briefer suit, the bikini, in 1946. The topless suit was introduced by Rudi Gernreich, California, in 1964.

Footwear was developed early in human history. Some extreme styles were developed during the Middle Ages and worn by dandies at various periods. The common people wore sandals, clogs, or slippers made of wood or sometimes of leather.[37] Some examples of footwear favored by men include sandals, curled toe slippers, and boots. A fine Egyptian woven papyrus sandal from 2000 B.C. has survived. The crackowe with a chain to hold up its extreme pointed toe was popular in the 1300s. Thigh-high, heavy cuffed jackboots were favored by gentlemen and soldiers from 1650 to 1775. Thomas Beard, America's first shoemaker, arrived in Salem, Massachusetts, in 1629.

Early shoes made little differentiation for the right or the left foot. Separate lasts (wooden forms) for each foot were used beginning in 1818. By 1851 John Nicols had adapted Elias Howe's sewing machine for shoes. Lyman Reed Blake, Abington, Massachusetts, invented a machine for sewing the soles to the upper shoe body in 1858. Standardized shoe sizes were introduced in the United States before the Civil War and were common by the 1880s.[38]

Further mechanization occurred during the second half of the 19th century. It was prompted in the United States in part by the Civil War because of the increased demand for boots and shoes. Most shoe manufactures were forced to install new machinery, including Blake's sewing machine. Jan Ernst Matzeliger invented a shoe lasting machine for shaping and fastening the leather over the shoe sole in 1882. By 1900 shoe manufacture in the U.S. was mechanized. By 1985 the laser was being used to develop custom orthopedic shoes.

Athletic shoes were being made near the end of 19th century and became popular as general footwear in the 1960s. The J. W. Foster Company, an English manufacturer that later introduced Reeboks in the 1980s, was established in 1895. They made the shoes for the British Olympic Track Team of 1924. By 1979 a U.S. business man was attracted to the shoes and arranged to sell them in the U.S. In 1983 brightly colored Reeboks were introduced to the U.S. A rival running shoe company, Adidas, was established in 1948 in West Germany. In the United States, Converse Chuck Taylor high-top athletic shoes were first marketed in 1923. Converse introduced its first athletic shoes, the All-Star, in 1917, the same year U.S. Rubber introduced its popular Keds sneakers. Initially introduced for basketball playing, the Converse Chuck Taylors were adopted by many others. More recently, NASA developed a high resiliency material, Dynacoil, that was incorporated into running shoes by KangaRoos USA (1985). These new materials lasted four times longer than older materials.

During the 1980s, business women in New York and elsewhere began wearing sneakers or jogging shoes on the street and changing into heels at work. High-heeled shoes for women were popularized by Catherine de

Médicis in 1533, who was short and needed the added height. By 1630 they were being imported to the United States from France. Massachusetts passed a law voiding a marriage where the woman used high heels to seduce the man. High heels went out of fashion until reintroduced in 1850 in a New Orleans brothel by a French prostitute. They gradually became a favorite form of footwear, but were decried by feminists and doctors in the late 19th century as damaging to the feet and, along with corsets, detrimental to health.[39] However, the stiletto heel gained popularity in the late 1950s and early 1960s, fading out of fashion only to reappear in the mid-1980s. During the 1980s women's shoes were available in a variety of styles and heel heights, including a variety of styles and colors of athletic shoes.

Rubber was applied to protective footwear as well as to shoes themselves. Goodyear advertised galoshes in 1824, while Aberdeen Rubber Sole and Heel Company, Scotland, introduced rubber heels for shoes in 1889, and Humphrey O'Sullivan of Massachusetts patented his rubber Safety Heel in 1899. Other shoe-related developments include Dr. Scholl's arch support patented in 1904 and Hush Puppies, which were first sold in 1958.

Socklike foot coverings were in use as early as 400 B.C. in Greece. Clergyman in Western Europe wore them from 400 A.D. as a sign of purity, and they became popular with noblemen by the 11th century. Early hose or stockings were made of cloth with two seams and tied to the doublet, vest, or upper garment. The French Stocking Knitters Guild was mentioned in 1527. The accounts of Sir Thomas l'Estrange recorded "four pair knytt hose" in 1533. The development of the stocking frame by William Lee in 1589 provided another means of producing knitted stockings. The first record of manufactured stockings was in 1657 in Paris, although stockings continued to be knitted in the home until 1900.

Stockings were initially made from natural fibers, particularly wool and silk. In the 20th century, manufactured fibers were used for stockings, and the first stockings made of a manufactured fiber, rayon, appeared in 1910. The first nylons were sold in 1940, and nylon panty hose appeared in 1958. Seam-free stockings were introduced in 1960, and control-top panty hose appeared around 1965. Nylon stretch yarn was used for men's socks in 1952.

As well as foot coverings, some early people, even in warm climates, wore hand coverings or gloves. In Egypt at the time of Tutankhamen (14th century B.C.), white linen gloves with drawstrings at the wrist were in use, as demonstrated by a pair recovered in his tomb. Gloves were also used by the Greeks and the Romans and were made in fabric and leather in medieval Europe for nobles and prelates. That 16th-century fashionplate Catherine de Médicis popularized women's gloves.[40] In 1834 Xavier Jouvin of Grenoble, France, invented a cutting die for gloves that allowed a precise fit for manufactured gloves, and by 1855 French glove-maker Paul-Albert Regnault invented metal snap fasteners for use in gloves.

Handkerchiefs were originally used to clean the face or, when heavily perfumed, to mask the effect of unpleasant smells. They were used by the Romans in the 4th century, but remained rare until the Middle Ages; the wife of Louis XI (ca. 1461) had only three. The pocket handkerchief was in use by the wealthy by 1503. To remove mucous from their noses, most people used their fingers or their sleeves. Frederick the Great of Prussia decreed soldiers were to have buttons on their sleeves to prevent them from wiping their noses with the sleeve.[41]

Handkerchiefs have largely been replaced for women and children by paper tissues. Kimberly Clark first introduced Celluwipes paper tissues, originally developed for use as gas mask filters in World War I, in 1924. The name was later changed to Kleenex. Initially introduced for the removal of cold cream, it was not until the 1930s that they were marketed as a handkerchief replacement. The public accepted them only gradually, but today they have almost entirely replaced women's handkerchiefs. Men's handkerchiefs are more readily available because men require larger tissues of greater strength. In 1987 Kimberly Clark was considering marketing for commercial use tissues impregnated with disinfectant to reduce the spread of colds and viruses.

Ostrich feathers, a fashion rage in the late 19th century for boas, hats, and trim, first made their appearance when a South African farmer sold the feathers in a local market in 1852. By 1913 ostrich feathers had become South Africa's fourth largest export, but the market collapsed soon after as fashions changed. The best feathers were used for decoration, hats, and boas, while the less attractive ones were made into feather dusters.

Hats or head coverings can be traced from friezes, buildings, and tomb and monument decorations for Egyptian, Assyrian, and other early civilizations. Ancient rural cultures of Asia Minor and Europe favored hats of straw, rush, or other vegetable fiber. Artisans of ancient Rome wore hats of felt, as did freed slaves.

Mature men and women wore head coverings indoors and out until the 19th century. Hats were made of a multitude of materials and in a variety of shapes: beaver hats became fashionable in Flanders in the 14th century, silk hats from Florence were worn in 1760, and the top hat was introduced by John Etherington of London in 1797. One impetus to the fur industry in North America was the popularity of beaver fur for hats and clothing, and by the end of the 19th century hunting had almost wiped out beavers in Canada and the United States.[42]

Initially a home task, hat making became a trade after the discovery of improved felt making in the 1400s. From soft caps, helmets, and various straw hats, hats evolved into more modern shapes by the 17th century. The broad brimmed Cavalier-style hat was worn in France from 1620 and by the mid-18th century had been replaced by the tricorne. In the United States,

Zadoc Benedict of Danbury, Connecticut, established the first U.S. hat factory in 1780.

A partial head covering, the earmuff, was devised by Chester Greenwood in 1873 and patented in 1877. During the 1970s they were difficult to find, but have again become popular in the 1980s.

Developments affecting men included the invention of jeans, zippers, suspenders, the tuxedo, and the starchless collar. Levi Strauss made his first pair of jeans in 1850 for sale to California gold miners. More recently designer jeans made them a fashion garment very different from the long-wearing, sturdy work pants they were originally intended to be. Zippers replaced buttons on jeans in 1926 but were not included in the design of men's trousers until 1935. Other developments related to menswear include India rubber suspenders first made by Thomas Hancock in 1820 and the modern type of suspender of cloth straps with rubber tabs offered by Ratier and Guibal, Paris, in 1840. Griswald Lorillard of Tuxedo Park, New York, first wore the tuxedo suit in 1886.

Collars have been worn throughout recorded history, sometimes as a separate decoration and other times as part of other garments including cloaks, dresses, and shirts. Detachable collars, popular in the 19th and early 20th centuries, were first devised by Hannah Lord Montague in 1825 and manufactured by Ebenezer Brown in 1829. A wipe-clean, paper version was patented in 1854 by Walter Hunt. Van Heusen introduced stiff, starchless collars for men's shirts in 1921, and Lacoste knit shirts, introduced in Europe in 1933, appeared in the United States in 1951.

By 1900, in Western societies, more and more ready-made clothing could be purchased and yet another major domestic task began to move outside the home to be purchased with wages earned from labor done outside the home. The transition did not occur evenly, and some households continued to make garments even into the 1930s.

Adornment

People have used a variety of aids to modify their body appearance throughout the ages. They have styled their hair and used jewelry, perfumes, and cosmetics. A number of tools, equipment, products, and methods were developed to accomplish such modifications. A few of these developments are discussed here.

Hair styling and dressing has had a long history: prehistoric combs and hairpins have been found by archaeologists. Much of what we know is available through the study of figures represented in carvings, statues, and paintings. The Egyptians shaved their heads and wore wigs with elaborate hair dressing. Wigs were also worn by both the Greeks and the Romans. The Romans admired the blond hair of Northern people and imported blond wigs from Germany as well as bleaching and dyeing hair.[43] Hair styles were simple during the early Middle Ages, but had become increasingly complex by the 17th century. Wigs and trains were fashionable in 1600.

As noted above, hair had been bleached and dyed early in history. Modern hair dyes using manufactured chemical ingredients were developed in the early 20th century and were initially only applied in hair salons by professional hairdressers. Products for home use were not developed until the late 1920s, and modern home coloring kits, requiring no separate processing to initiate and stop dye action, were developed in 1953. Eugene Schueller of France successfully tested hair dye in 1909 and founded the French Harmless Hair Dye Company, which later became l'Oreal, in 1911. Imedia hair dye, which was developed in 1927, offered more shades. Home hair-coloring kits were introduced in 1929, replacing the henna, powder, pomades, and other dyes used in earlier ages. Imedia also developed a one-step hair dye in 1953. Earlier kits required use of hydrogen peroxide to oxidize the dye and to penetrate the hair. Some modern hair colors can be shampooed in and do not require a separate process to set the color.

A variety of substances, including fats, soaps, and lacquer, have been used to hold hair in various elaborate hair styles. The Celtic people on the European continent used soap to shape hair into elaborate styles worn during battle. The Japanese women used lacquer to maintain the shape of their stylized coiffures. More recently, L'Oreal developed Elnett Hair Spray in 1960.

Other developments include permanent waving and electrical hair styling tools. The marcel wave, named for its inventor, a French hairdresser, was devised in 1872, and a permanent-wave machine was demonstrated in London in 1906 by Karl L. Nessler. L'Oreal patented the "cold" permanent wave in 1945. Roman women had used heated irons to curl their hair. Such irons were heated in a fire or, in later times, on a stove. Electrically heated curling irons were used in Berlin in 1887. Carmen hair curlers, electrically heated, were invented in Denmark in the early 1960s and imported to the United States by Clairol in 1967. Rene Lelievre and Roger Lemoine invented a heated combing iron in 1959, while the Shavex styling comb was imported from Switzerland in 1965.

The hair dryer first appeared in 1920, although the Pneumatic Cleaner Company had tactfully suggested in advertisements that its vacuum cleaner exhaust could be used to dry "mi'lady's tresses."[44] By 1951 Sears was offering the bonnet hair dryer to its catalog customers.

Perfumes were a necessity in much of human history since frequent bathing, though common in Roman times, was frowned on in United States and English society until near the end of the 19th century.[45] Florida Water, a popular toilet water, was introduced in 1808 and was used by Scarlett O'Hara in *Gone with the Wind*. Once the province of the wealthy, modern chemistry significantly lowered the cost of fragrances and provided more options for aromas. This change was led by the invention of eau de cologne, a weaker and less expensive substance than perfume, by Jean Marie Farina of Cologne in 1850. Bruno Court, a French firm, registered a patent for solid stick perfume in 1944. Notable expensive fragrances include Coco Chanel's Chanel No. 5, introduced in 1920, followed in 1925 by Shalimar, by Miss Dior and L'Air du Temps in 1947, by Azzaro Femme in 1970, by Giorgio in 1981, and by Christian Dior's Poison in 1985.

Other cosmetic aids include rouge, cold cream, and lipstick. Cosmetics have long been used: Egyptian women used kohl to darken their lashes and emphasize their eyes, while Roman women often painted their faces with white lead. Many natural substances, including berry juice, were used to redden lips or cheeks and to darken eyebrows. Around the turn of the century women used petroleum jelly and cucumbers as a face cream. Modern makeup packaging began with Boujois in 1890, with the offering of makeup in cartons. Boujois also developed small compacts of rouge in 1912. Beiersdorf, Germany, invented Nivea cold cream for removing makeup in 1911, and it was followed by lipstick in a metal tube in 1915 offered by Maurice Levy of the United States.

Table 5.5. Apparel and Adornment

Dates	Event
B.C.	
2000	Woven papryus sandals worn in Egypt (EB).
1766-1122	Silk twill damask fragments left in China that required advanced weaving techniques (EB).
1350	Assyrian king depicted in frieze shaded by a long-handled parasol (RM).
25	Collapsible umbrella buried in Korean tomb (RM).
A.D.	
4th c	Romans use muscinium for mucus (handkerchief) (RM).
400	Clergy wore stockings as a sign of purity (EB).
767	Pope Paul I gave Pepin the Short a jeweled, short-handled umbrella (RM).
>1150	Drawers became a male undergarment (WC).
>1300	Crackowe, pointed shoe with a chain to hold up the toe, worn (WB).
>1436	Wife of Louis XI had three handkerchiefs (RM).
1503	Pocket handkerchief used (TH).
1527	Reference made to French Stocking Knitters Guild (1ST).
1533	Four pair "knytt" hose recorded in the accounts of Sir Thomas L'Estrange (1ST).

	Petite Catherine de Médicis popularized high heels in Paris (CDIS).
>1533	Catherine de Médicis popularized drawers, long trouserlike underwear (GBL).
>1560	Leloir described "les calecons," drawer-type undergarment worn by French ladies (WC).
1600	Wigs and dress trains were fashionable (TH).
1605-17	Fynes Morrison in his *Itinerary* noted that Italian women wore drawers as underwear (WC).
1629	Thomas Beard, the first U.S. shoemaker, came to Salem, Massachusetts (WB).
>1630	High heels imported from France to the United States (CDIS).
1637	Waterproof umbrella listed in the inventory of King Louis XIII, France (1ST).
1640	Massachusetts law voided marriage if woman used high-heeled shoes to induce marriage (CDIS).
1657	First stockings manufactured in Paris (TH).
1660	Croatian regiment marched in Paris with bright handkerchiefs around their necks; *cravat* is French for *Croatian* (RM).
1696	Jonathan Swift mentioned an umbrella in *A Tale of a Tub* (1ST).
1715	Collapsible shaft, folding umbrella, manufactured by Marius, Paris (1ST).
1748	François Fresneau described a waterproof overcoat and boots coated with latex sap (WABI).
1760	First silk hats imported from Florence (TH).
>1770	Women in Anglo-Saxon countries wore pantaloons as underwear (GBL).
1650-1775	Jackboots, heavy, thigh-high cuffed boots, worn by gentlemen and soldiers (WB).
1780	Zadoc Benedict, Danbury, Connecticut, opened the first U.S. hat factory (WB).
1785	Gosselin, Amiens, constructed a fold-up, pocket umbrella (RM).
1793	Hannah Wilkinson, Pawtucket, R.I., manufactured cotton sewing thread that was superior to previously used linen thread (FFF).
1797	John Etherington, London, introduced the top hat (USAIR).
1808	Florida Water, fragrant toilet water, was introduced (UNITD).

1818	Right and left shoe lasts, wooden forms for shaping the shoe, developed (EB).
1820	Braces were first made of India rubber by T. Hancock, London (DID).
1821	First manufactured raincoat offered for sale by G. Fox, London (1ST).
1824	Rubber galoshes advertised by W. Goodyear, Boston (1ST).
1825	Hannah Lord Montague, Troy, New York, devised the detachable collar (FFF).
1829	Ebenezer Brown, Troy, New York, began manufacturing detachable collars (FFF).
1830	First elastic manufactured by Rattier et Guibal, Paris (1ST).
1834	Xavier Jouvain, Grenoble, France, invented a die for cutting gloves (EB).
1840	Rattier and Guibal, Rouen, France, invented suspenders (see also 1820, Hancock) (WABI).
1843	Thomas Hancock used Charles Goodyear's vulcanization process to improve Macintosh's waterproof cloth (1ST).
1846	Crinoline (woven horsehair) petticoats first worn (WABI).
1848	First large-scale department store, Marble Dry Goods Palace, New York, occupying one city block, opened (1ST).
1850	Levi Strauss made jeans (1ST).
	Eau de Cologne invented by Jean Marie Farina, Cologne (DID).
>1850	High-heeled shoes reappeared in New Orleans brothel worn by a French prostitute (CDIS).
1851	Amelia Bloomer tried to popularize Turkish pantaloons for women (EB).
	John Nicols, Lynn, Massachusetts, adaped Elias Howe's sewing machine for sewing shoes (WB).
>1852	South African farmer sold ostrich feathers in local market, leading to eventual export of feathers (WJ).
1854	Walter Hunt, New York City, patented the (detachable) paper collar (FFF).
1856	Crinoline petticoats were replaced by the steel-framed underskirt invented by Tavernier, France (WABI).
1858	Lyman Reed Blake, Abington, Massachusetts, invented a machine for sewing the soles to shoes (WB).
1872	Marcel, French hairdresser, invented the marcel wave (WB).
1873	Chester Greenwood, Farmington, Maine, devised the first earmuffs (WJ).
1877	Chester Greenwood patented earmuffs (WJ).

1882	Jan Ernst Matzeliger invented a machine for shaping and fastening the leather over the shoe sole (WB).
1884	Bust improvers advertised (WC).
1885	Lady's suit invented by Redfern of England (WABI).
1886	Griswold Lorillard, Tuxedo Park, New York, devised the Tuxedo, evening wear for men (WJ).
1887	Electrically heated curling tongs available in Berlin according to *Scientific American* (EL).
1889	Rubber heels introduced by the Aberdeen Rubber Sole & Heel Company, Scotland (1ST).
1890	French firm Bourjois launched makeup packaged in an elegant carton (WABI).
1895	J. W. Foster and Sons, England, oldest athletic shoe company, established (see 1979, manufacture of Reeboks for export to the U.S.) (BUSWK).
	Montgomery Ward catalog depicted "breast support," early bra (MONTW).
1899	Safety heel, of rubber, patented by Humphrey O'Sullivan, Lowell, Massachusetts (patent 618128) (FFF).
1900	Ready-made clothing was sold (RC).
	Women used vaseline and cucumber as a face cream (HS).
1902	British newspaper advertisement offered "patent bust improvers" (DBE).
1904	Dr. Scholl, former shoe salesman and Illinois medical school graduate, patented the arch support (TT).
1905	Elastic rubber replaced traditional whalebone and lacing in foundation garments (TT).
1906	First demonstration of permanent wave machine developed by Karl L. Nessler, London (1ST).
1907	Annette Kellerman wore a one-piece, skirtless bathing suit at Revere Beach, Boston (HS).
1909	Brassiere for dressy occasions mentioned in *Vogue* (1ST).
	Eugene Schueller successfully tested hair dye (WABI).
1910	Eugene Schueller founded the French Harmless Hair Dye Co. (later L'Oreal) (WABI).
1911	Beiersdorf, Germany, invented Nivea cold cream (WABI).
1912	Girdle introduced (DBE).
	Otto Titzling invented the brassiere but failed to patent it (PA).
	French firm Bourjois created pastel cheek makeup, the first mass-produced rouge (WABI).
	Army and Navy Stores catalog advertised "golf-shape knickers" (WC).

1913	Ostrich feathers became South Africa's fourth largest export, the market collapsed shortly after (WJ).
1914	Brassiere patented by Mary Phelps Jacob (TT).
	Eugene Sutter developed a hair dryer as part of a permanent wave machine (WABI).
	Short panties, petticoats with legs, introduced (GBL).
1915	Lipstick in a metal tube was manufactured by Maurice Levy, U.S. (1ST).
1917	Converse Shoe Company introduced the All-Star athletic shoe (ER).
1920	Coco Chanel introduced No. 5 perfume (TT).
>1920	Race electric hair dryer introduced by Racine Universal Motor Co. (EL).
1921	Artificial, or cultured, pearls produced (TT).
	Van Heusen's stiff, starchless collars introduced (TT).
1923	Converse Shoe Company introduced the high-top Chuck Taylor basketball shoe (ER).
1924	Kleenex paper tissues introduced in U.S. (1ST).
1925	Pierre and Jacques Guerlain created Shalimar (WABI).
1927	Imedia hair dye developed offering more shades and using organic materials for better coverage (WABI).
1929	D. J. Kennedy, U.K., patented the padded bra (PA).
	Phillipe de Brassiere, France, used Titzling brassiere design in dressy fabrics and became famous (PA).
	First do-it-yourself hair coloring introduced (TT).
	Panty girdle introduced (COLLR).
1935	Cooper & Sons, Wisconsin, introduced Jockey Briefs based on a French men's swimsuit (EOOET).
1938	Jockey shorts for men manufactured by Lyle and Scott Ltd., U.K. (WC).
1939	Warner Co., Bridgeport, Connecticut, sold bras with cup sizes (HS).
1940	First nylon stockings marketed using Dupont nylon yarn (1ST).
1944	Bruno Court, French Company, registered solid stick perfume (WABI).
1945	L'Oreal Laboratories, France, patented the cold permanent wave (WABI).
1946	First bikini swimsuit designed by French couturier Louis Reard (1ST).
1947	Christian Dior introduced the "New Look" with long skirts (WABI).
	Robert Ricci marketed L'Air du Temps perfume (WABI).

Christian Dior marketed Miss Dior perfume (WABI).

1948   Adidas, athletic shoe manufacturer, established in West Germany (WJ).

>1950   Bikini panties patterned on the bikini swimsuit introduced (GBL).

1951   Sears offered "bonnet" hair dryer in its catalogs (EL).

Lacoste shirts first sold in the U.S. (HS).

1953   One-step hair dye, Imedia Creme, developed (WABI).

1958   Hush Puppies entered the U.S. shoe market (EP).

Panty hose sold (DBE).

1959   Kleenex dispenser box introduced (RM).

Rene Lelievre and Roger Lemoine invented a heating-combing iron, Babyliss (WABI).

1960   L'Oreal developed Elnett hair spray (WABI).

Seamfree women's stockings introduced (JM).

>1960   Aren Byhjing Pederson, Denmark, invented Carmen, electrically heated curlers (EL).

1964   Topless swimsuit designed by Rudi Gernreich, California (1ST).

1965   Shavex electric styling comb imported to the U.S. from Switzerland (EL).

Mary Quant, England, created the miniskirt (WABI).

>1965   Control panty hose introduced (JM).

1967   Clairol imported Carmen electric curlers to U.S. (EL).

1970   Azzaro Femme perfume marketed by Loris Azzaro (WABI).

1979   Paul Fireman discovered Reeboks in England and planned introduction to the U.S. (BUSWK).

1979-80   Leg warmers for women were fashion wear (JM).

1980   Minnetonka bought Calvin Klein Cosmetics and Obsession perfume (BUSWK).

1981   Giorgio perfume launched by Fred and Gale Hayman (CDIS).

1983   Brightly colored Reeboks introduced to the U.S. (BUSWK).

1985   Wacoal Corporation, Japan, introduced a machine-washable bra using "shape-memory" alloy wire (BUSWK).

NASA developed Dynacoil, material used for high-resiliency running shoe by KangaRoos USA, St. Louis (BUSWK).

Laser used to develop custom orthopedic shoes (IEEE).

Christian Dior's Poison perfume launched (CDIS).

1987   Kimberly-Clark considered marketing germicidal tissues to corporate customers (BR).

Stores

Markets or places to exchange goods have existed almost as long as people. Stores related to food retailing were discussed in chapter 3 and overlap the development of department and variety stores. General stores and trading posts in North America carried all types of merchandise and some newer stores are combining clothing, hardware, household goods, and food again. Large department stores and shopping malls are a relatively recent phenomenon and are associated with the development of urban centers. Dorothy Davis in *A History of Shopping* provides an interesting overview of the history of shopping and the evolution of markets.[46]

Department stores began in the early 19th century and flourished in the latter half of the same century. Belle Jardiniere, a Paris department store, opened in 1824. In America, Adam Gimbel, who later developed a large department store chain, opened his first store in Indiana in 1842. The first large-scale department store, occupying one city block, the Marble Dry Goods Palace, opened in New York in 1848. Marshall Field established his first store in Chicago in 1852, and the Crystal Palace Bazaar, London's first department store, opened in 1858. Woolworth's, the first variety chain store, began operation in Utica, New York, in 1879.

Many other familiar retailing names began operating stores in the 19th and early 20th centuries. A number started as small dry goods or men's stores. These include Henry Sands Brooks, who opened his first store, later Brooks Brothers, in 1818 in New York; Simon Lazarus opened his first Lazarus Store in Columbus, Ohio, in 1851; William Filene opened his first store in Lynn, Massachusetts, in 1856 and his first Boston stores in the 1870s; Rowland Hussey Macy opened the first Macy's in 1858; Morris Rich opened Rich's Atlanta store in 1867 just after the Civil War; and I. Magnin, San Francisco, opened a store in 1876. Early 20th-century openings included J. C. Penney's of Kemmerer, Ohio, in 1902 and Neiman-Marcus of Dallas and Bullock's of Los Angeles in 1907.

Catalog shopping dates to 1844, when Orvis Company issued its first catalog for fishing equipment, followed in 1845 by a Tiffany Company catalog. General catalog merchandising, however, really started in 1872 with the establishment of Montgomery Ward, which offered a wide variety of home furnishings and equipment and clothing through its catalogs. Montgomery Ward's closest competitor, Sears, offered only watches in its first catalog issued in 1888. Sears's first full-line catalog was issued in 1894, and by 1909 Sears even offered houses together with materials and plans. Neiman-Marcus offered its first Christmas catalog in 1960.

Among the more recent developments in catalog shopping are catalog show rooms and a proliferation of catalogs aimed at up-scale customers. During the 1960s a number of catalog show rooms were opened by Best Products, Service Merchandise, Consumers Distributors, and others. These stores provide displays of each product sold, but stock is kept in a separate area, and customers place orders with a clerk and generally received the merchandise immediately from on-site storage. More recently discount warehouses, large stores with open shelves, allow the customers to pick merchandise directly from open shelves instead of having clerks retrieve stock. Such stores have few fixtures or amenities. Changes in retailing practices reflect such pressures as more competitive pricing from department stores and discounters, increased competition from traditional catalog suppliers, and the introduction of wholesale clubs that have all made catalog showrooms less popular.[47] Catalog retailing has moved away from the original aim of serving remote and rural populations and now focuses on more up-scale, urban or suburban clients.[48]

A new innovation currently being tested is electronic shopping. R. R. Donnelley & Company, a large commercial printer, has developed the Electronicstore kiosk where a customer can select and order merchandise using a credit card for payment. The kiosk has been installed in shopping malls and some individual stores (1985). Sears and Penney's have moved closer to this concept with their telephone ordering. Customers, having identified the desired item in a catalog, call, place an order, and charge it to a credit card. Another electronic service is Compuserve's Electronic Mall, first tested in 1984. It also used catalogs and a customer inquiry service relying on Compuserve's electronic communication system.

A fixture in medium-sized communities is the shopping mall. Shopping centers began in 1916 with Market Square in Lake Forest, Illinois, the first planned shopping district. Country Club Plaza in Kansas City opened in 1922, expanding the concept to include a group of buildings developed and managed as a unit, a special parking area, and a street exclusively for pedestrians. In the Highland Park Shopping Village of Dallas, which opened in 1931, the stores faced each other, and in the early 1950s the Northland Mall near Detroit pioneered multistory construction. The first enclosed, climate-controlled shopping mall was the Southdale Center near Minneapolis, which opened in 1956. The cold Minnesota winters were a strong impetus to enclosing the mall. There are roughly 3,000 enclosed malls in the U.S. and 100 in Canada, and about 27,000 shopping centers of all types in North America.[49]

Other retailing concepts include the renovation of old buildings, often former factory or warehouse space: in 1964, Ghirardelli Square, an old San Francisco chocolate factory, was renovated, as was the Cannery, and both offer a multilevel variety of boutiques and restaurants. Faneuil Hall, Boston, was originally built in 1742 as a market and meeting place, and was renovated in the 1970s. It was reopened in 1976 as a cluster of shops, boutiques, and restaurants.

Table 5.6. Stores

Dates A.D.	Event
1615	Merchant Adventurers granted a monopoly for the export of English cloth (TH).
1742	Faneuil Hall, Boston, was built as a market and meeting place (WK).
1818	Henry Sands Brooks opened a clothing store in New York (later Brooks Brothers) (HS).
1824	First department store, Belle Jardiniere, Paris, was founded by Pierre Parissot (WABI).
1842	Adam Gimbel, Indiana, opened his first store (RMRB).
1844	Orvis issued their first catalog offering fishing equipment (ACTB).
1845	Tiffany issued their first catalog (ACTB).
1851	Brigantine *La Ville de Paris* arrived in San Francisco loaded with luxury goods (NB).
	Simon Lazarus opened a men's clothing store in Columbus, Ohio (MP).
1852	Marshall Field established a store in Chicago (RC).
1854	City of Paris Department Store opened in San Francisco by Felix and Emile Verdeu of Paris (NB).
1856	William Filene opened his first store in Salem, Massachusetts (MP).
1858	Crystal Palace Bazar, first London department store, opened (WABI).
	Rowland Hussey Macy opened the first Macy's (HS).
1864	William Filene opened a wholesale business in New York City (MP).
1867	Morris Rich opened his first Atlanta store (MP).
1868	William Filene financially ruined in Jay Gould's Black Friday (MP).
>1868	William Filene opened a new store in Lynn, Massachusetts (MP).
1872	Montgomery Ward, Chicago, began catalog sales (WABI).
1876	First I. Magnin store was opened in San Franciso (NB).
1879	First variety chain store, Woolworth's, Utica, New York, opened (1ST).

< 1880	William Filene opened two Boston stores (MP).
1888	First Sears catalog issued, offering only watches (TIME).
1894	First full line Sears catalog offered (TIME).
> 1900	L. L. Bean issued its first catalog (ACTB).
1902	J. C. Penney opened his first store in Kemmerer, Ohio (HS).
1907	Bullock's Department Store, Los Angeles, opened by John Bullock (NB).
	Al Neiman and Herbert Marcus opened their first Dallas store (MP).
1909	New City of Paris store built to replace the store destroyed in the 1906 earthquake in San Francisco (NB).
1916	Market Square, the first planned shopping district in the U.S., opened in Lake Forest, Illinois (WK).
1922	Country Club Plaza, Kansas City, included a group of buildings developed and managed as unit; it included parking (WK).
1931	Highland Park Shopping Village, Dallas, shopping center with stores facing each other instead of the road (WK).
1956	Southdale Center, near Minneapolis, was opened as the first enclosed shopping mall (WK).
1960	First Nieman-Marcus Christmas catalog (CDIS).
> 1960	Catalog showrooms began operation (BUSWK).
1964	Ghirardelli Square, San Francisco, renovated chocolate factory converted to shops and restaurants (WK).
1972	City of Paris Department Store closed permanently (NB).
1976	Renovated Faneuil Hall opened containing stores, restaurants, and boutiques (WK).
1981	R. R. Donnelley began investigations of electronic shopping (MKTG).
1983	R. R. Donnelley built a prototype Electronicstore kiosk for merchandise ordering (MKTG).
1984	CompuServe tested the Electronic Mail, an on-line purchasing service (INFO).
1985	R. R. Donnelley made Electronicstore kiosks available for shopping malls and department store use (MKTG).

Our lives have been changed significantly by cloth and clothing technology. Few of us make our own clothes or do so only for special garments. Today we buy ready-made garments at relatively low cost, particularly if calculated in terms of hours we must work to purchase them. At one time it took two years to process flax fiber to produce thread or yarn; with modern technology it can be done in ten days. Formerly women and children had to card, spin, weave, knit, and sew throughout the year to clothe the family. While the wealthy could afford to employ tailors or dressmakers, most people could not. Today we can go to the shopping mall and choose from a large variety of suits, dresses, and coats, purchasing them with wages earned outside the home. If we sew, it is by choice, not by necessity.

In a sense, with the development of the 20th-century suburban shopping center or mall we have come full circle. Like the markets of old, the malls offer a variety of goods, entertainment, and food stuffs, and a gathering place for people.

6
Cleaning

Throughout the ages we have striven to improve our lot and our living conditions. Even animals seem driven to a certain level of cleanliness: witness the grooming habits of such diverse animals as birds, cats, and some breeds of dogs. Standards of cleanliness and attitudes toward cleaning have varied through the ages, from society to society, and from geographic region to region. The Cretans and the Romans appear to have had standards of cleanliness and bathing similar to our own. Large Roman public baths survive in a number of places and even gave a name to Bath, England. The Romans used water lavishly; they used half again as much per citizen as New Yorkers today.[1] Until recently, cleaning efforts often focused on making things look clean. According to Caroline Davidson's overview of housework in England, much of the cleaning dealt more with surface appearances than reality.[2] Lice and bugs were perennial problems, even in well-to-do households until the 20th century. Today we have a variety of cleaning products and tools that were not available to our ancestors. For much of history many cleaning tools or substances were home-made or gathered from nature. The Egyptians used natron, a naturally occurring form of carbonate of soda, while the Romans used fuller's earth for cleaning clothing.[3] Sand was used for cleaning dishes, floors, and even skin. Soap was made from wood ashes and animal fats when available. Fats were used for a variety of purposes including cooking, lubrication, leather working, and lighting. Their use for making soap was not the most important. Other barriers to soap and manufactured cleaning substance use were taxation, guilds, or other monopolies prohibiting or restricting home production of some cleaning substances.[4] For a long time soap products were taxed in Britain, and as a consequence, the poor often used lye made from ashes, urine, and even dung to clean clothes.

Because of their pervasiveness, soaps and cleaning products are discussed first, followed by laundry equipment, household cleaning, and personal hygiene. Useful sources include Caroline Davidson's *A Woman's Work Is Never Done*,[5] the American Home Laundry Manufacturers Association conferences for washing machines,[6] Loris Russell's *Handy Things to Have around the House*[7] for various small items, and Earle Lifshey's *Housewares Story*[8] for electric irons and ironing boards.

Soaps and Cleaning Products

Soaps and synthetic detergents are the most common cleaning products, but other cleaning products, such as bleach and ammonia, are also important. The chronology of their development is given in table 6.1. By definition soap as a cleaning agent, is also a detergent, but use of the term *detergent* is here confined to synthetic detergents. Soap may be made by a variety of processes, but usually involves animal fat in combination with an alkali such as certain types of wood ashes or lye made from them. Some synthetic detergents also contain animal fats, other do not. Generally, synthetic detergents will clean in cold as well as hot water and not leave residual deposits on clothes or washing machines, whereas soaps work best in hot water. Soap, one of the first cleaning products, is reviewed first, followed by synthetic detergents, developed during World War I, and then by other selected cleaning products.

Soaps

Both the Sumerians and the Phoenicians made soap, but the Phoenicians used it primarily as a salve. The Sumerians used alkalies in 3000 B.C., while the Phoenicians in 600 B.C. used goat tallow and wood ashes. Pliny the Elder, in 77 A.D., recorded that the Phoenicians traded soap with the Celts in 600 B.C., but the Celts also learned to make their own soap and later traded with the Romans. Soap in Rome was also used as a medication and as a hair dressing, and Pliny indicated that the Romans preferred German soap to Gallic soap. While the Romans used fuller's earth and sand for laundry, Galen of Pergamun, in 200 A.D., mentioned the use of soap for cleaning. By the 8th century, soap was in use throughout southern Europe. Latif, an Arab philosopher, described soap in 1200, and by 1259 soap manufacture was recorded at Sophar Lane in London. Despite its early discovery and use for some purposes, soap was not used extensively during much of the Middle Ages and even during the early Renaissance. People in Western societies believed that bathing was detrimental to health and considered Queen Elizabeth I eccentric for her "frequent"--monthly--bathing.[9] In addition to superstitious beliefs, the difficulty of obtaining water and soap contributed to its restricted usage. Most water had to be transported from a fountain, well, or stream and then heated. Soaps were harsh and animal fats were also needed for many other products besides soaps. As the water supply systems improved, soap usage increased.[10] Eastern societies had different attitudes toward cleanliness; the Japanese made bathing a social occasion, and ritual bathing was important in India. A major event that made soap less expensive and more readily available was the development by Nicolas LeBlanc of a process for making sodium carbonate from salt in 1791. His process provided a cheap form of soda for cleaning and for the manufacture of soap, and was used in 1814 at a plant in St. Helen's, Lancashire. Within a few decades soda, which was less dependent on hot water for cleaning power, became a fixture in most homes and soap usage also increased.[11] A second development leading to improved soaps was the isolation of fatty acids by M. Chevreul, France, in 1809. Gradually, the improved soaps, available at lower prices, replaced other cleaning products such as dung, urine, and lye.

Soap has been made and used in both liquid and solid form for many years, but powdered soap was a 19th-century invention. The solid form is easier to transport and takes less room to store, but the liquid form is better suited to some types of cleaning tasks. The first U.S. patented liquid soap was registered by William Sheppard, New York, in 1865; soap in liquid form was also produced at home and was used in Appalachia into the 20th century.[12] The first soap powder, Babbitt's Best Soap, was introduced in New York in 1843.

Improvements in soaps and soap making continued and brands now familiar began to appear in the late 18th and throughout the 19th century. In 1789 Andrew Pears began making Pears' Transparent Soap. Ivory Soap, first called White Soap, was marketed by Procter & Gamble in 1878. The "soap that floats," Ivory, was created accidentally when a workman whipped too much air into a batch of soap, but the customers liked it and kept asking for the floating soap. The first wrapped soap, Old Brown London Soap, was sold by John Atkinson in 1892. In 1907 Frank Giffin Carter invented an abrasive cleaning soap using marble flour (stone dust). Soap is still used for personal cleansing, but most laundry is now done with synthetic detergents.

Synthetic Detergents

Synthetic detergents were developed largely in response to a shortage of fats for soap manufacture in wartime Germany, although earlier work in the late 19th century made rapid development possible. A. Krafft of Germany discovered substances that had cleaning properties but were not soaps in

1886, and made further related discoveries in 1896. A. Richler, Belgium, developed some synthetic chemical detergents in 1913. Nekal, the first commercial synthetic detergent used in the textile industry, was marketed in Germany in 1917. It was followed in 1925 by Igepin. Fluorescent brightening powders were discovered in 1929 and patented by I. G. Farbenindustrie in 1941. They were added to washing powders in 1945.

Figure 6.1. Soaps and Cleaning Products

YEAR

1987		
	Tide with bleach and enzymes, Liquid Tide	
	In dryer fabric softeners, Bounce, Cling-Free	
1975		
	Biodegradable detergents, enzymes in detergents	
	Liquid fabric softeners, Sta-Puf, NuSoft	
1950		
	Built detergents, Tide	
	Nekal, Igepin	Dreft, Drene shampoo
1900	Abrasive soap	Triclorethylene
	Detergent properties	Ivory soap, carbon tetrachloride
	Liquid soap	Soap powder, turpentine
1800	Fatty acids	
	Leblanc soda process	Bleaching powder
	Pears soap	Sulphuric bleach
1600		Bluing
1200	Soap manufacture (London)	
400		
	Soap used for cleaning (Rome)	
	Soap (Phoenicia)	
-1200		
-4400	Soap (Sumeria)	Natron (Egypt)

The 1930s marked the beginning of modern U.S. synthetic detergents, many of which are still popular laundry products. Dreft, a synthetic detergent for wool and silk, was introduced in 1933. Tide, introduced in 1946, was the first "built" detergent, combining phosphates with detergent surfactants. Biodegradable detergents appeared in 1966, and in 1967 Procter & Gamble introduced detergents with enzymes that "digest" dirt and organic matter, thereby improving the detergent's cleaning effectiveness. Liquid Tide was introduced in 1984, and an improved powdered Tide with enzymes and bleach was marketed in 1985.

Other Cleaning Products

In the search for cleanliness, soap and synthetic detergents were not the only substances used. Bluing, ammonia, bleach, and even urine have all been used. Bluing, obtained from powdered glass of cobalt blue (smalts) or from ultramarine extracted from lapis lazuli, was used to tone down the yellowish tinge left in cotton and linen fabrics by repeated washings. Bluing is now made from Prussian blue or coal-tar blue. It was introduced into England from Holland in 1500. Detergents with fluorescent brighteners have now replaced bluing.

Ammonia has been used for cleaning and laundry for a long time, first as a naturally occurring substance in urine and later as a manufactured substance. Raymond Lully, a Majorcan philosopher, discovered ammonia in 1315 and was, according to Carter, stoned to death by angry citizens for his discovery. While Lully (or Llull) was stoned to death, it is uncertain whether it was for his science or his religious beliefs.[13] In 1677 Johann Kunchel described an aqueous solution of ammonia, by 1774 Joseph Priestly had isolated ammonia gas, and in 1787 van Marcum and van Troostwijk liquefied it. In 1908 Fritz Haber was able to synthesize it. Much of the interest in ammonia related to its use in fertilizers and gunpowder, not as a household cleaner. Stale urine contains ammonia and in part explains its successful use by the poor in producing whiter laundry. Dung has a similar property, but was also useful as a fertilizer or fuel and, as such, too valuable for extended use in laundering. Clothes cleaned with either substance required thorough rinsing in clear water. Dung and urine were used mainly by the poor and by rural inhabitants. Caroline Davidson discusses at some length the use of these latter materials for cleaning.[14]

Bleaching products, like ammonia, were of interest for industrial and commercial uses particularly in the textile industry. In 1746 John Roebuck invented a method for mass production of sulfuric acid for use in bleach manufacturing, and in 1756 dilute sulfuric acid was used by Hulme as a bleach. In 1785 C. L. Berthollet invented Eau de Javal, a chemical bleach, which the Lille bleachers used in their work. C. W. Steele discovered chlorine in 1774, and Faraday liquefied it in 1823. In 1789 Berthollet invented a chlorine and alkali bleaching liquid. Charles Tennant, Glasgow, improved Berthollet's process; he invented bleaching powder in 1799 and began manufacturing it in 1800.

Dry cleaning uses special chemicals instead of soap or detergent and water to clean garments. J. B. Jolly, Paris, established the first dry cleaners in 1855 using turpentine as a cleaning fluid. Unfortunately, turpentine was also highly flammable. Louis Anthelin, Germany, discovered the cleaning properties of carbon tetrachloride in 1897. Carbon tetrachloride, while less flammable than turpentine, is a toxic chemical and was replaced in 1918 by trichlorethylene,which was found to have similar cleaning properties but to be less toxic.

Table 6.1. Soaps and Cleaning Products

Dates	Event
B.C.	
3000	Sumerians made soap from alkalies and fats (DBE).
600	Phoenicians prepared soap from goat tallow and wood ashes and used it as a salve (DBE).
A.D.	
77	Pliny indicated Romans preferred German to Gallic soap (prepared with natural soda and fats) (DID).
>200	Galen mentioned the use of soap for cleaning (DBE).
>700	Soap used in most of southern Europe (EDE).
>1200	Latif, an Arab, described soap (DID).
1259	Soap manufacture done at Sopar Lane, London (1ST).
1300	Alum discovered at Rocca, Syria (DID).
1315	Majorcan philosopher Raymond Lully identified ammonia (DID).
1500	Bluing for bleaching introduced into England from Holland (DID).
1524	Soap first made commercially in London (it had been made earlier at Bristol) (DID).
1564	Starch introduced into England from Flanders by Dinghen van der Plasse (CD).
1677	Johann Kunchel described aqueous solution of ammonia (DID).
1746	John Roebuck invented a method of mass producing sulphuric acid (used in bleach manufacture) (EB).
1756	Dilute sulphuric acid used for bleaching by Hulme (DID).
1774	C. W. Scheele discovered chlorine (DID).
	Joseph Priestly isolated ammonia gas (DID).
1785	C. L. Berthollet invented chemical bleach, Eau de Javal, which was adopted by the Lille bleachers (TH).
	Hulme, U.K., used dilute sulphuric acid as a bleach (DID).

1787	M. van Marcum and A. Paeto van Troostwijk liquefied ammonia (DID).
1789	Andrew Pears made Pears' Transparent Soap (CD).
1791	Nicolas LeBlanc, France, patented a process for the manufacture of sodium carbonate (used in soap manufacturing) from salt (EB).
1799	Charles Tennant of Glasgow improved Berthollet process by inventing bleaching powder (DID).
1809	M. Chevreul, France, isolated fatty acids--olive, palmitic, and stearic acids--leading to improved soap (EB).
1814	LeBlanc's soda process was first used at a plant in St. Helen's, Lancashire (CD).
1823	Faraday succeeded in liquefying chlorine (1ST).
1829	First wrapped soap, Old Brown London Soap, sold by John Atkinson (1ST).
1830	Uniform cakes of soap for resale by grocers were sold in Newburgh, New York; previously grocers cut soap from large blocks as needed (FFF).
1843	First soap powder marketed by Babbitt's Best Soap, New York (1ST).
1865	Soap in liquid form patented by William Sheppard, New York (FFF).
1878	Ivory Soap (first called White Soap) marketed by Procter & Gamble, Cincinnati, Ohio (FFF).
1886	Krafft, Germany, discovered detergent properties (further discoveries were made in 1896) (DBE).
1889	Lysol disinfectant first introduced for cleaning (CDIS).
1897	Ludwig Anthelin, Germany, discovered carbon tetrachloride for cleaning (WABI).
1907	First household powdered soap, Persil, marketed by Henkel & Cie, Dusseldorf (1ST).
	Frank Giffin Carter invented inlaid soap with insert of abrasive marble flour (DID).
1908	Fritz Haber synthesized ammonia (TH).
1913	A. Richler, Belgium, discovered synthetic chemical detergents (TT).
1917	Shortage of soap in Germany led to marketing of Nekal, first commercial detergent for textile industry (TT).
1918	Because of toxic fumes, carbon tetrachloride was replaced by trichlorethylene for dry cleaning (WABI).
>1925	Igepon detergents invented by Platz and Daimler (DBE).
1929	German chemist discovered fluorescent brightening agents, later used in washing powders (TT).

1932   Calgon detergent developed (WABI).

1933   Dreft, synthetic detergent for wool and silk, marketed for home use by Procter & Gamble (FFF).

        First soapless shampoo, Drene, introduced by Procter & Gamble in U.S. (1ST).

1941   Fluorescent brightening agents patented by I. G. Farbenindustrie (TT).

1945   Brightening agents added to washing powders (TT).

1946   Procter & Gamble introduced Tide, first "built" detergent (TWA).

1966   Biodegradable liquid detergents introduced (TT).

1967   Procter & Gamble introduced detergents with enzymes that digested dirt (TT).

1984   Liquid Tide introduced (WJ).

1985   Tide with enzymes and bleach introduced, Procter & Gamble noted bleach worked for wider range of temperatures (WJ).

Laundry Equipment

Originally people washed their clothes in the water of a stream or lake and pounded them with rocks or a board to loosen the dirt. Much washing in Britain, as elsewhere, was done along streams or at other convenient water sources. To get hot water, fires were built and pots or kettles of water placed on them or stones were heated and dropped into containers of water to heat the water. Many poor people washed with cold water because of the inconvenience or the high cost of fuel for heating water.[15] Laundry is still done this way in remote parts of the world, including parts of Kenya, China, and India. Later, various substances including soap were found to make the job easier, but soap generally needed hot water to produce satisfactory results. In addition to the cleaning substances discussed above, laundry equipment also made cleaning easier. Table 6.2 outlines developments in laundry equipment.

Washing Machines

The first recorded mention of a mechanical device for washing clothes was Robert Hook's 1677 description of Sir John Hoskins's method of washing fine linen in a bag fastened at one end and strained by a wheel and cylinder at the other. The invention of the first washing machine is unrecorded, however, although a variety of mechanical contrivances existed early in the 18th century. When William Bailey invented his washing machine in 1758, he described it as an improvement on the then existing "machine washing-tubs."

His machine, operated by one person, could perform the work of three washerwomen. Manual devices for home use were also produced, including plungers and washboards. A hand plunger for manipulating the wash dates from 1770. James McCalla, Galton, Ohio, patented a clothes pounder with multiple cones in 1883. The washboard was patented in 1833.

Figure 6.2. Laundry Equipment

YEAR

1987		
	Last Maytag wringer washer	Cordless iron
	Programmed washers	
1975		
	Hoover twin tub washer	
	Bendix washer/dryer combined	Steam and dry irons
1950		
	Bendix automatic washer	Hamilton automatic dryer
	Spin dryer	Thermostatic and steam irons
1900	Electric/agitator, wringer	P&O dryer, electric iron
	Self-reversing washer	Steam-heated and gas irons
	Wash board	
1800		
	Washing machine (Bailey)	Pochon clothes dryer
1600		Mangle

A number of inventors developed various mechanized washing machines. Noah Cushing, Quebec City, patented a washing and fulling machine in 1824. In 1849 Charles Mattee designed a crude wooden wash wheel for the Davis Laundry in Oakland, California. This was followed by Joel Houghton's cylindrical, hand-operated, washing machine with agitated water, patented in 1850. Samuel Yost, Connorsville, Indiana, patented a rotary washboard washing machine in 1855, and in 1858 Hamilton Smith, Philadelphia, patented a mechanical washing machine with paddles to manipulate the wash. Smith patented a second, improved machine in 1863 with self-reversing motion.

Most washing machines were produced for institutional laundries, not for home use. Human labor or water provided the power. Water power required a swift stream nearby. Too, with the many industrial uses for water power, it was not readily available to run laundry equipment. An exception to this were the various Shaker communities, which used both water and steam power for their laundries, including driving washing mills (washing machines) and steam-heated centrifugal dryers.

199

The real breakthrough in washing machines occurred with the application of electric power. In 1907 Alva J. Fisher designed the Thor electric washing machine manufactured by the Hurley Machine Corporation. This was the first washing machine that did not require an operator to crank a handle to perform the washing action, although clothes still had to be fed by hand through a wringer or mangle to remove excess water. In 1909 the P&O Line fitted a passenger ship with two steam-heated electric washing machines and two electrical centrifugal dryers.

In the U.S. washing machines operated on piped hot water from a hot water heater (hot water and water supplies are discussed in chapter 7). In other countries the washing machine, like those on the P&O liner, included a mechanism to heat water. Early washers required the operator to remove or drain wash and rinse water, whereas modern machines automatically pump waste water into the drainage system.

A number of developments followed in the next thirty years. Maytag manufactured their first wringer washer in 1907; Barlow and Seelig, later incorporated as Speed Queen, offered one in 1911; and John Fisher produced an agitator machine in 1915. By 1924 the Savage Arms Corporation, Utica, New York, had manufactured a spin dryer to remove excess water from clothes as part of the washing process. Hoover incorporated the spin dryer in the popular twin tub washer in 1957. Bendix introduced an automatic washer in 1937, programmed washers were introduced in 1978, and by 1983 the last Maytag wringer washer was manufactured--supplanted by the easier to use spin dryers and automatic washers. Clothes dryers also had a similar extended pattern of development. An early clothes dryer, developed by Pochan, France, in 1799, consisted of a perforated metal drum that was heated externally and turned by hand. Inattention could result in unevenly dried or burned clothes. The first modern dryers were the two centrifugal electrically heated dryers installed in the P&O line in 1909. In the United States, Hamilton developed, and was marketing, an automatic dryer by 1939. By 1953 Bendix developed the Duomatic, a combination washer and dryer. Tumbler dryers for home use were marketed in the United Kingdom by Parnell in 1958 after successful use in laundrettes there.

Laundries existed in Rome during the Roman Empire but disappeared during the aftermath. The first public laundry in England opened on Frederick Street, Salford, in 1841. Liverpool opened public wash houses in 1844. The first commercial laundry in the United States was opened around 1849 by Charles Davis, a failed miner, in Oakland, California, to provide service to the gold prospectors of California. The first United States laundrette, the Washateria, opened in Fort Worth, Texas, in 1934. Use of its washing machines were charged for by the hour.

When considering why a particular technology is used, misused, or not used at all, common sense and an understanding of human nature may provide answers. Caroline Davidson, in her overview of English housework, expressed surprise that while washerwomen were hired to help with household laundry, few households made use of commercial laundries.[16] The answer may lie in basic human nature, and may have some commonality with the response to the question of why women clean the house before the cleaning lady comes; they do not want to be judged as lazy or as a careless housekeeper. In the case of laundry, outer garments are generally made of sturdier fabrics than undergarments, and while such delicate fabrics of the latter may require special care, woolens and dyed fabrics also require separate processing. In addition to care requirements, because of the cost and labor required, relatively few garments were owned, and any delays in laundering could be inconvenient. If the laundry was done at home, the householder was assured it was handled properly, in a timely fashion, and would not be misplaced.

Irons

Ironing tools developed earlier than washing machines. The Greeks used a "goffering" iron, a cylindrical bar, for pleating linen garments, and by the 15th century, some Europeans were using "hot box" irons with compartments for coals or a heated brick. Mangles--devices for smoothing table linens or removing excess water from wet wash--were in use by the 17th century. Ann, duchess of Hamilton, ordered a mangle from Edinburgh in 1696. In Great Britain Hugh Oxenham patented an upright mangle in 1774, while George Jee patented an improved mangle in 1797 requiring less strength and fewer operators than earlier models. In the United States, R. A. Stratter, Philadelphia, patented a mangle in 1857. Heated ironers were available by the end of the 19th century. Montgomery Ward offered, in their 1895 catalog, a gas- or steam-heated home ironer for table linens and sheets. An industrial ironer, the Ducoredun, was used circa 1900. Domestic electric ironers were available by 1926, and a folding model was offered by Earle Ludgin and Company of Chicago in 1946.

A number of patents were issued in the 19th century for hand irons. Until the development of electric power, irons were heated in the fireplace or on stoves, or hot coals were placed in receptacles in the irons. The thickness and mass of metal determined how long the iron had to be heated and how long it retained heat. It was impossible with an external heat source to maintain a constant temperature. Among some of the irons patented was one by Nicolas Taliaferro and Augusta and William Cummings of Murphyville, Kentucky, for a self-heating iron in 1852, and another by Mary F. Potts, Ottumwa, Iowa, for a double-pointed sad iron--a heavy flatiron-- with a spring-loaded, clamp-on handle in 1871. A gas iron was patented in 1874. A number of machines were used and some patented for pleating and fluting and at least one was issued to a woman: Susan K. Knox patented a fluting machine, used for pleating ruffles, in 1886. Such irons had grooves for pressing fabric into pleats or ruffles.

The iron became a more easily used tool with the advent of electricity and a means of maintaining a constant temperature. Henry W. Seely, New York, patented the electric iron in 1882, but the first thermostatically controlled one was not sold in the United States until 1934. The development of thermostats for electric appliances was discussed in chapter 4. They were first used in appliances in the early 1930s. General Electric introduced the first travel iron in 1909, and Winsted Hardware marketed the first folding travel iron in 1937. Nocord introduced the first cordless iron in 1922, but it did not become popular; West Bend introduced their cordless model as a new innovation in 1984. The steam iron was developed in the late 1920s and by the early 1960s combination irons were marketed. The first steam iron was imported to the United States in 1927 and the first steam iron manufactured in the U.S. was the Steem-Electric in 1937. General Electric introduced the steam and dry iron to the U.S. in 1957 and Hoover introduced a similar model to Britain in 1963. Variations on these models were developed by iron manufacturers: Sunbeam introduced the spray-mist iron in 1962, a Shot of Steam iron in 1969; General Electric added a Teflon soleplate for smoother ironing in 1964 and a self cleaning steam, spray, and dry iron in 1972. Both General Electric and Sunbeam were selling irons that automatically shut themselves off in 1984.

Laundry Aids

Laundry aids include clothespins or clothes pegs, clothes hoists, fabric softeners, and sponges. It is uncertain exactly when and where the clothespin first appeared. Since clothing was washed infrequently until well into the 19th century,[17] and clothes were spread on bushes or grass, or hung over lines or other objects to dry, there was little need for large numbers of clothespins until larger, more frequent washings became common. Wood clothes pegs were carved by the men of the household as needed to hold the wash. The spring-type clothespin was patented by D. M. Smith, Springfield, Vermont, in 1853. The clothesline evolved from a simple rope to more complex devices. The rotary clothes hoist was invented in Australia in 1946. This allowed the householder to lower the line to hang and take down wash, to raise it after the laundry was hung to catch the breeze, and also provided maximum hanging space in a very compact arrangement. Lines were nested in the shape of a square around the center pole. The development and rapid adoption of dryers in the United States led to the abandonment of most clotheslines. A resurgence of interest, spurred by potential energy savings, occurred in the 1970s, but the realities of winter climates encouraged continued reliance on dryers, at least during cold weather.

One consequence of artificial drying was the phenomena of overdrying and the resultant static cling. Liquid products, added to the final rinse cycle of the wash, were introduced in the 1950s and 1960s: Sta-Puf was first introduced in 1957, followed by NuSoft and Downey in 1961. Products for use in the dryer, rather than in the washing machine, were introduced in the mid-1970s. Procter & Gamble marketed Bounce for use in dryers in 1976, and it was quickly followed by Calgon's Cling Free. Some detergents also incorporated fabric softeners.

Rags and sponges have long been standard cleaning equipment. In earlier times, fabric was valuable, even in the form of rags, and was carefully hoarded and reused. Cuffs were turned or trimmed to hide wear, rugs or quilts were made from rags, and finally, if no other use was possible, they were used for cleaning. A sign of affluent times and a mobile, throw-away society is the disappearance of rags for cleaning. Colgate-Palmolive, while searching for innovative new products, found an alternative use for Johnson & Johnson's sanitary napkin covering material. They introduced Handiwipes, disposable but reusable, nonwoven cleaning cloths, in 1966.

Scott Paper Company introduced Sani-Towels, the first consumer paper towels, in 1907 for drying the hands of school children and cleaning. Later the name was changed to ScotTowels. Scott manufactured toilet paper, but received a shipment of paper too wrinkled for that use and developed paper towels as another way to use the paper. In 1984 Scott introduced ScotTowel Jr., a cheaper, smaller sized towel for bathroom use.

Natural sponges generally require warm water to grow; consequently, they were only readily available in a limited geographic area. Transportation costs and relative scarcity made them expensive until the development of artificial sponges. Sponge rubber was patented by Charles Hancock in 1846. The first synthetic sponges were manufactured by Novacel, France, in 1932.

Dishwashing

Many of the products and laundry aids described above can also be used for dishwashing, including soaps, synthetic detergents, sponges, and rags. Other products were developed specifically for dishwashing. Calgon dishwashing detergent was marketed in 1932. Brillo steel wool pads were introduced in 1913 for cleaning pots and pans; they were followed in 1917 by S.O.S. soap-impregnated pads developed by Edwin Cox of San Francisco. Lem-me, a lemon-scented dish scrubber, was introduced in 1985 by Swiss-Tex. Polyethylene dishpans were available in 1948. The first dishwashing machine was invented by Mrs. W. A. Cockran, Shelbyville, Indiana, in 1889. She built both domestic and hotel models, with the latter using a steam-driven engine.

Her first domestic-use model was introduced in 1914, but required such large quantities of hot water that it was not successful. It was not until 1932, however, that the first practical model for domestic use was marketed in the United States. The automatic dishwasher was marketed in 1940; Europe imported dishwashers from the United States in 1960. In 1962 Kelvinator demonstrated an ultrasonic dishwasher in Seattle, but did not market it because it was too costly for normal use. Kelvinator used sound waves beyond the range of human hearing to loosen and remove the soil.

Table 6.2. Laundry Equipment and Aids

Dates	Event
B.C.	
4th c	Greeks used a "goffering" iron, a cylindrical heated iron bar, to produce pleats in linen garments (EOOET).
A.D.	
15th c	Wealthy Europeans used "hot box" irons (irons had a compartment for hot coals or a hot brick) (EOOET).
1677	Robert Hooke recorded Sir John Hoskins's way of washing fine linen in a bag strained by a wheel (DBE).
1696	Ann, duchess of Hamilton, ordered a mangle from Edinburgh (CD).
1758	William Bailey invented a washing machine, an improvement on "machine washing-tubs" then in use (CD).
1770	Hand plunger designed specifically for manipulating wash sold (WB).
1774	Hugh Oxenham, England, patented a mangle (CD).
1797	Nathaniel Briggs, New Hampshire, patented a washing machine (FFF).
	George Jee devised an improved mangle, reducing the number of operators and strength to operate (CD).
1799	Pochan, France, invented a clothes dryer (WABI).
1824	Noah Cushing, Quebec City, patented a washing and fulling machine (LR).
1833	Washboard patented in U.S. (EL).
1841	Public laundry opened on Frederick Street, Salford, England (1ST).
1844	First public baths and wash houses opened in Liverpool, England (TH).
1846	Sponge rubber patented by Charles Hancock, London (1ST).

1849	First commercial laundry in U.S. started by a failed miner, Davis, for other miners in California (WB).
	Dry cleaning process discovered by M. Jolly-Bellin, Paris (1ST).
>1849	Charles Mattee designed a crude wooden wash wheel for Davis's laundry (WB).
1850	Cylindrical hand-operated washing machine with agitated water invented by Joel Houghton, U.S. (DID).
1852	Nicolas Taliaferro, Augusta, and William Cummings of Murphyville, Kentucky, patented a self-heating iron (LR).
1853	D. M. Smith, Springfield, Vermont, patented the spring clothespin (LR).
1855	Samuel Yost, Connorsville, Indiana, developed a rotary washboard washing machine (LR).
	J. B. Jolly, Paris, established the first dry cleaners using turpentine as a cleaning fluid (WABI).
1857	R. A. Stratter, Philadelphia, patented a mangle (LR).
1858	First rotary type, mechanical washing machine was patented by Hamilton E. Smith (DID).
1863	Hamilton E. Smith invented the first self-reversible washing machine, an improvement on his 1858 machine (DID).
1869	Shaker elder recorded use of a steam-operated laundry, including a Parker wash mill (washing machine) and centrifugal dryer (HDI).
1871	Mary F. Potts, Ottumwa, Iowa, patented a double-pointed sad-iron with spring-loaded clamp-on handle (LR).
	B. F. Goodrich, Akron, Ohio, manufactured rubber clothes wringers (EOOET).
1874	First gas iron patents were issued in the U.S. (EL).
1882	Electric iron patented by Henry W. Seely, New York (1ST).
1883	Clothes pounder with multiple cones for manipulating wash patented by James McCalla, Galton, Ohio (LR).
1886	Susan K. Knox patented a fluting machine (LR).
1889	Mrs. W. A. Cockran perfected the first dishwashing machine (1ST).
1891	Stevens dishwasher, Cleveland, advertised (FHTC).
1892	Electric iron demonstrated by J. J. Dowsing at Crystal Palace (1ST).
1895	Montgomery Ward's catalog offered a steam or gas heated ironer (MONTW).
1898	J. R. Clark Co., Minneapolis, introduced the first ironing board with legs (EL).

Forerunner of Whirlpool-Seegar offered a reciprocating tub washing machine (AHLMA).

1900 Ducoudum Ironer, French industrial ironer, used (SG).

1903 Earl Richardson, California, developed an iron with a "hot point," which later became the Hotpoint iron (EL).

General Electric introduced the detachable cord for irons (EL).

1905 Portable vacuum cleaner for domestic use was marketed by Chapman and Skinner, San Francisco (1ST).

1907 Thor, an electric washing machine, designed by Alva J. Fisher, manufactured by Hurley Machine Corp., U.S. (1ST).

Scott Paper Company introduced Sani-Towels; they later became ScotTowels (UNITD).

First Maytag wringer washing machine manufactured (FORB).

1908 Blackstone offered an improved agitator washing machine (AHLMA).

Forerunner of Whirlpool-Seegar offered the Dolly washing machine (AHLMA).

1909 P & O Line fitted with two steam heated electric washing machines and two electrical centrifugal dryers (DID).

General Electric introduced the first travel iron (EL).

1910 First stamped, nickel-plated steel shell introduced for General Electric irons (EL).

1911 Barlow and Seeling, later Speed Queen, offered a wringer washing machine (AHLMA).

1912 Electric dishwasher marketed in U.S. (WABI).

1913 Brillo pads sold for cleaning pots and pans (TT).

Maytag made a hand-operated washing machine (AHLMA).

1914 Mrs. Cochran's company introduced a domestic dishwashing machine (EOOET).

1915 John Fisher, U.S., developed a prototype of the agitator washing machine (TT).

Upton Machine Company, later Whirlpool-Seegar, was marketing the Cataract Rocking Tub washing machine (AHLMA).

1917 S.O.S., soap-impregnated steel wool pads, were devised by Edwin W. Cox, San Francisco (EOOET).

1920 Barlow-Seeling, later Speed Queen, offered a vacuum washing machine (AHLMA).

1921 First iron with embedded Calrod heating element introduced (EL).

1922	Nocord Electric Company introduced a cordless electric iron (EL).
1924	Spin dryer used for washing clothes made by Savage Arms Corp., Utica, New York (TT).
	Westinghouse introduced nonadjustable, automatic iron featuring Klixon disc thermostat (EL).
1926	Eldec Co., New York City, introduced "steeming iron" imported from Switzerland (EL).
	Ironer on wheels offered in mail-order catalog (SG).
	Speed Queen manufactured the Vacuum Dasher washing machine (AHLMA).
1927	Liberty Gauge and Instrument Company, Cleveland, introduced adjustable-temperature automatic iron (EL).
1929	General Electric introduced an iron with Button Nook to facilitate ironing around buttons (EL).
1931	General Electric introduced a cast-aluminum soleplate iron (EL).
	Synthetic sponge manufactured by Novacel, France (1ST).
1932	First practical dishwasher marketed in U.S. (TT).
1934	First thermostatically controlled iron sold in U.S. by Sunbeam; it used double-pole automatic thermostat (EL).
	First launderette was the Washateria, opened at Fort Worth, Texas, by J. F. Cantrell (1ST).
1937	Durabilt, first folding travel iron, introduced by Winsted Hardware, Winsted, Connecticut (EL).
	Steem-Electric Company, St. Louis, introduced Steem-Electric, first U.S.-made steam iron (EL).
	Bendix manufactured the automatic washer (AHLMA).
1939	Met-L-Top, all metal ironing table with adjustable legs, introduced (EL).
	Hamilton, U.S., marketed an automatic clothes dryer (AHLMA).
1940	Automatic dishwasher developed in the U.S. (WABI).
1946	Rotary clothes hoist invented in Australia (TT).
	Folding ironer marketed by Earle Ludgin & Company, Chicago (SG).
1948	Dishpans made of polyethylene introduced (1ST).
1953	General Electric introduced the first steam travel iron (EL).
	Bendix introduced the Duomatic, a combination washer and dryer (AHLMA).
1954	General Electric introduced the first steam/dry iron (EL).

1957   Hoover made the spin dryer part of its twin-tub washing
        machine (TT).
        First steam/spray iron introduced by General Electric (EL).
        StaPuf liquid fabric softener introduced (WHIRL).
        NuSoft fabric softener introduced (WHIRL).
1958   Domestic tumble dryers marketed by Parnell after
        successful use in coin-operated laundrettes (TT).
1960   Europe imported the automatic dishwasher from the U.S.
        (WABI).
1961   Downey liquid fabric softener marketed (WHIRL).
1962   Kelvinator, Seattle, demonstrated an ultrasonic dishwasher,
        but it was too costly for normal use (TT).
        Sunbeam introduced a spray-mist steam iron (EL).
1963   Hoover steam and dry electric iron introduced in Britain
        (TT).
1965   General Electric introduced the Teflon-coated soleplate
        iron (EL).
1966   Colgate-Palmolive introduced Handiwipes, developed from
        Johnson and Johnson's nonwoven sanitary napkin fabric
        (CP).
1969   Shot-of-Steam iron with trigger-controlled ejector
        introduced by Sunbeam (EL).
1972   General Electric introduced "self-cleaning" steam, spray,
        and dry iron (EL).
1976   Bounce and Cling-Free in-dryer fabric softeners marketed
        (CREPT).
1978   Programmed washing machine introduced (TT).
1983   Last Maytag wringer washing machine manufactured
        (FORB).
1984   General Electric and Sunbeam marketed irons that shut
        themselves off automatically (WD).
        Scott Paper Company introduced ScotTowel Jr., smaller-
        sized paper towel (WJ).
1985   Lem-me, dish washer scrubber, introduced by Swiss-tex
        (MKTGN).

Household Cleaning

Major household cleaning chores include dusting and cleaning floors. Table
6.3 provides a summary of household cleaning developments. Brooms and
mops, the primary means of cleaning floors, were originally homemade. A
patent was issued to Cyrus T. Moore, Concord, New Hampshire, for an

improved corn straw broom in 1852. The self-wring mop, a flat sponge with a metal plate, was patented in the United States in 1961.

A number of carpet sweepers were patented, including a mechanical sweeper patented in 1811 by James Hume. Joseph Whitworth patented a street-cleaning machine in 1841 and later, after modifying his design, patented a carpet sweeper in 1859. Hiram H. Herrick, East Boston, Massachusetts, also patented a carpet sweeper, but it would only sweep in one direction. The first practical carpet sweeper was patented in 1876 by Melville R. Bissell, Grand Rapids, Michigan. He developed his machine to combat his own dust allergy. Carpet sweepers are still marketed and used for quick cleaning by energy-conscious users.

The vacuum cleaner first appeared in the late 19th century, but did not become practical until the early 20th century and the development of electrically powered models. The Whirlwind, 1869, appears to have been the first vacuum cleaner sold. It was hand-cranked to create suction. The first vacuum cleaner patent was filed by John S. Thurman, who was granted a patent for a "pneumatic carpet renovator" in 1899. Hubert Cecil Booth devised his vacuum cleaner in 1901, and a prototype was produced in 1902. Chapman and Skinner, San Francisco, were marketing a portable vacuum cleaner for domestic use in 1905. The unit weighed ninety pounds! Compared to a weight of less than fifteen pounds for a modern vacuum cleaner, this precursor would not be considered portable today. Modern electric vacuum cleaners evolved from Murray Spangler's design, sold to W. Hoover in 1907. Hoover began selling vacuum cleaners for home use in 1908. Lux, Sweden, produced a cleaner suitable for cleaning curtains, cars, and upholstery in 1918. Hoover introduced the hand-held vacuum in 1930 and the upright Hoover, "it beats as it sweeps, as it cleans", in 1936.

Table 6.3. Household Cleaning

Dates A.D.	Event
1699	Edmund Heming patented a device for sweeping the streets of London (EL).
1811	James Hume patented a mechanical sweeper (EDE).
1841	Street cleaning machine invented by James Whitworth (1ST).
1852	Patent issued to Cyrus T. Moore, Concord, New Hampshire, on an improved corn straw broom (LR).
1858	Hiram H. Herrick, East Boston, Massachusetts, patented a carpet sweeper that swept in only one direction (LR). Lucius Bigelow patented an improved carpet sweeper (EL).

1859	Sir Joseph Whitworth invented a carpet sweeper based on his 1842 road sweeper (DBE).
1865	Carpet sweeper in general use (TH).
1869	Whirlwind, the first vacuum cleaner sold, used a hand crank to create suction (WJ).
1876	Melville R. Bissell, Grand Rapids, Michigan, patented the carpet sweeper (1ST).
1899	John S. Thurman, U.S., patented the pneumatic carpet renovator, a vacuum cleaner (PA).
1901	Vacuum cleaner devised by Hubert Cecil Booth (1ST).
1902	Vacuum Cleaner Co. Ltd. produced a prototype based on Hubert Booth's design (1ST).
1907	Murray Spangler sold the rights to his domestic upright vacuum cleaner to Hoover (TT).
	David E. Kenney, New York, granted a patent on a suction cleaner originally applied for in 1901 (EL).
1908	W. Hoover introduced the vacuum cleaner for home use (TT).
1918	Lux, Sweden, produced a vacuum suitable for cleaning curtains, cars, and upholstery (TT).
1930	Hoover introduced the first small, hand-held vacuum (TT).
1936	Upright Hoover--"beats as it sweeps as it cleans"--introduced (TT).
1961	Self-wringing mop, flat sponge with metal plate, was patented in the U.S. (TT).

Personal Hygiene

The availability of soap, the influence of inadequate water supplies, and the lack of hot water were mentioned earlier. This final section is devoted to a brief history of the toothbrush, dentifrices and the razor. The date of origin of the toothbrush uncertain. The Chinese claim to have invented one in 1498. In 1649 a correspondent of Sir Ralph Varney asked the latter to purchase a toothbrush in Paris. By 1690 toothbrushes were for sale in London. A refillable toothbrush with removable bristles was introduced in 1931, but it was not successful. The first nylon toothbrush was marketed in 1938, and the first electric toothbrush was introduced by Squibb in 1961. Disposable toothbrushes, molded from one piece of plastic, were made in Milan, Italy, in 1979. Other devices for cleaning teeth include the Water-Pik, developed in 1966; an acoustic wand developed by Joseph Heyman, NASA, in 1984; and a new type of electric toothbrush developed by the Japanese in 1986, which included a semiconductor in the handle that when exposed to light emitted electrons through the toothbrush bristles. These electrons neutralized the acidic film produced by plaque-causing bacteria so it could easily be brushed away.

In addition to these tools, people have used a number of substances for cleaning teeth and sweetening breath. Among the more unusual was the Roman use of urine to whiten teeth.[18] Soda and salt are still reliable standbys. Toothpaste was first sold in a collapsible tube by Dr. Washington Sheffield, New London, Connecticut, in 1892. Work on stannous fluoride for tooth decay prevention was started by Dr. Joseph C. Mulder at Indiana University in 1945. Procter & Gamble provided sponsorship in 1949 and clinical trials of the toothpaste began in 1952. Crest toothpaste with stannous fluoride was introduced in 1956 and was endorsed by the American Dental Association in 1960. After 1960, Crest was followed by a number of imitators and some improved toothpastes. Gels, packaging, and most recently tartar or plaque control toothpastes have been urged on consumers. Colgate-Palmolive introduced gel toothpaste in 1981 and followed with the introduction of a pump dispenser toothpaste in 1984 to compete with Minnetonka's Check-up pump dispenser introduced earlier. Minnetonka used a German pump dispenser. Such pumps had been used in Europe since the late 1970s; and Procter & Gamble had also tested a pump dispenser for Crest, but manufacturing problems had delayed widespread availability in the U.S. during 1985. Vipont introduced an antiplaque toothpaste in 1983, while Procter & Gamble responded with a new Crest for controlling tartar in 1985. Dr. Joseph Lister popularized the use of carbolic acid as a surgical antiseptic after discovering its effectiveness in 1965, and wrote and lectured to convince others. In 1876 he visited Philadelphia and was heard by a Missouri physician, Joseph Lawrence. Impressed with Dr. Lister, Lawrence developed an antibacterial mouthwash he named for Lister and began marketing in 1880 as Listerine.

Man has shaved for much of his existence. Cave drawings indicate the use of clam shells, flint, and other sharp tools for shaving. Knives have been, and are still, used by some for shaving. Razors were developed early and were found in Egyptian tombs of 3000 B.C.; the Romans were shaving by 499 B.C.

Even safety razors had a period of development before King Gillette marketed his successful model. M. Moreau devised a "safety razor" using a toothed guard in 1762. The forerunner of the modern safety razor appeared in 1828 in Sheffield. Another toothed guarded razor was developed by William S. Henson, Chard, Somerset, in 1847, and the Star Safety Razor with a fixed blade was patented in 1880. Gillette invented the modern safety razor with a removable, replaceable blade in 1895 and patented it in 1901. He created a new industry with replaceable razor blades when he began manufacture and sale of his razor in 1903. Wilkinson Sword introduced the long-lasting, stainless steel razor blade in 1955. Disposable razors were introduced in Europe by Bic in 1974; both Bic and Gillette introduced the disposable razor to the U.S. market in 1976. The new razors were inexpensive and could be used for up to five shaves. In 1985 Gillette introduced Lubra-Smooth, a twin cartridge disposable razor with a water-soluble plastic resin on the blade edge to reduce friction.

Electric razors were developed in 1900 in the United States, but the first successful electric razor was developed and patented by Jacob Schick in 1923 and first manufactured in 1931. By 1937 Philips of Holland had introduced a razor with a circular cutting head.

Figure 6.3. Personal Hygiene

YEAR

YEAR		
1987	Semiconductor toothbrush	Antiplaque toothpastes
		Gel toothpaste
1975	Disposable razors	Water-Pik
	Stainless steel razor blades	
	Electric toothbrush	Crest toothpaste
		Stannous floride
1950		
	Electric razor	Nylon toothbrushes
1900	Modern safety razor (U.S.)	
		Toothpaste in a tube
	Early safety razor (U.K.)	
1800		
	Early safety razor (France)	
		Toothbrush (Paris)
1600		
		Toothbrush (China)
1200		
400		
	Razors (Rome)	
-1200		
	Razors (Egypt)	
-4400		

Table 6.4. Personal Hygiene

Dates	*Event*
A.D.	
1498	Chinese invented the toothbrush (1ST).
1649	Correspondent of Sir Ralph Varney asked the latter to purchase a toothbrush in Paris (1ST).
1690	Toothbrushes sold in London (1ST).
1762	M. Moreau devised a safety razor that used a toothed guard (CAD).
1828	Forerunner of the modern safety razor developed in Sheffield, England, and in the U.S. (EB).
1847	Safety razor with comb teeth guard invented by William Samuel Henson, Chard, Somerset (DID).
1880	Listerine antiseptic introduced (HS).
	Kampfe Brothers, New York, developed the Star Safety Razor using a fixed (not removable) blade (FFF).
1892	Dr. Washington Sheffield, New London, Connecticut, introduced toothpaste in a tube (1ST).
	Beecham's Tooth Paste, U.K., was marketed in a tube (1ST).
1895	King C. Gillette invented the successful safety razor (TH).
1900	Unsuccessful electric razors were devised in the U.S. (EB).
1901	Gillete patented the safety razor (1ST).
1903	Gillette began manufacture and sale of his razor at American Safety Razor (1ST).
1923	Jacob Schick patented the electric razor (TH).
1931	Refillable toothbrush with removable bristles introduced (TT).
	First electric razor manufactured by Schick Incorporated, Stamford, Connecticut (1ST).
1937	Philips, Holland, introduced an electric razor with a circular cutting head (TT).
1938	First nylon toothbrush marketed in U.S. (1ST).
1945	Joseph C. Mulder, Indiana University, investigated stannous fluoride for preventing tooth decay (NY).
1949	Procter & Gamble began sponsorship of Dr. Mulder's research on a fluoride toothpaste (NY).
1952	Clinical tests began with Crest toothpaste (NY).
1955	Wilkinson Sword introduced stainless steel razor blades (TT).

1956	Crest toothpaste with stannous fluoride introduced (NY).
1960	American Dental Association endorsed Crest as an effective means of protecting teeth (NY).
1961	First electric toothbrush manufactured by Squibb Company, New York (1ST).
1966	Water-Pik developed by Dr. Gerald M. Moyer (EL).
1974	Bic introduced the disposable razor in Europe (NEWSW).
<1976	Europeans developed pump dispensers for toothpaste (WJ).
1976	Bic and Gillette introduced disposable razors in the U.S. (NEWSW).
1979	Throw-away toothbrushes molded from one piece of plastic made in Milan, Italy (TT).
1981	Colgate-Palmolive introduced gel toothpaste (WJ).
1983	Vipont marketed an antiplaque toothpaste (WJ).
1984	Acoustic wand for cleaning teeth devised by Joseph S. Heyman, NASA (IEEE).
	Colgate-Palmolive introduced pump dispenser for toothpaste to compete with Minnetonka's Check-up (WJ).
1985	Gillette introduced a twin-blade cartridge razor with water-soluble strip of plastic resin (WJ).
	Procter & Gamble tested tartar-control formula Crest toothpaste (WJ).
1986	Japanese scientists developed a new electric toothbrush that neutralized plaque film (SMR).

Summary

Cleaning products have changed and improved over time. Our standards of cleanliness have changed and are more demanding than those of our ancestors. We are less tolerant of filth; we have learned to associate disease and illness with dirt. Some, such as Ruth Cowan, have said that our standards may be too high.[19] Doctors are now advising us to wash less frequently in winter months to avoid dry skin.[20]

Like other technology improvements, we have the choice of how much we clean and when we do it. The development of electrically powered washing machines was one of the most significant developments for households. Previously washing could take several days to complete and even longer to dry fully in rainy weather. Now it can be done in the space of a few hours. Programmed washers and dryers allow the householder to do other tasks while the laundry is being washed and dried. The householder can choose when to do laundry and is not dependent on others to help with the task. The elderly and the infirm can manage modern washing equipment. This was not true of traditional washing techniques.

Although the ultrasonic dishwasher was too expensive for widespread use, this technology may yet be developed in the future, making cleaning chores even simpler and less labor intensive. NASA scientists have developed an acoustic wand for cleaning teeth. The same techniques can be applied to other cleaning tasks. Some day the ultrasonic shower may be a reality, but one wonders whether it will be as satisfying as the flow of warm water cascading over a tired body.

This chapter has reviewed the cleaning tools available for our homes and our bodies. Effective use of these requires an adequate supply of clean water and means of waste disposal. These topics, along with the development of the bathroom, are discussed more fully in the next chapter.

7
Water and Waste Disposal

The state and sophistication of a civilization or society can most readily be judged by its water supply and its waste disposal methods. We have made major strides in both, but worry about being engulfed by our own industrial and chemical wastes. Even some primitive societies dealt with waste disposal, as evidenced by latrines with primitive drains found in Neolithic stone huts in the Orkney Islands dating from 3200 B.C. (the *Britannica* cites this date as 8000 B.C.) and ruins in Crete, Rome, and Mohenjo-daro.[1] Some modern societies still do not meet these earlier standards. Suburbs of Sydney, Australia, where rapid expansion and rocky ground have prevented the use of septic systems, still used night soil collection well into the 1970s. Night soil service uses trucks that periodically pump out holding tanks where modern sewers are unavailable. The state of sanitation in some Third World countries is more primitive.

Sanitation and waste disposal become a problem when people live together in increasing numbers and when they remain in the same place. Agricultural societies found many uses for waste products. Garbage and solid organic matter were used as fertilizers or even fuel. Liquid wastes such as urine, rich in ammonia, were used for laundry, as long as the clothes were rinsed thoroughly in clean water afterwards, as was the practice in ancient Rome and in England until the advent of hot water systems and inexpensive cleaning products.[2] Such waste materials, because of their ammonia content, were effective in whitening and cleaning fabrics.

With some of the recycling facilities proposed recently, we may come full circle to our agricultural ancestors. Waste matter can be used to power

217

our homes and liquids can be purified and reused. In fact, plans for space stations incorporate such features. The space shuttle uses water generated as a by-product from the operation of its fuel cells.

This overview considers, first, water supply, then sewage and waste disposal, and, finally bathrooms, which depend on both a source of water and a means of waste disposal. Two excellent sources for this chapter were Gimpel's *The Medieval Machine*[3] and Lawrence Wright's *Clean and Decent*.[4]

Figure 7.1. Water Supplies and Waste Disposal

YEAR

1987		
	Fuel cells create hot water and drinking water	
1975		
1950	Desalinization plant (Kuwait)	
		Fluoridation of water
	Sink disposal	Progas water heater
1900	Porcelain bathtubs	Chlorination of water
	Garbage cans	Solar water heating
	London sewers modernized	
1800		
	Modern toilet	Bidet
	Ball valve	Pumped, piped water
1600		
		Bath with hot and cold water
	Early toilet	
1200		
	Monasteries used sophisticated water supply and disposal	
	Moors built waterworks (Spain)	
400		
	Julius Caesar desalted water	
	Heated baths	
	Roman aqueducts	
-1200	Jerusalem channeled water (underground)	
		Bathrooms, piped water, sewers (Mohenjo-daro, Knossos)
-4400		
	Latrines (Orkney)	
-10800		

Water Supply

Modern water supplies include a means of delivering water to individual residences and the means of ensuring the purity of the supply. Moving water from its source to where people live is no small task. It requires a means of moving the water and something to move it in. The Romans used gravity to make the water flow and stone and cement to contain it. Wood was seldom used because it was prone to rotting and splintering. Some lead and pottery pipes were also used, but could contain only limited pressure. Effective water systems required pumps and metal pipes.

Among the earliest examples of piped water, roughly 2000 B.C., are the baths and homes in Mohenjo-daro and the piped water system for the palace of Knossos, Crete. The Indians used brick to channel water and wastes. The Cretans used tapered, terra-cotta pipes with projecting "handles" used to lash the pipes together to prevent them from being displaced either by land movement or by the force of the flow. The tapered shape prevented sediment from forming and blocking water flow.[5] Other early examples of piped water supplies were Jerusalem, Athens, and Rome. The water supply for Jerusalem was carried in subterranean tunnels around 1000 B.C., while Athens was using piped water by 600 B.C.

The Romans are famous for their aqueducts carrying water to Rome and other cities, but most of their water was carried below ground. The Roman Aqua Appia aqueduct was built in 312 B.C., and by 305 B.C. the Romans had fourteen aqueducts covering 359 miles and with fifty stone arches. Roman aqueducts were capable of carrying 300 million gallons of water a day. By contrast, New York's water system, serving eight times the population, carries only five times the volume of water.[6] De Camp, in *The Ancient Engineers*, points out that much of the time one or more of the aqueducts was under repair and normal water losses would reduce the amount delivered to something comparable to modern usage.[7]

Most aqueducts relied on gravity to carry water from one place to another. One exception was a pressure aqueduct built at Pergamon in 200 B.C., with the reservoir at 1,200 feet and the cistern at 391 feet. In order to contain the force of the water, Roman pressure aqueducts required metal pipes that were expensive, so the Romans generally used stone bridges except where the height was too great for their bridge construction techniques. Considerable care was taken to plan a route that used gravity to force the water flow and to avoid routes that would require costly metal piping or could not be bridged.[8] The Moors carried on the Roman tradition, building a waterworks in Cordova, Spain, in the 9th century.

During the Middle Ages the most sophisticated water systems in Europe appear to have been in various cathedrals and monasteries. The cathedral priory of Canterbury had a piped water and sanitation system in

1167. Water power was used to run monastery factories and to provide a clean water supply. Excellent discussions of the use of water powɛr in the Middle Ages can be found in Gimpel[9] and in Reynolds.[10]

For much of its history, London relied on wells and on the Thames River for most of its water. The well-documented history of London's water supply was paralleled by developments on the Continent and in other English cities. In 1236 Gilbert Sandford built the first public water conduit from Tyburn. By 1582 Peter Morice had installed a water-powered pump at London Bridge for the London water supply, and in 1613 Hugh Myddleton constructed the "New River" to bring added water to London. In 1619 pipes were laid to individual homes.

Pumps were developed by Samuel Morland, Thomas Savery, and John Desaguliers. By 1675 Sir Samuel Morland's water pump was in commercial production and he was using it to supply water in his own home. In 1693 he erected a public drinking fountain in Hammersmith. Thomas Savery patented the first working steam engine for raising water in 1698; originally devised for mine work, it was also used in several homes. In 1717 Desaguliers improved Savery's engine, and by 1761 steam pumping was used to supply London's water.

Another significant advance for water supplies was the development of iron pipe. Pipe made of wood rotted and had to be replaced every few years, whereas iron pipe lasted for decades and in addition could withstand high pressure. The Germans were using cast iron pipe in the 15th century. In 1796 an iron aqueduct, designed by Thomas Telford, was built at Longden upon Tern, Shropshire. In some locations, piped water was not available to private homes until well into the 20th century and is still not available in many lesser developed countries.

Piping water to individual homes under sufficient pressure or to where it could be easily pumped relieved the householder of the burden of queuing up for water or hauling it long distances. In Colombia, one household saved two hours a day formerly required to haul water for household use once they were able to obtain piped water from a recently installed aqueduct. Water for washing or cooking could now be obtained by turning on a faucet.[11]

While delivery of water was important, so was its quality. Pollution became a significant problem as cities and towns grew. Isolated dwellings or small communities that took care to site their water intake upstream and their sewage outlets downstream had few problems unless a significant source of pollution existed upstream.

Various methods for purifying water were developed. Atheneaos and Pliny described water purification in the 1st century A.D. using filtration and natural substances to remove salt. In 1300 the Chinese used sandstone and unglazed porcelain as filters. In 1790 Mrs. Johanna Hempel patented the first English filter, and in 1791 James Peacock developed the sand-rising water filter. From then through mid-century a series of filters was devised including vegetable-charcoal, animal-charcoal, and multiple sand. In 1804 the water supply in Paisley, Scotland, was filtered, and in 1827 James Simpson constructed a sand filter for the London water supply. Simpson's filter was installed in 1829, and by 1855 filtration was made compulsory for all river water supplies for London.

Water conditioning included other processes in addition to filtration. Thomas Clark patented a soft-water process in 1841, and in 1854 a water-softening plant was installed at Plumstead Waterworks, England. In the United States, home water softening was introduced by Emmett Culligan of St. Paul, Minnesota, in 1924. Chlorination was introduced into a public water supply in Pola, Italy, in 1896; hypochlorite of lime was used to disinfect English water supplies in 1897; and in 1914 liquid chlorine was introduced to water supplies. Fluoride to prevent tooth decay was added in 1945 to the Grand Rapids, Michigan, water supply system and subsequently elsewhere.

Desalinization has a long history beginning with Aristotle's experiments in 350 B.C. for removing salt from sea water. Julius Caesar was supposed to have applied Aristotle's process to supply drinking water to the legions in 49 B.C. during his campaigns in Egypt. Although distillation was known by the Moors, who introduced it to Europe in the 10th century, it was not practical or economic on a large scale. In 1855 a water desalinization plant was described by S. Sidey, instructor at the War Office Revictualling Station, Heligoland, and the first English patent for a desalinization process was granted shortly after in 1869. The English built a desalinization plant at Aden on the Red Sea that same year to supply British ships with water. Large-scale facilities were not built until the 20th century. The first large still for commercial and industrial purposes was built at Aruba near Venezuela in 1930. The first large land-based, seawater desalting plant was built in Kuwait in 1948 with a capacity of 1.2 million gallons/day, which was increased to five million in 1958. Desalting processes require the use of energy sources and chemicals to provide potable water, and the expense cannot be justified unless natural sources of fresh water are unavailable.

A more cost-effective method for producing fresh water may be the use of fuel cells. Water for the space shuttle is generated by fuel cells that yield water for drinking and rehydrating foods as a by-product of energy creation. In 1985 a San Francisco building was also using fuel cells to generate power and to provide hot water.

Table 7.1. Water Supplies

Dates	Event
B.C.	
1700-1450	Covered drains for waste water and clay pipes used for water in Knossos, Crete (EB).
1000-900	Water supply in Jerusalem carried in subterranean tunnels (TH).
600-500	Athens used piped water (TH).
350	Aristotle recorded an experiment for removing salt from sea water (DID).
312	Aqua Appia Aqueduct of Rome built (EB).
305	Fourteen Roman aqueducts existed, consisting of 359 miles with fifty stone arches (EB).
200	Pressure aqueduct used at Pergamon with a reservoir at 1,200 feet and cistern at 391 feet (EB).
	Chinese used natural gas to evaporate salt from brine (EB).
49	Julius Caesar used Aristotle's process of desalting water to supply drinking water to legions (DID).
A.D.	
1st c	Vitruvius's *De Architectura,* ten volumes on building materials, hydraulics, and town planning, published (EB).
50	Athenaeos, Attalia, *Book on the Purification of Water,* discussed filtration and natural desalting (RF).
77	Pliny described the purification of water (RF).
756	Arabs began systematic irrigation of Spanish river valleys (RF).
9th c	Moors built waterworks in Cordova, Spain (EB).
1150	Moors introduced water distillation into Europe (DID).
1167	Water supply and sanitation system of the cathedral priory of Canterbury used piped water (MM).
1236	Gilbert Sandford built the first London public water conduit from Tyburn (RF).
1300	Chinese used sandstone and unglazed porcelain as filters (DID).
1455	Cast-iron water pipes used in Dillenburg Castle, Germany (RF).
1582	Peter Morice installed a water-powered pump at London Bridge for London water supply (EB).
1613	Hugh Myddleton constructed the New River, cut to bring water to London (TH).

1619	Water pipes laid bringing water to individual homes in London (EB).
1675	Samuel Morland's water pump was in commercial production and used in his own London home (CD).
1693	Public drinking fountain erected at Hammersmith, London, by Sir Samuel Morland (1ST).
1717	John Desaguliers improved Thomas Savery's engine (CD).
1761	Steam pumping used for the London water supply (EB).
1790	Mrs. Johanna Hempel patented the first English filter (DID).
1791	Sand-rising water filter invented by James Peacock (DID).
1796	Iron aqueduct built at Longden upon Tern, Shropshire, England (1ST).
1802	Vegetable-charcoal filter introduced (DID).
1804	Filtered water supply used in Paisley, Scotland (1ST).
1812	Cylinder filter invented by Paul, Geneva (DID).
1814	Multiple sand filter invented by Ducommon, France (DID).
1815	Pressure filter invented by Count Real, France (DID).
1818	Animal-charcoal filter first used (DID).
1819	Compressed air filter invented by Hoffman, Liepzig (DID).
1824	Bag or stocking filter invented by Cleland (DID).
1827	James Simpson constructed a sand filter for the purification of London's water supply (TH).
1829	Water filtration introduced by James Simpson, Chelsea and Lambeth (EB).
1831	Earthenware ascending filter invented by Lelage, France (DID).
1834	Solid carbon-block filter invented (DID).
1841	Soft-water process patented by Thomas Clark in Great Britain (1ST).
1845	Howard invented the linen filter (DID).
1854	Water-softening plant was installed at Plumstead Waterworks, England (1ST).
1855	Filtration of all river-water supplies for London made compulsory (EB).
	Water desalinization plant described by S. Sidey, instructor at War Office Revictualling Station (1ST).
1869	English patent for desalinization granted (EB).
	English built a desalinization plant at Aden (EB).
1886	Henri Moissan isolated fluorine (WB).
1896	Chlorinated water supply introduced in Pola, Italy (1ST).

1897	Hypochlorite of lime used to disinfect water supplies in England (EB).
1914	Liquid chlorine used in water supply systems (EB).
1930	Large still for commercial and industrial use built at Ariba (EB).
1945	Fluoridated water supply introduced in Grand Rapids, Michigan (1ST).
1948	First large, land-based desalting plant built in Kuwait with capacity of 1.2 million gallons/day (EB).
1958	Capacity of Kuwaiti desalting plant increased to five million gallons/day (EB).
1983	Space shuttle produced its water supply from fuel cells (NASA).
>1983	Ray Ward, Mesa, Arizona, adapted shuttle water filter for home use and marketed it commercially (EB).

Sewage and Waste Disposal

What goes in must go out, so water brought to the home and used for a variety of purposes has to be removed. Again Mohenjo-daro, Crete, and Rome made provision for removing waste water in pipes or underground sewers. By 600 B.C. part of the Cloaca Maxima, Rome's main sewer, was vaulted, reducing the smell and improving land use. Rome's sewers carried rainwater, used bathwater, and latrine waste.

In contrast to the complex and sophisticated aqueduct and sewer systems in ancient Rome, the waste disposal facilities in much of Europe through the Middle Ages were poor. One exception appears to have been the monasteries, which had well-developed water supply and disposal systems: fresh water was used for food preparation and laundry and the waste water was used to carry away sewage, which was then used by one Cistercian monastery in Italy to fertilize nearby meadows.[12] In most cities, however, the streets were the sewers, and when it rained filth was everywhere. In an effort to reduce the stench and waste disposal problems, the French Parliament issued a decree in 1366 forcing Paris butchers to slaughter animals and clean the carcasses by running streams.

In addition to the odors created by poor waste disposal habits, serious health risks developed. After repeated outbreaks of cholera in London, John Snow, in 1854, traced the source of the disease to contaminated wells. Snow had published a book, *On the Mode of Communication of Cholera*, in 1849, postulating sewage contamination of water supplies, but his ideas had only limited acceptance at that time. London was riddled with cesspools and privy vaults. In some instances old vaults were forgotten and new ones created. Little notice was taken of their proximity to nearby wells. Snow's evidence forced the city fathers to change policies and to promote the use of sewers instead of other disposal means. In 1855 the London sewers were modernized, and the sewer system was completed in 1875.

Sewage cannot merely be drained away: it must also be sanitized and disposed of, or used in some way. In 1914 the first major sewer works to use bacteria in decomposition of waste was opened at Manchester, England, and in 1985 the United States Navy was using strains of bacteria to clean ship sewage tanks. In the period 1912 to 1915 the British developed the activated sludge process for sewage treatment. This process combines the use of compressed air and recycling of the sludge for improved efficiency. Because the sludge is high in bacteria and other life forms, it was called *activated sludge*. A more passive method was proposed in 1975 by the Imperial Chemical Industries for the use of deep shaft sewage treatment, that is, dumping wastes into deep shafts. Such techniques had been used in Colorado at Rocky Mountains Flats but generated an undesirable side effect, namely, earthquakes, and put a damper on such techniques. The liquid wastes disposed of by this method at Rocky Mountain Flats apparently lubricated earth faults that caused movement.[13]

Pollutants such as chemical and oil spills and metals present special problems. Phosphates, first used in laundry detergents in the 1940s and into the 1960s, are also used in fertilizers and promote the rapid growth of algae and other vegetation. This led to sluggish streams and diminished oxygen content in lakes and streams for fish and other aquatic life. It also led to state legislation and eventually to changes in laundry detergents, reducing, or even removing, phosphates from some products. In 1970 George Levin developed Phostrip, a process for removing phosphates from water, while in 1985 Ralph Portier, Louisiana State University, developed fourteen strains of yeast and bacteria that eat environmental pollutants, including PCB (polychlorinated biphenyls) and PCP (phencyclidines). In 1987 Margalith Galun, an Israeli botanist, identified a metal-eating fungus that could be used to remove metals from waste waters. Waste shellfish shells from shrimp and crab can be processed to remove their protein, which can then be used in many ways, including water purification. NASA scientists also discovered that plants make effective air purifiers. In 1985 they announced the spider plant was especially effective at removing formaldehydes from the air, a discovery that should be a boon to allergy sufferers and to astronauts.

Several methods are available for aiding with the disposal of household garbage, including garbage cans, sink disposals, and trash compactors. The portable garbage container was introduced by Eugene Poubelle, the Paris prefect of police, in 1883, and the French word for garbage can, *poubelle*, was derived from his name. A sink waste disposal appeared in 1929, and by 1935 General Electric had introduced the electric sink waste disposal unit, reducing the trash disposal in some communities to mainly inorganic materials such as paper, cans, and glass. In 1983 a hand-powered trash compactor originally developed for use in space was adapted for home and recreational use.

Space waste disposal involves either jettisoning waste into space so that it will remain in orbit or eventually burn up during reentry to the Earth's atmosphere, or bringing it back for disposal on Earth. North American Aerospace Defense Command, NORAD, estimates six tons of debris are orbiting the Earth, including sewage and food containers from spacecraft.[14] The Russians cosmonauts package their solid wastes in plastic bags that are packed in aluminum cans and jettisoned. The United States has, since Gemini, brought back solid waste including used food containers.

Increasingly, governments and private companies are seeking ways to use wastes to create energy: the Getty Synthetics Fuel Company was extracting methane gas from landfill in 1975; Denmark provides 3 percent of that country's heating requirements from burning solid wastes; and Columbus, Ohio, has a trash-burning power plant generating electricity. Other means of recycling waste materials include the Dutch use of waste to create hills for recreational areas; the Japanese use of solid waste to reclaim land from the sea; new paper created from waste paper; the use in steel production of some recycled metals; and the sale by zoos of animal manure as fertilizer to gardeners. The U.S. Geological Survey has developed a process for the removal of metallic contaminants from sludge and for processing the sludge into a solid fuel similar to coal. Recycling, land reclamation (or landfill), or energy production are three of the most productive ways of managing waste disposal.

Table 7.2. Sewage and Waste Disposal

Dates	Event
B.C.	
2800	Latrines with crude drains used in Neolithic stone huts on the Orkney Islands (EB).
2500	Harrappas at Mohenjo-daro used brick-lined wells and sanitary drainage system of brick (EB).
1700-1450	Pedestrian walks built in the center of cobblestone streets in Knossos, Crete (EB).
600	Part of Cloaca Maxima, Rome's main sewer, was vaulted (EB).
A.D.	
1st c	Pompeii used a drainage system for waste water, rain, and public baths (EB).
74	Public conveniences in Rome used running water (RF).

1150	Cistercians, Italy, used refuse water to fertilize meadow (RF).
1366	French parliamentary decree forced Paris butchers to slaughter and clean near a running stream (MM).
1854	John Snow traced cholera outbreak to contamination of well water by privy vault forcing use of sewer (EB).
1855	London sewers modernized (TH).
1875	London's main sewerage system completed (TH).
1883	Galvanized portable garbage container introduced by Eugene Poubelle, Paris prefect of police (1ST).
1897	Motorized garbage trucks used by Cheswick Vestry (1ST).
1912-15	British developed activated sludge process for sewage treatment (EB).
1914	First major sewage works used bacteria in decomposition of waste opened at Manchester, England (TT).
1929	Sink waste disposal unit developed (WABI).
1935	Sink waste disposal unit marketed by General Electric Corporation, Bridgeport, Connecticut (1ST).
1970	Gilbert Levin, U.S., developed a process, PhoStrip, for removing phosphorus from waste water (FORT).
1975	Imperial Chemical Industries (ICI) promoted use of "deep shaft" sewage treatment (TT).
	Getty Synthetic Fuels extracted landfill gas as a fuel source (CDIS).
1983	Hand-powered, rachet-driven trash compactor developed by Nelson and Johnson Engineering adapted for home and camp (EB).
1985	NASA scientists announced that the spider plant was effective in removing formaldehyde from air (RD).
	Ralph Portier, Louisiana State University, developed fourteen strains of bacteria and yeasts that ate pollutants (RD).
	Bacteria used to clean Navy sewage tanks (CDIS).

Bathrooms

Once a water supply was established and waste disposal systems were available, the modern bathroom with its facilities became feasible. An excellent discussion of the bathroom and how it developed is given by Lawrence Wright in *Clean and Decent*.[15] The three areas reviewed here are toilets, hot water supplies, and the bathroom itself.

Toilets

Facilities for disposing of human wastes have varied from holes in the ground and the use of running streams to latrines and more complex flushing toilets. Latrines with primitive drains have been found in the Orkney Islands dating from roughly 3200 B.C. Both the palace of Knossos and the cities at Mohenjo-daro had toilet facilities. The fixture at Knossos may have used a water reservoir for flushing wastes. Sargon II's palace at Dur Sharrukin, Assyria, 717-707 B.C., had jars of water and dippers next to latrines for flushing wastes. Streams of water were supplied to the public conveniences of Rome in 74 A.D. and to medieval monasteries' *necessaria* (lavatories) to remove waste.

Most facilities in the Middle Ages remained primitive: the castles of wealthy nobility had garderobes either clustered around a deep shaft or projecting outside the walls to fall below to the ground, into a stream of water, or the moat. Families in London found living on London Bridge convenient for this purpose; the Thames took care of the disposal problem until it virtually became a cesspool itself.

As noted earlier, the monasteries had excellent plans for sanitation. Their *necessaria* were located over running water to drain away the wastes and smell. Most had a complex system with incoming fresh water used for cooking and washing, and waste bath or kitchen water used to power various water mills; remaining water was channeled under the latrines to remove waste products. This sanitation system is one reason many of the monasteries were free of plague.[16]

The toilet, or water closet as the English call it, had several false starts. Thomas Brightfield, St. Martin's Parish, developed one in 1449 that was flushed by piped water from a cistern. Sir John Harrington, a man ahead of his time, installed a toilet of his own design in his country house at Kelston near Bath and one for Queen Elizabeth at Richmond Palace in 1586. Harrington, who bathed every day, was considered a bit eccentric and his invention did not become popular.[17] Toilets were imported to England from France in 1660, but only for the rich. Sir John Carew, Beddington, Surrey, 1678, had an automatic flushing lavatory, but because of limited water supplies and plumbing technology, such facilities did not become broadly available until well into the 19th century.

The bidet, as implied by its name, was a French innovation. The first recorded use of a bidet is in 1710 by Mme de Prie, a noted Parisian beauty. A portable model was available in 1726 for use by army officers on campaign.

Lack of adequate piped water supplies and suitable plumbing fixtures held progress back. In London, water was only pumped on three days of the week, and householders had to fill their tanks during the two to three hours on those days when water was actually available. In 1770 Mr. Melmouth,

Bath, was forced by the water supply company to remove a toilet he had installed because of the heavier requirement its use placed on the limited water supply.[18]

The development of the ball valve and its introduction in 1748 diminished these problems, and in 1775 Alexander Cummings patented the elements of the modern toilet. Joseph Bramah, Yorkshire, constructed an improved model in 1778 and in 1979 patented his design. In 1870 Hellyer's Optimus, improved valve toilet was introduced, and in 1884 Jennings exhibited his award winning modern pedestal toilet with an oval portrait-frame seat. In 1889 D. Bostel, Brighton, introduced the wash-down toilet. Public lavatories with toilets were opened in London in 1852 by the Society of Arts, and in 1855 the London Municipal lavatories were opened. The 19th-century inventors and plumbers developed a variety of toilets each claiming to be the best and each exhibiting a different approach to the aesthetics of the design. Wright, in *Clean and Decent*, provides copious illustrations of the different models available.[19]

Special paper for use with toilets was first made by Joseph Gayetty of New York in 1857. Previously a variety of materials had been used, including moss and rags. Once paper was available the need for appropriate packaging and dispensing was apparent. Seth Wheeler of New York developed the toilet roll in 1871.

The limited space, weight, and self-contained nature of space vehicles have imposed new restrictions on waste disposal systems. Space-age sanitation has changed from the limited waste removal of water only on Mercury to the fecal mittens, or plastic bags, on Gemini and Apollo, to the galley and toilet on the shuttle. The plastic bags were fastened to the buttocks with pressure sensitive adhesives and removed and sealed after use. Waste matter was jettisoned after removal of any required samples. Imagine the potential waste disposal problem for future space sanitation crews picking up used fecal mittens. Maybe they can be used to create a new form of space fuel. People in early societies (and in some remote areas today) burned animal dung for fuel.

NASA developed a more sophisticated system for the shuttle, but it failed to work properly. Waste matter was forced downward by air flow, freeze dried, and stored for disposal after return to Earth. The equipment suffered numerous malfunctions and NASA settled for storage without drying. The Russians also use sealed plastic bags stored in an aluminum canister and jettisoned into space before returning to Earth. Space teams of both countries jettison urine and waste water.

Hot Water Supplies

The Romans and the Cretans heated bath water, but it is only in the 20th century that this convenience has been easily available for most people, at least in the developed countries. In 1523 Pope Clement VII had hot and cold water taps and hot air heat in his bath. In 1700 a bathroom at Chatsworth, Derbyshire, had hot and cold running water. The *Magazine of Science and School of the Arts* recorded bathtubs with a unit for heating bath water in 1842, and an 1850 issue of the *Journal of Gas Lighting* described Defries Magic Heater for the same purpose. Also in 1850, a Boston plumber's advertisement described a servant-operated, pumped system for hot and cold water.

The application of cast iron to kitchen ranges also included the development of cast iron boilers as part of the range. This was followed by the use of gas and electric power leading to more efficient, more convenient methods for rapidly heating water. Benjamin Maugham produced the first gas "geyser" for quickly heating water in 1886, while the multipoint pressure geyser, Califont, was developed by Ewart in 1899, and a dual water immersion heater for bath or sink was introduced by Hotpoint, Bocker, and Bray in 1932. The first domestic water heater to work efficiently, the Progas Instantaneous Water Heater, was introduced in 1930. Continuous hot water was available at last, the only limitation being the fuel source available, the frequency of the heating cycle, and the size of the heater tank.

Although solar energy had been used to distill water from sea water in ancient Rome, it was not used for hot water systems until the late 19th century and into the early 20th century. Solar water heating became practical with the invention of the Climax solar water heater patented by Clarence M. Kemp of Baltimore, Maryland, in 1891. He and others improved on the design, and in 1895 he sold the California marketing rights to E. F. Brooks and W. H. Congers of Pasadena.

Until the discovery of its oil and gas reserves, California had to import coal and other fuels, making the cost of heating water expensive. With its extended sunshine, it was an ideal place for solar water heating, and many of the early developments in practical solar water heaters took place there. Frank Walker of Los Angeles patented an improved solar water heater with a unit that was recessed into the roof in 1908, while William J. Bailey introduced his Day and Night solar water heater in 1909. Bailey's unit incorporated an insulated storage tank to retain water heat overnight and remained popular in California until the discovery and development of natural gas in the Los Angeles Basin in the 1920s. With the energy crisis in the 1970s, there was a resurgence of interest in solar water heating.[20]

Solar-heated hot water systems are now available and used to supplement conventional hot water systems. GMB Company of France developed individual portable solar water heaters in 1977, and by 1978 Dave Little of Australia had developed Suntrac, a solar-water heating system.

Bathing

After Roman times, separate bathrooms containing all the functions of the present bathroom did not exist. Castles or seats of nobility did have separate areas for latrines as did the monasteries, but most homes were small, often consisting of one or two rooms. Baths were taken, if at all, in streams during mild weather or by the fire, usually in the kitchen of smaller homes. Public baths did exist for a brief period during the Middle Ages, but after repeated attacks by the clergy who believed that public bathing encouraged immorality, and after several occurrences of the plague, they disappeared. Until fairly recently in the United States the tradition was the Saturday night bath.

Much of the development of the bathroom occurred in the 19th and 20th centuries. The first bathroom with hot and cold running water in Great Britain was located at Chatsworth, Derbyshire, in 1700. The first hotel bathroom, located on the ground floor, was at Tremont House, Boston, Massachusetts, in 1829, the same year the St. George Municipal Baths, Liverpool also opened. The first hotel to offer private baths was the Mount Vernon Hotel, Cape May, New Jersey (1853). This was followed shortly after by George Vanderbilt's installation of a bathroom in his New York residence in 1855. In 1888 the first working-class homes with bathrooms were built at Port Sunlight, Cheshire, England. By 1915 enameled bathtubs began to replace the earlier cast-iron tubs.

The bathroom had arrived: hot and cold running water was available the instant the tap was turned and in sufficient quantities for a leisurely bath or a quick shower. No longer did servants or the householder have to haul water in and out or fill and empty the bath. There was no longer a need to empty the slops bucket; the push of a handle took care of it all. Winter no longer meant a trip outside, the risk of frostbite, or the disposal problems to be faced with the ground frozen. We take so much of our present existence for granted that it only becomes apparent how much when we are faced with a lesser standard such as in a primitive camp ground without showers, hot water, and electricity, or when we visit a country with different standards and capabilities than our own. How far have we come? At the 1985 National Association of Home Builders Convention, Houston, exhibitors showed a steam shower by Thermasol and the Bathwomb by Water Jet Corporation, a covered tub with nine massage jets, facial mist, four water-resistant stereo speakers, a pillow headrest, speaker phone, clock, control panel, and folding tray table.

Table 7.3. Bathrooms, Toilets, and Hot Water

Dates	Event
B.C.	
2800	Latrines with crude drains used in Neolithic stone huts on the Orkney Islands (EB).
2500-1700	Bathrooms used in private homes in India (EB).
717-707	Assyrian palace at Dur Sharrukin had jugs of water at each latrine for flushing (RM).
27	First large Roman public bath built by Agrippa (RM).
A.D.	
1449	Thomas Brightfield, St. Martin's Parish, built a water closet flushed by water piped from a cistern (CAD).
1523	Pope Clement VII had a bath with hot and cold water taps and hot air heat (CAD).
1586	Sir John Harrington designed and installed the first water closet in his country home at Kelston near Bath (1ST).
1660	Water closets imported from France to England (TH).
1678	Sir John Carew, Beddington, Surrey, had an automatic flushing lavatory (1ST).
1700	Bathroom with hot and cold running water installed at Chatsworth, Derbyshire (1ST).
1710	Bidet recorded as used by Mme de Prie, noted Parisian beauty (1ST).
1726	Portable bidet for officers on campaign offered in Paris (CAD).
1748	Ball valve for water supply systems was introduced, essential for modern plumbing and control (1ST).
1775	Alexander Cummings patented elements of the modern valve water closet (CAD).
1778	Improved water closet constructed by Joseph Bramah, Yorkshire (TH).
1797	Joseph Bramah patented his improved water closet design (1ST).
1829	First hotel bathroom installed at Tremont House, Boston, Massachusetts (1ST).
	Municipal baths opened, St. George's Baths, Liverpool (1ST).
1842	*The Magazine of Science and School of the Arts* recorded baths with a unit for heating water (CAD).
1850	Plumber's advertisement, Boston, described servant-pumped hot and cold water system for the bath (CAD).

The Journal of Gas Lighting described Defries Magic Heater for heating bath water (CAD).

1852	Public lavatories with water closets opened in London (1ST).
1853	First hotel to offer private baths was the Mount Vernon Hotel, Cape May, New Jersey (1ST).
1855	London municipal lavatories opened (1ST).
	George Vanderbilt installed the first bathroom in a private home in the U.S. in his New York home (HS).
1857	First toilet paper made by Joseph Gayetty, New York, with unbleached pearl-colored manila hemp paper (RM).
1860	The Reverend Henry Moule, U.K., developed the earth closet toilet (CAD).
1870	Hellyer's Optimus, improved valve water closet, introduced (CAD).
1871	Toilet roll devised by Seth Wheeler, New York (1ST).
1884	Modern pedestal water closet with more efficient flushing mechanism exhibited by Jennings (1ST).
1886	Benjamin Waddy Maugham produced the first gas "geyser" for rapid heating of water (CAD).
1888	Working-class dwellings were built with bathrooms, Port Sunlight, Cheshire, England (1ST).
1889	D. T. Bostel, Brighton, introduced the wash down water closet (CAD).
1891	Clarence M. Kemp, Baltimore, Maryland, patented the first Climax solar water heater (BP).
1895	E. F. Brooks and W. H. Congers, Pasadena, California, bought marketing rights to Clarence Kemp's Climax heater (BP).
1898	Frank Walker, Los Angeles, patented a recessed model of a roof-top solar water heater (BP).
1899	Califont, a multipoint pressure geyser, was developed by Ewart (CAD).
1908	Willsie and Boyle built a solar-powered water pump, but it was uneconomic (TT).
1909	William J. Bailey introduced the Day and Night insulated solar water heater (BP).
1915	Enameled baths began to replace cast iron baths (TT).
1916	Double-shelled enameled bathtubs were mass produced in the U.S., replacing cast-iron tubs (ACTB).
1930	First domestic water heater to work efficiently was the Progas instantaneous water heater (TT).
1932	Dual water immersion heaters for bath or sink were introduced by Hotpoint, Bocker, and Bray (TT).

1965-72	Fecal mitten used on Gemini and Apollo (EB).
1977	GMB Co., France, developed an individual portable solar water heater (WABI).
1978	Dave Little, Australia, invented the Suntrac solar water heater (WABI).
1983	NASA shuttle used a toilet facility with freeze drying of solid waste matter (NASA).
1985	NASA abandoned freeze drying of toilet wastes for simpler storage system (CDIS).
	California State office building, San Francisco, got five percent of its power and most of its hot water from fuel cells (USA).
	Bathwomb introduced by Water Jet Corporation with nine massage jets, facial mist, four stereo speakers, and phone (CDIS).

We have succeeded in controlling our environment and creating, at least within our homes, comfort and relative ease. Some of the means we have used are discussed more fully in the next chapter.

8
Heating and Housing

Our penchant for controlling our environment is clearly demonstrated in our approach to heating and housing. Through the types of homes we build, the way we provide heating or cooling, and even by the means we choose to decorate them, we influence not only the aesthetics of our homes, but also their effectiveness in providing shelter and comfort. Warmth and shelter are among the basic human needs. People in early societies used wood for fire, as did their descendants, until the wood supply all but disappeared in certain areas. The competing needs for wood as a construction material and wood as fuel presented real problems to governments as recently as the time of Queen Elizabeth I. By the 17th century, Ireland had largely been stripped of trees. In Great Britain, the use of coal and its attendant air pollution problems was one of the consequences. Cheap fuel sources are still a problem for African and other Third World countries.[1]

Most of this chapter is devoted to heating, with a smaller section on cooling and ventilation, followed by a discussion of construction materials, and, last, a brief section on decoration and furnishings. Sources for this chapter have been particularly diverse and it is not possible to cite one or two that cover the entire chapter; however, Caroline Davidson's *A Woman's Work Is Never Done*[2] and Butti and Perlin's *A Golden Thread*[3] were among the most used.

Heating

Humans discovered fire some 500,000 years ago. They used it to warm themselves, to cook, to drive animals away or to slaughter, and for light. Once pottery and metal technology developed there were even more uses for fire and fuel. Its application to cooking was described in chapter 4 and its use for lighting is discussed in chapter 9. The properties of the fuels available influenced both the type of heating device used and its effectiveness.

Energy Sources

Wood, where available, seems to have been the energy source of choice; bones, grasses, peat, and animal dung have all been used, particularly where wood was scarce. Wood, used indoors, was often first processed into charcoal, which gave off considerable heat but little smoke. Most homes did not have chimneys until the late Middle Ages, so smokeless fuel was highly desirable.

By 1797 Sir Frederick Eden reported that agricultural laborers in southern England had only enough fuel for one fire a day. They relied on the village baker to bake their bread and Sunday roast. Some laborers received part of their wages in wood or lumber.[4] Gradually, in England and other areas with access to supplies, coal became the fuel of necessity, first for the poor and then somewhat reluctantly for the more well-to-do. When improperly burned in inadequate fireplaces, coal gives off fumes, and since many people relied on open fireplaces, coal was not their first choice. Peat was also used for fuel, but primarily in areas where it was readily available, such as Ireland, Scotland and parts of England.

Wood shortages are still a problem for developing countries. Clearing land for agriculture has reduced the supply of wood for fuel. As wood sources decline, more effort has to be spent in seeking further afield for supplies. The people also turn to other sources such as cattle dung and crop residues, reducing the availability of these as crop fertilizers. In many developing countries, people do not have money to buy fuel and rely on scavenging to supply their needs. In most of these areas, coal or other fuel sources are not readily available or are too costly.[5]

Coal was used in some Stone Age cultures and in early Middle Eastern, Greek, and Roman settlements. Coal cinders have been found among funeral pyres in Wales dating from 1500 B.C. Aristotle and his pupil Theophrastus mentioned coal, and the Romans in Britain used coal to heat their homes and bath water. The design of the hypocaust (a Roman hot air furnace, 80 B.C.) kept smoke and fumes outside: a fire built in the cellar under the house, or at the outside edge of the house, generated hot air, which was channeled through pipes or open ducts under the floor and sometimes up into the walls to provide hot air heating.

Figure 8.1. Heating

YEAR		
1987		
1975	Iceland uses hot springs for heating	
	Russians use volcanic steam	
1950		
	MIT Solar house	Round thermostat
	Heat pump	Electric clock thermostat
1900	Central gas heat (Missouri)	Electric radiator
	Burton electric heating	Butz Damper-Flapper
	Bunsen burner adapted	Heat pump principles
1800	Argand burner adapted	Thermostatic principles
	Rumford fireplace principles	Heat responsive element
1600		
	Stoves (Germany)	
1200		
400		
	Hypocaust (Rome)	
	Stoves (China)	
-1200		
-4400		
-10800		
-23600		
-59200		
-110400		
-212800		
-407600		
	Fire	

In some regions, coal was readily accessible without mining. Outcroppings of coal occur in Britain in Northumberland and near Fife. Some submarine outcroppings were broken off and chunks washed ashore by the sea where they could be easily collected. As a result the British call coal "sea-coal." Coal was discovered near Liege in 1049, and the records of the priory of Saint Sauveur-en-Reu, Forez, mentioned coal in 1095. In 1234 the freemen of Newcastle were granted a charter to mine coal.

Coal was also used in other areas. In the Americas, the Hopi Indians used coal for heating and cooking in the 12th century. Marco Polo reported widespread use of coal by the Chinese during the 13th century. The Chinese had used stoves from 600 B.C.

While oil was known and used from ancient times, it was not used intensively as a heating fuel until the 19th century. The first oil refinery in Great Britain was established by James Young in 1848 using a petroleum spring to make lubricants and paraffin wax. Elsewhere, the Rumanians were producing 2,000 barrels of oil in 1857 from hand-dug wells, and James Miller Williams, Oil Springs, Ontario, dug a well and established a refinery at Oil Springs in 1857.

The discovery of oil at Titusville, Pennsylvania, in 1859 is regarded by most historians as the real birth of the oil industry. Initially oil was used as a lubricant and for lighting. Heating applications had to await the development of appropriate burners.

Attempts have been made to use volcanoes, hot springs, and the sun as sources of energy. Initial, but unsuccessful, efforts to apply steam heat from volcanic activity were made at Lardarello, Tuscany, Italy, in 1904. The Russians were using volcanic steam to generate electricity in 1973, while by 1974 half of Iceland's population used hot springs heat piped to home and offices.

Early people learned how to adapt their dwellings to use the sun for heating. The lower angle of the sun in winter was used to penetrate into dwelling areas that in summer, when the sun was higher, were shaded by overhanging eaves. Courtyards with southward facing rooms were used by the Chinese, Greeks, and Romans. Fuel shortages in Greece and Rome encouraged such use of passive solar heating.[6] The Romans even advocated the use of a sublayer of pottery and ash under a dark floor to retain heat after the sun set.

The Romans also protected solar access by law in the 2d century A.D.; they prohibited the construction of buildings that blocked access to sunlight, and such protection was incorporated into the 6th-century Justinian Code.[7] While in Greece solar heating was used by most people, it was primarily used by the wealthy in Rome. The Romans also introduced the use of transparent coverings of windows or wall openings using mica or glass to increase the effectiveness of passive solar heating.[8]

In America passive solar heating was used by Indians of the Southwest and by early colonists. By the mid- to late-19th century, however, with increased urban growth, solar design principles were ignored. In 1910

William Atkinson, a Boston architect, experimented with a "sun box" and in 1912 published *The Orientation of Buildings, or Planning for Sunlight* that described solar heating's potential. Atkinson's work went unheeded for more than twenty years.

In the 1930s, encouraged by European developments, American architects again began looking at passive solar heating. The Royal Institute of British Architects completed a study of building orientation in 1931-32 and devised a helidon for experimenting with model building designs and solar exposure. C. D. Haven, U.S., devised thermopane glass in 1930, which Libby Owens Corning developed into thermopane windows in 1935. By reducing the heat loss from windows, thermopanes made the use of windows in passive solar heating more effective. George Keck designed and built a number of homes incorporating thermopane windows and relying on passive solar heating to reduce overall heating costs.

In 1938, Henry N. Wright published the results of his study on the sun's heating effects on buildings in the journal *Architectural Forum*, making his results widely available. This was the same year the Massachusetts Institute of Technology began their two-decade study of solar heating. By 1940 their first solar water-heated house was operating. It was followed in 1944 by Dr. George Lof's solar air-heating system used to supplement heat in his home in Boulder, Colorado. Dr. Maria Telkas, also of MIT, designed yet another solar heating system (1948), using metallic salts to store heat instead of water or stone. Her system used no auxiliary heating and was designed to accommodate up to seven days of cloudy weather. Unfortunately, the Massachusetts climate was uncooperative and produced periods of clouds exceeding two weeks, so her system was eventually supplemented with electric heaters.

We have continued to develop new ways to use or supplement energy sources that generate both heat and power. One of the major older tools in the use of fuels for heat was the development of the stove.

Stoves and Furnaces

Although the open hearth is attractive, it is an inefficient means of heating and cooking--most of the heat goes up the chimney. Lacedaemonions of Greece, relying on the principle that hot air rises, piped it in flues under the floor to heat the Great Temple at Ephesus in 350 B.C. Even earlier, before 1200 B.C., ruins in Anatolia reveal ducts under the floor that may have been used for heating. Gaius Sergius Orata, 80 B.C., also applied rising hot air to his reinvention of central heating with the hypocaust. He used fire to heat bath water and piped the hot air under the floor to heat the baths. It was also applied to homes. While heating was not essential in southern Italy, it was important elsewhere in Europe and particularly in the British Isles.

After the decline of the Roman Empire and its technology, heating reverted to the more primitive but reliable technique of the open fire. Until the late Middle Ages most homes were heated by a fire in the middle of the floor with an opening in the roof for smoke. The first mention of a chimney is in 1347 when records indicate that several were destroyed by an earthquake in Venice.

The Chinese used stoves for heating in 600 B.C. The stove traveled to Russia and to Germany and by 1350 German tile makers were making stove tiles.[9] The stove was translated by Benjamin Franklin into the Franklin stove for more efficient heating in 1740. Franklin's stove or "Philadelphia fireplace" could be used in existing fireplaces or as a free-standing unit connected to a flue. It was not a closed stove; wood was burned on a grate and fed air by a draft for more efficient burning. Count Rumford outlined the basic principles for the design of fireplaces and chimneys for effective heating in 1796. In 1849 Elisha Foote, U.S., patented automatic control of the damper to improve burning.

Slower to evolve was central heating. While used by the Romans, it was not rediscovered until the 18th century. The Chateau de Pecq had central heating installed in 1777 by the architect M. Bonnemain. He also used a lead rod as a primitive thermostat to control air flow to the boiler.[10] Steam heating was first suggested by Sir Hugh Platt in 1594 for heating greenhouses, but was not applied to home heating until 1745 by Colonel Cook. Matthew Boulton also heated a friend's home by steam heat in 1795.

Steam heat became popular in the United States with a steam plant supplying heat to a number of buildings. The system installed in Lockport, New York, in the 1880s was a model for its thorough insulation of steam pipes. Steam heat, however, created a dry environment because of the hot piping that is essentially at one temperature, that of steam.

The development of hot water heating systems reduced the dryness problems associated with steam heating and allowed greater temperature variation and more control.[11] Jacob Perkins, England, patented a high pressure hot water heating system in 1831. It was compact and efficient, pumping water through coils in a furnace and then to the area to be heated.

Gas and oil heating followed the adoption of coal. However, the Chinese burned the natural gas they encountered when drilling for salt to evaporate brine, circa 200 B.C. Oil did not become practical until after the discovery and exploitation of major oil fields in the United States and Canada. Many of the gas and oil burners were adapted from lamps, which had been the major consumers of oil and oil by-products. In 1799 Phillipe Lebon patented a combined lamp/heating unit, and in 1808 Samuel Clegg adapted the Argand lamp originally developed in 1784 as a gas burner. The first domestic gas fire was produced in 1853. Bunsen developed his gas burner in 1855, and it was adapted by Pettit and Smith for the domestic heating for the commercial market in 1856. The gas ring fire was introduced in 1860. By 1904 the Laclede Gas Light Company began experiments in central house heating in St. Louis, Missouri. This was the same year gas was used to power central heating and large-scale hot water supply in Clampton, England. An automatic electrically ignited oil burner for home heat was introduced in 1917, and in 1972 the Japanese developed an ultrasonic kerosene heater.

During the 1950s and 1960s many American homes were converted first from coal to oil and later from both to natural gas. This eliminated the ash disposal problem, the coal dust, and the task of stoking the furnace. Oil and gas were clean and easy to use. Furthermore, gas did not require on-site storage of fuel and was, for a time, considerably cheaper than coal or oil.

After the oil shortage of the 1970s and the rise in both oil and natural gas prices, home owners had second thoughts. Some turned to wood stoves, and many improved models with more efficient heating were marketed; others turned to electric heat. In colder climates, insulation and energy efficient buildings gained in popularity. Most home owners saw heating bills rise significantly.

Electricity was first used for lighting, but was soon applied to heating as well. An electric heating system was patented by Dr. W. Leigh Burton in 1887 and introduced commercially by the Burton Electric Company, Richmond, Virginia, in 1889. Electric central heating was offered by the American Electric Heating Company in 1893. R. E. Bell Crompton and J. H. Dowsing of England patented an electric radiator in 1892 using a cast iron plate.

Effective electric heating required durable, efficient heating elements and carriers for such elements. Albert March of Lake Country, Illinois, invented a nickel chrome alloy in 1906 that was ideal for electrical heating as well as burner elements in cooking stoves. R. Belling, Enfield, England, developed a fire-proof clay for cores, wrapped with Marsh's nickel-chromium wire; he used these materials in his 1912 Standard Radiator.

A series of electric heaters was developed in the early 20th century, including a convection heater in 1910, one using incandescent bulbs in 1911, and the Globar heater in 1926. The first commercial convection heater equipped with an electric fan was introduced in the United States by Belling in 1936.

Heat pumps were first envisioned in the mid-19th century, but were not fully developed until the 20th century. The principles for the heat pump were first developed by Lord Kelvin in 1851. These were applied by T. G. N. Haldane of England to build the first successful heat pump in 1927. It was used to heat his office in London; a second one heated his house in Scotland.

Crucial to all efficient home heating systems is a reliable method of temperature control. For many modern heating systems, the thermostat performs this function. Modern thermostats rest on a number of scientific principles including temperature measurement or sensing and feedback control to act on changes in temperature. The first known use of a bimetallic temperature responsive device was the "grid-iron" pendulum built in England in 1726 and used to improve the accuracy of a clock. Thermostatic control principles were used by French architect Bonnemain in 1777. He used a lead rod to control the boiler air intake and thus the temperature in a heating system.[12] Two important principles relating to thermostatic control were discovered in 1821. Sir Humphrey Davy of England discovered that the electrical resistance of a metal depends on the temperature; and Thomas Johann Seeback of Germany discovered that dissimilar metals expand at different rates and generate an electrical current in a circuit with dissimilar junctions held at different temperatures. The first use of the word thermostat was by Andrew Ure, a Scottish chemist, in 1830 when he patented a "heat-responsive element." Elisha Foote of the United States patented an automatic damper control to improve combustion in a stove in 1849, and by 1879 Julian Bradford, Portland, Maine, had devised an electric heat and vapor governor to control the humidity levels in spinning and weaving factories.

The major advance for home heating began in 1883 when Alfred Butz, Minneapolis, Minnesota, applied Bradford's work to the development of the flapper damper. In 1885 Butz founded the Thermo-Electric Regulator Company, which later became Honeywell. Butz's flapper damper incorporated a thermostat with an electric circuit that operated a spring motor for opening or closing the furnace dampers and was patented in 1886. By 1905 the first clock-controlled thermostat had been developed, and by 1915 a model that could lower the temperature and then reset it in the morning had been introduced. An electric clock thermostat was marketed in 1933, eliminating the chore of winding. Other developments included a newer sleek design in 1935 styled by Henry Dreyfuss, an industrial designer, and the familiar round thermostat first introduced in 1942, reintroduced in 1952, and still in use today. Wartime conditions were such that little attention was paid to thermostats for home use. More recently computerized models have been introduced, and computer-controlled home heating systems are now available.

Table 8.1. Heating

Dates	Event
B.C.	
500,000	Fire used (HEP).
1500	Coal used in funeral pyres in Wales (EB).
600	Chinese used heating stoves (EB).

371	Aristotle and his pupil Theophrastus mentioned coal use (EBHFB).
350	Lacedaemonians, Greece, centrally heated floor of Great Temple, Ephesus, with flues laid under floor (EB).
80	Gaius Sergius Orata devised Roman central heating, the hypocaust (AE).
A.D.	
100	Romans in Britain used coal for heating (EB).
4th c	House bellows described by Decimus Magnus Ausonius (DID).
1049	Coal discovered near Liege (HFB).
1095	Records of the priory of Saint Sauveur-en-Rue, Forez, mentioned coal (MM).
12th c	Hopi Indians used coal for heating and cooking (EB).
1234	Freemen of Newcastle were granted a charter to mine coal (HFB).
1306	Marco Polo reported widespread use of coal in China in 13th century (EB).
1347	Chimneys destroyed in Venetian earthquake (HFB).
1594	Sir Hugh Platt suggested steam heat for greenhouses (HFB).
1726	First bimetallic temperature responsive device, a "grid-iron" pendulum built in England for a clock (EA).
1740	Franklin stove developed for more efficient heating (RC).
1745	Colonel Cook applied steam heating to heat a home (DBE).
1777	Central heating installed at Chateau de Pecq by the architect M. Bonnemain (1ST).
1795	Matthew Boulton used steam heat to heat a friend's home (HFB).
1796	Count Rumford developed the basic principles for fireplaces and chimneys for effective heating (HFB).
1799	Combined lamp/heating unit patented by Philippe Lebon, France (1ST).
1808	Samuel Clegg used the Argand lamp (invented 1784) as a gas burner (DID).
1821	Thomas Johann Seebeck, Germany, discovered that junctions of dissimilar metals generate an electric current (EB).
	Sir Humphrey Davy, England, discovered that the electrical resistance of a metal depends on temperature (EB).
1830	Andrew Ure, Scottish professor of chemistry, was issued a patent on a "heat-responsive element" (EA).
1831	Jacob Perkins, England, patented a high-pressure hot-water system (HFB).

1848	First oil refinery in Great Britain established by James Young, Riddings, Derby, England (1ST).
1849	Elisha Foote, U.S., patented a stove with a thermostatically controlled damper (HFB).
1850	Joseph Glass invented chimney sweeping tools (DID).
1851	Lord Kelvin completed development of his principles for heat pumps (EB).
1853	First domestic gas-fire produced (DID).
1856	Pettit & Smith adapted the Bunsen burner (1855) for domestic heating and commercially marketed it (1ST).
1857	Rumanians produced 2,000 barrels of oil from hand dug wells (WB).
	James Miller Williams, Canada, dug an oil well and established a refinery at Oil Springs, Ontario (WB).
1859	First commercial oil well at Titusville, Pennsylvania, dug (WB).
1860	Gas-ring fire introduced (DID).
1865	First oil pipeline laid in Pennsylvania (TH).
1871	Sir William Siemens, Germany, proposed an electrical resistance thermometer (EB).
1879	Julien Bradford, Portland, Maine, devised an electric heat and vapor governor for spinning shops (HONEY).
1880	Holly, Lockport, New York, installed a well-insulated steam heating system (HFB).
1882	Edward S. Moore, Salem, Massachusetts, installed a solar "hot box" at the Peabody Museum (BP).
1883	Albert Butz, Minneapolis, began working on a thermo-electric regulator for the coal furnace (HONEY).
1885	Albert Butz formed the Butz Therm-Electric Regulator Company, later renamed Honeywell (HONEY).
1886	Butz patented the thermostatically controlled damper/flapper (HONEY).
1887	Electric heating system patented by Dr. W. Leigh Burton (1ST).
	Coin-in-the-slot gas meter patented by R. W. Brown, Birmingham (1ST).
1889	Electric heating system was introduced commercially by Burton Electric Company, Richmond, Virginia (1ST).
1892	R. E. Bell Crompton and J. H. Dowsing, England, patented an electric radiator using a cast iron plate (WABI).
1893	Electric central heating introduced by American Electric Heating Corp. (1ST).

1901	First coin-in-slot electric meter approved for use in Britain (TT).
1904	Gas used for first time to power central heating and large-scale hot water supply, Clampton, England (TT).
	Laclede Gas Light Co. began experiments in central house heating in St. Louis, Missouri (TT).
	Initial efforts to harness steam from volcanic region made at Lardarello, Tuscany, Italy (TT).
1905	First clock-controlled thermostat to turn down heat introduced (HONEY).
1910	New convection electric heater developed that was easier and safer to install and use (TT).
	William Atkinson, Boston architect, experimented with a "sun box" (BP).
1911	Electric heater based on H. J. Dowsing's patent using incandescent lamps marketed by General Electric Co., Britain (TT).
1912	William Atkinson, Boston architect, published *The Orientation of Buildings or Planning for Sunlight* (BP).
	R. Belling, Enfield, England, developed a fireproof clay for radiator cores and built the Standard (WABI).
1915	First clock-controlled thermostat to both lower and restore furnace temperature introduced (HONEY).
1917	M. Hammer and C. Lewis, U.S., produced a fully automatic, electrically ignited oil burner for home heat (TT).
1920	Magicoal, imitation coal fire, introduced by H. H. Berry (TT).
1926	Globar electric heater introduced by Carborundum Co.; that used a silicon-carbide rod (TT).
1927	T. G. N. Haldane, England, built a heat pump and used it in his office in London and in a home in Scotland (WABI).
1931	Mercury switches introduced into thermostats for heating (HONEY).
1931-1932	Royal Institute of British Architects devised the helidon for aiding proper solar orientation (BP).
1933	First electric clock incorporated in heating thermostat (HONEY).
1935	First thermostat styled by industrial designer introduced by Honeywell (HONEY).
1936	First commercial convection heater equipped with electric fan introduced in the U.S. by Belling (TT).
1938	Massachusetts Institute of Technology staff began two decades of research into solar heating (BP).

Henry W. Wright published the results of his study of the sun's heating effects (BP).

1940   Massachusetts Institute of Technology staff built a solar house (BP).

1942   First round thermostat for home heat introduced, reintroduced after the end of the war in 1952 (HONEY).

1944   Dr. George Lof, Boulder, Colorado, used solar heated air for home heating (BP).

1948   Solar house built by Dr. Maria Telkes using glass covered, black painted metal plates to heat air (BP).

1952   Round thermostat reintroduced by Honeywell for home heating systems (HONEY).

1972   Matsushita Electric and Osaka Industrial Research Institute developed an ultrasonic kerosene heater (TT).

1973   Russian power station used volcanic steam (TT).

1974   Half of Iceland population used hot springs heat piped to homes and offices (TT).

Ventilation

We expend energy to keep our environment relatively stable all year round, heating our homes in the winter and cooling them in summer. Until the 20th century, heating and cooling were limited by the available power sources and the then existing technology. Nonetheless, people have tried from earliest times to control the temperature of their living space. The Egyptians, Greeks, Romans, and Indians, living in relatively low humidity climates, used wet mats to cool indoor air. One of the first mechanized devices was the rotary ventilating fan developed in China by 180 A.D. Most ventilation was accomplished by the use of fires drawing air upward. This principle was used in mines and public buildings.[13]

Leonardo da Vinci built the first mechanical ventilating fan in Europe in 1500. It was powered by water. The first rotary fan for mines was built in 1533 by the English, and a centrifugal mine fan was patented by George Lloyd in 1848, the same year Guibal invented and produced a 17-foot diameter fan.

The development of electricity placed power at the disposal of the individual home owner, but widespread use in the United States did not occur until after the end of the 19th century. The electric fan for commercial production was developed by Dr. Schuyler Wheeler, New York, in 1882, and the first oscillating electric fan in the United States was produced in 1908 by Eck Dynamo and Electric Company.

Air conditioning, other than the use of evaporation in low humidity climates, was not practical until 1844 when John Gorrie installed mechanical air conditioning at the American Hospital for Tropical Fevers, Apalachicola, Florida. Joseph McCreery, Toledo, Ohio, patented a spray mechanism for air conditioning in 1897. Further improvements occurred in the 20th century. Alfred Wolff designed air conditioning systems for Carnegie Hall and other New York City buildings in 1902, the same year Willis Carrier designed the first scientifically based system for air conditioning using water to both cool and condition the air (since at lower temperatures the air retains less humidity). Carrier founded the Carrier Corporation in 1915 to market his design. Room air conditioners were available for home use in 1932, but economic constraints arising from the depression and the World War II, together with the uneven spread of electrical power distribution systems, prevented widespread use until after the war. By 1981 57 percent of U.S. homes had air conditioning.

Places like Houston and New Orleans would be unpleasant to live in during the summer without air conditioning. Even Midwestern states such as Kansas, Michigan, and Ohio can be unpleasant for sleeping during summer months. As recently as the early 1960s some people used front porches for summer sleeping. Now we push a button or flip a switch and quickly adjust the internal temperature to more comfortable levels. Some energy conscious home owners rely instead on electric fans, natural breezes, or insulated construction. We, unlike people in less technologically advanced societies, can live where we choose and adjust our environment to suit our comfort, limited only by our ability to pay for the equipment or energy consumption required.

Table 8.2. Ventilation

Dates A.D.	*Event*
180	Rotary ventilating fan used in China (DID).
1500	Da Vinci built the first mechanical fan for ventilation using water power (WB).
1533	English built rotary fans to ventilate mines (WB).
1837	William Fourness, Leeds, invented an exhaust fan (DID).
1844	Air conditioning installed by John Gorrie at American Hospital for Tropical Fevers, Apalachicola, Florida (1ST).
1848	George Lloyd patented a centrifugal mine fan (DID). Guibal invented a rotary fan and built a seventeen-foot diameter one (DID).
1882	Electric fan for commercial production developed by Dr. Schuyler Wheeler, Crocker & Curtis, New York (1ST).

1897	Joseph McCreery, Toledo, Ohio, patented an air-conditioning spray process (WB).
1902	Alfred R. Wolff designed air conditioning for Carnegie Hall and other New York City buildings (WB).
	Willis H. Carrier designed a scientifically based air conditioning system (WB).
1908	First oscillating electric fan produced in U.S. by Eck Dynamo & Electric Co. (1ST).
1915	Willis Carrier founded Carrier Corporation to market his air conditioner design (ACTB).
1931	Patent request for window air conditioner filed (WABI).
1932	Room air conditioners available (WB).
1981	57 percent of U.S. homes had air conditioners (FUTUR).

Housing Materials

Part of the control of our environment involves the homes we live in. We have many choices of the types of units available: apartments, condominiums, houses, and high-rise co-ops are some of the options. Materials and construction techniques have evolved throughout human history. Table 8.3 provides a historical overview of some of the materials, devices, and processes used. Materials reviewed below include cement, glass, metal, and paint. Devices described include locks and security systems, and electrical plugs and outlets.

Materials for housing vary with locality. Wood, thatch, stone, brick, bone, and mud have all been used. Wood and stone were not available in some places. Paleolithic Russians used mammoth bones to construct shelters.[14] In dry areas where wood was scarce, mud or mud brick construction was used. This method was common in Mesopotamia as well as in the American Southwest.

The Romans are credited with discovering cement, or at least an excellent lime mortar that could be used in the same way. The Romans made lime mortar at Pozzuoli in 150 B.C. and used it for public buildings, aqueducts, and homes. Their secret was lost until rediscovered in 1796 by James Parker. A number of different types of cement were developed in the 19th century. Portland cement was patented in 1824 by Joseph Aspdin, and L. Vicat invented a cement made from chalk and clay in the same year. In 1844 I. Johnson made artificial hydraulic cement. By 1867 reinforced concrete was patented by Joseph Monier, cellular concrete was used in Holland to provide low-cost housing in 1923, and glass-reinforced cement was developed in 1971.

Fasteners are useful in wooden construction and nails, a commonly used fastener, were made by hand until the 18th century. Ezekiel Reed,

United States, invented a nail-making machine in 1786. Sarah Babbitt, a Shaker sister, developed cut nails in 1810. A wire nail making machine was made by Adolph Felix, New York, in 1851.

Glass was used for windows in Rome in 14 B.C., but not in private English homes until 1180. Early homes had few windows, and these were covered initially with wooden shutters. Later, thin mica or oiled, parchmentlike skins were used. The Chinese used rice paper on some windows. Shutters remained as added protection and to keep out cold in northern climates. Sash windows were installed by Inigo Jones at Raynham Hall, Norfolk, England,in 1630,and by 1688 the first plate glass was cast. Hand-made glass was uneven in quality and expensive. Most glass windows were made from blown glass that was flattened, creating a bump in the middle, hence the name *Crown glass*. Later, glass was formed in cylinders that were cut and flattened.

Machine-made glass reduced the cost of glass and increased the quality. High quality sheet glass was made using the Fourcault machine in 1913. A major advance occurred when the Pilkington Brothers, Britain, in 1959 developed a cheaper process for producing high-quality glass using the float process, which provided the qualities of plate glass, but at a significantly lower cost. Liquid glass is poured onto molten tin in a carefully controlled atmosphere producing a flat, high quality glass. Alistair Pilkington also developed, in 1980, Kappafloat glass, which traps solar heat.

A variety of materials has been used for roofs. Stone Age huts had roofs of thatched straw. Later societies used wood, slate, or pottery tiles. Roofing tiles replaced thatched roofs in London in 1212. Organic materials such as thatch had to be renewed periodically. Slate and tile, while long lasting once installed, had to be handled with care while installing the roof, were heavy, and could be difficult to patch or replace. Corrugated iron, available in 1832, was relatively cheap and durable, but was hot in the summer and noisy in rainy weather. Galvanized metal was developed in 1837 by Sorel of France, and Carpentier patented a process for ribbing it in 1851.

Paint has been used both for its protective and decorative qualities. Lead had been used as a pigment in white paint from Roman times because it covered well and added to the water-shedding quality of the paint, but only in recent times have the hazards posed by such paints been fully recognized. Initially pigments were added to a paint base by professional painters or householders before the paints were applied. Some manufacturers began supplying premixed paints for commercial use by 1867, including Averill Paint Company of New York. It was not until 1880 that Henry Sherwin and Edward Williams began offering Sherwin-Williams standardized premixed paints for the domestic use market. Previously, premixed paints varied greatly from batch to batch. Paints were further improved in 1918 when it was discovered that titanium oxide added to white paint improved color and coverage with reduced toxicity; it began to replace lead for white paint pigments. Alkyd resins were developed from petroleum the same year to produce better paints. The first acrylic polymer was produced commercially in 1927 for use as lacquer and for glass. Modern latex or water-based paints were first available in the late 1940s.

People have been concerned about personal security and the safety of their property throughout much of history. Wooden weighted tumbler locks were used on Egyptian mummy cases. The Romans introduced metal locks and made them in many different sizes and shapes. Lock technology advanced significantly during the late 18th century and in the 19th century. Robert Barron, England, patented a double-acting tumbler lock in 1778, and Joseph Bramah developed the Bramah lock using a small metal tube for a key in 1784.

Modern locks owe much to Jeremiah Chubb and Linus Yale. Chubb developed his lock in 1818, and Linus Yale patented his in 1851. Yale's lock was made of metal and, though more precise in design, incorporated the same principles used in the Egyptian locks. Yale's son added further improvements in the 1860s, developing the familiar Yale cylinder lock.

Other means were also used to ensure security. Some homes in Scandinavia and northern Europe during the Middle Ages had small doors, forcing persons entering to stoop or bend over to enter. This put their heads at a convenient angle for those inside to decapitate them if they were enemies. Small doorways also reduced the entry of cold air and helped keep what heat there was inside.

We have been using deadlocks on our doors since 1924 and now use even more sophisticated electronic monitoring and security systems. Visitors can be screened first on closed circuit television. We can decide before opening the door whether to admit we are home. Electronic locks were used in 1972 in high security buildings, are used in hotels, and may in the future may be used in homes.

In 1979 X-10, Northvale, New Jersey, introduced its X-10 home security system using the house power lines for controlling appliances, lighting, heating, and cooling. In 1985 General Electric added easy-to-use graphics software to the X-10 system with control via a television set with a hand-held remote control device. The system was sold for about $450. Mitsubishi also introduced its Housekeeping System which required special wiring for $2,000 to $3,000. The Mitsubishi system combined data, video, and audio in one system and provided more functions. The General Electric system could be added easily to an existing home, while the Mitsubishi system was easier to install in a new home as it was built. A number of cable system operators also offer home security systems.

Other devices also can be used to protect the home. Ben Franklin installed a lightning conductor in 1752. The first electric bell was developed by Joseph Henry, Albany, New York, in 1831. Safer, more convenient access to electric power was provided by the electric switch socket designed by Sir David Salomons in 1888 reducing the risk of frayed and tangled cords and possible electrical fire.

Electric power was first used for lighting and later for appliances and heating. Lighting was the first, and for some years, the only use, for electricity; homes were originally wired for light fixtures in the ceiling or high on the wall. There were no baseboard outlets of the type now used. Appliances were developed with plugs that screwed into lighting fixtures, which led to festoons of cords. The first U.S. patent for a concealed floor socket was issued in 1890; some English wall sockets carry the date 1893. A ceramic-copper socket was patented by Frederick A. Chapman of Philadelphia in 1895. A major advance in plug and socket design occurred when Harvey Hubbell developed a two-part plug and socket in 1904 with a screw portion that could be left permanently in a light socket and a modern pronged plug for the appliance cord. In 1915 he developed the modern duplex wall outlet that could accommodate modern pronged plugs as well as parallel or tandem plugs by the use of t-shaped slots. During the same year, the National Electric Lighting Association met in New York to seek some standardization in outlets and appliance plugs. By 1917 six manufacturers had agreed on a standard, and by 1931 virtually all appliance manufacturers had adopted the standard. Houses were then wired for both lighting and power and householders no longer had to stretch and strain to plug appliances into ceiling lighting fixtures. It also meant the end of dangling cords and plugs that would not fit the available socket. The most recent advance was the addition in 1962 of a third opening in outlets for a rounded ground prong that provided increased safety, although many small appliances still come with only two-pronged plugs.

Along with changes in electric power distribution and home wiring, home construction has also changed. Entire factory fabricated homes can be delivered for on-site assembly. Hodgson Houses Incorporated was established in 1892 to build prefabricated home for African missionaries. Hodgson also made prefabricated summer cottages for well-to-do East Coast families. Sears sold mail-order homes from 1909 until 1937. All construction materials and detailed plans for electric wires were included. The price for the precut homes varied from $595 to $5,000. This put quality housing within the reach of many people. The depression and the attendant high repossession rate led Sears to cease distribution of the homes.

Prefabricated home kits are still available from a number of small firms. A few large firms offer both home kits and parts to builders. Many traditionally built homes involve some premanufactured parts, such as roof trusses, doors, or windows. Some communities' building codes prohibit homes built from prefabricated kits for a variety reasons, including fear of cheap, sub-standard construction, the influence of construction unions, and the impact of a sudden influx of new, rapidly built homes straining community resources. The homes are offered in a wide variety of styles and prices, including log cabins, geodesic domes, and Cape Cod and Tudor styles. Prices depend on the style, materials used, shipping distance, and manufacturer. The main advantages to builders are the reduced need for skilled craft workers and reduced construction time.[15]

Table 8.3. Housing Materials

Dates	Event
B.C.	
2000	Weighted tumbler locks used on Egyptian mummy cases (DID).
150	Lime mortar made by Romans at Pozzuoli (DID).
27-14	Glass used for windows in Rome (EB).
A.D.	
50	Window glass manufactured in Rome (WB).
1180	Glass windows used in English private homes (TH).
1212	Tiles replaced thatched roofs in London (TH).
1572	Hand fan introduced into England from France (DID).
1608	First U.S. glass made at Jamestown, Virginia (WB).
1630	Sash windows installed at Raynham Hall, Norfolk, England, by Inigo Jones (1ST).
1688	Plate glass cast for the first time by Louis Lucas, France, and used for mirrors (WB).
1739	Caspar Wistar reestablished U.S. glassmaking at Salem, New Jersey (WB).
1752	Ben Franklin installed the first lightning conductor, Philadelphia (1ST).
1778	Robert Barron, England, patented a double-acting tumbler lock (EB).
1784	Joseph Bramah developed the Bramah lock using a small metal tube for a key (EB).
1796	Count Rumford developed the basic principles for fireplaces and chimneys for effective heating (HFB).
	James Parker rediscovered puzzolina hydraulic cement, patented by John Smeaton as Roman (hydraulic) cement (DID).
1818	Jeremiah Chubb, Portsmouth, England, developed his pick-proof lock (EB).
1824	Portland cement patented by Joseph Aspdin, Yorkshire (EI).

L. J. Vicat, France, invented cement made from chalk and clay (DID).

1831   First electric bell devised by Joseph Henry, Albany, New York (1ST).

1832   Corrugated iron manufactured by John Walker, Rotherhithe, England (1ST).

1837   Sorel, France, developed a process for galvanizing metal (WABI).

1844   I. C. Johnson made artificial hydraulic cement (DID).

1851   Linus Yale patented his Yale lock (1ST).

Robert Newell, New York, displayed his parautopic lock at the Great Exhibition, London (EB).

Wire-nail-making machine was made by Adolph Felix Browne, New York (DID).

Pierre Carpentier, France, patented a process for ribbing galvanized metal sheets (WABI).

1853   Dr. Stonehouse invented a charcoal air filter (DID).

>1860   Linus Yale, Jr., devised the Yale cylinder lock (EB).

1867   Reinforced concrete patented by Joseph Monier, France (EI).

Averill Paint Company, New York, offered premixed paints based on patent 66773 granted to D. R. Averill of Newburg, Ohio (FFF).

1873   James Sargent, Rochester, New York, devised a lock based on a Scottish lock patent for a time-lock (EB).

1880   Sherwin-Williams Company, Cleveland, Ohio, began manufacturing standarized premixed paints (EOOET).

1888   Electric switch socket designed by Sir David Salomons, Tunbridge Wells, England (1ST).

1890   U.S. patent 421802 issued for a concealed floor socket for two-pronged electric plug (TAC).

1893   Wire reinforced glass fabricated by Leon Appert, France, for use in casings and windows (WABI).

English electric wall socket manufactured (TAC).

1895   Frederick A. Chapman, Philadelphia, patented a cut-out block and box for electric power outlets in walls and ceilings (TAC).

1902   Emile Fourcault, Belgium, developed a machine to draw a continuous sheet of glass (WB).

1903   Laminated glass invented by Edouard Benedictus, France (WB).

1904   Harvey Hubbell invented a separable plug for electrical appliances with a screw part for light socket (TAC).

1908-17   Sheet window glass drawing machine invented by I. W. Colburn and developed by Libby Owens Sheet Glass Co. (WB).

1908   Titanium dioxide extracted by two U.S. chemists made possible purer white products, especially paints (TT).

1909   Sears offered low-cost homes in its catalog; the program lasted until 1937 and included plans, paint, building materials, and nails (WJ).

Edouard Benedictus, France, patented laminated glass (WABI).

1913   Sheet glass of high quality produced as Fourcault machine started commercial operation (TT).

1915   National Electric Light Association met in New York City to establish some uniformity in appliance plugs and outlets (TAC).

1917   Six U.S. manufacturers agreed on a standard appliance plug and receptacle for electric power (TAC).

1918   Alkyd resins developed and manufactured on a large scale from petroleum to produce better paints (TT).

Titanium oxide introduced into white paint for improved color, coverage, and nontoxicity (TT).

1923   Cellular concrete used to build low-cost housing at Scheveninger, Holland (TT).

1924   Dead-bolt locks used for homes (TT).

1926   Machine to draw glass directly from the furnace developed by Pittsburgh Plate Glass (WB).

Mass production of safety glass was developed by Libby Owens Glass Company (WB).

1927   First acrylic polymer produced commercially by Rohm and Haas, Germany; it was used as lacquer and for glass (TT).

1930   Thermopane, insulated window glass, invented by C. D. Haven, U.S. (WB).

1931   Glass building bricks introduced (DID).

Most U.S. electric appliances had converted to standardized plug first adopted in 1917 (TAC).

1935   Libby-Owens Corning manufactured thermopane windows (BP).

1938   Fiberglass developed by Owens-Illinois Glass Co. (WB).

1942   Foam glass developed by Pittsburgh Corning (WB).

1959   Pilkington Brothers, Britain, developed float glass, a cheaper process for high-quality glass (TT).

Hollow clay bricks used for thermal insulation (TT).

1962	Standard electric power receptacles had a third hole added for grounding (TAC).
1971	Light, fireproof glass-reinforced cement invented by the Building Research Establishment, England (TT).
1972	Electronic lock and card key system used in high security buildings (TT).
1979	X-10, Northvale, New Jersey, marketed a home security system (EW).
1980	Alistair Pilkington, England, invented Kappafloat, glass with the property of trapping solar heat (WABI).
1984	General Electric Home Minder home security system based on X-10 components introduced (EW).
1985	Mitsubishi introduced the "wired home," a housekeeping system, to monitor appliances, heating, and entertainment devices (EW).
	Ivarsons, Minneapolis, offered prefabricated, energy-saving Swedish homes for the U.S. market (CDIS).

Decorating and Furnishings

Home decoration has had a long history. Although experts are uncertain of the precise function of Stone Age cave paintings and have speculated on various reasons for them, including hunting magic, religion, or decoration, many favor hunting magic as the explanation. In later ages, painting served a decorative function rather than a solely ritualistic function. The Egyptians, Assyrians, and Sumerians used painted and carved walls to decorate places and buildings. The Romans used mosaics and frescoes to decorate their dwellings. The medieval castles used tapestries as much to reduce drafts as for decoration. An overview of home decoration and furnishings is provided in table 8.4. Wallpapers, floor coverings, and furniture and furnishings are discussed briefly below.

Wallpapers were first used by those unable to afford tapestries and hangings and were often imitations of them. Wallpaper for decorating has been traced back to the early 16th century. Wallpaper was printed and sold by Herman Schinkel of Delft in 1508 and paper made by Hugo Goes was used at the Master's Lodging, Christ College, Cambridge, in 1509. The earliest flocked wallpaper, dating from 1680, was found at Worcester.

As printing and papermaking technologies improved, these were applied to the manufacture of wallpapers. In 1785 Christophe-Phillippe Oberkampf invented the first machine for printing wallpaper, and this was closely followed by Nicolas-Louis Robert's 1798 process for manufacturing continuous rolls.

The 19th century saw a number of changes in designs and manufacture. Jean-Baptiste Reveillan and Joseph Dufour were two popular designers of French wallpapers early in the 19th century. French wallpapers depicted country landscapes and also contained certain architectural motifs such as columns, capitals, and moldings. The first English machine-printed wallpapers were produced in 1840 by a Lancashire printer, and the first wallpaper designed by William Morris was produced in 1862. Few advances took place in wallpaper manufacture for the next hundred years.

Figure 8.2. Furnishings

YEAR	FLOOR COVERINGS	FURNITURE
1987		
1975		
	Epoxy floors	Water bed
	Polypropylene carpet fiber	Blow chair
1950		
	Vinyl asbestos tiles	
	Asphalt tiles, patterned linoleum	
1900		Murphy bed
	Linoleum	Rolltop desk
	Kamptulion, Bigelow rug loom	Lady's chair
1800		Metal springs
		Bureau à cylindre, secretaire, tambour desk
	Axminster, Wilton charters	Rocking chair, commode
1600	Gobelin factory (France)	Upholstery
		Drawers, chests
1200	Spanish carpets	
		Beds (France, Germany)
400		
	Carpets (Altair)	
		Couches
-1200		
	Tapestry (Egypt)	
		Tables, boxes
		Chairs, beds (Egypt)
-4400		

In the 1950s and 1960s photogravure and high-speed printing techniques improved the quality and lowered the costs of fine wallpapers.

Cleaning of wallpaper was made easier by the development of washable wallpaper by Dupont in 1968.

Early domestic floors were earthen. Public buildings employed stone or wood for floors, but after Roman times most dwellings again reverted to dirt. Primitive people used furs, skins, and woven mats to ease the chill of earthen or stone floors. Rushes were used as a floor covering in many dwellings during the Middle Ages.

The first tapestry--probably used as a wall hanging--appeared in Egypt around 1480 B.C., but evidence of the first carpets dates to more recent times. Remnants of early carpets were found in royal graves in the Altai Mountains from 499 B.C. Carpets next appeared from Egypt and Syria during the 12th century A.D. Later, Eleanor of Castile introduced Spanish carpets to England in 1255, and carpets were recorded in the inventory of the duke of Bedford in 1482. During the 16th century pile carpets were made in England. Pile carpets have a raised texture from looped threads, unlike the flat surface of a woven carpet.

Industrialization of carpet making began to occur in the 17th century. King Henry IV, France, set up 200 workmen from Flanders to make tapestries in the Gobelin Paris factory, thus starting the French carpet and tapestry industry in 1601. Peter Dupont and Simon Laudet started a manufactorum (factory) for carpets in 1620. In 1701 royal charters were issued to the weavers of Axminster and Wilton in England. Because of the hand labor and materials used, carpets were expensive and used only by the wealthy. Rag rugs were popular with colonial settlers because they were easy to make, materials were available, and they were relatively easy to clean.

As yarn and loom technology developed, carpets became less expensive. E. B. Bigelow constructed a power loom for carpet manufacture in 1845, thereby reducing the cost of producing carpets. The development of manufactured fibers aided in further reducing the cost of carpets and rugs. In 1954 polypropylene was invented by Dr. Natter as a less costly substitute for nylon in carpets, and it was commercially produced in 1957.

During the 19th century, several artificial floor coverings were developed. One of the first such was Kamptulicon, patented in 1844. Although durable, it was expensive and was superseded by less expensive cork linoleum invented by Frederick Walton in 1860. The first linoleum coverings were plain, and decorative linoleum was not developed until the 1930s. Another popular flooring material, asphalt tile, was developed in the United States during the 1920s from a mixture of asphalt, asbestos fiber, and mineral fillers. The British called these tiles *thermo-plastic* tiles. In 1933 vinyl asbestos tiles were introduced at the Chicago World's Fair, but because of resin shortages they were not generally available until 1948. Seamless epoxy floors were developed during the 1960s.[16]

Although carpets were a luxury for much of human history, furniture evolved early in the civilized world. The Egyptians used beds, chairs, and stools. Some of the surviving royal furniture is elegant and resembles modern furniture in the simplicity of line. In addition to the beds, chairs, and stools of early Egypt, the Mesopotamians used tables and boxes. The Greeks used couches instead of beds and passed the tradition to the Romans. The Romans developed a veritable passion for tables.[17]

Furniture in Europe after the fall of Rome was sparse and usually portable; it was taken by the wealthy from residence to residence as they traveled.[18] The unrest of the times also meant that things that could be easily packed and moved would be taken with the owner and not left to marauding bands, invading armies, or lax caretakers. The best examples were found in the monasteries. By 750 beds were popular in France and Germany. During the late Middle Ages chests and drawers came into use. By 1700 the commode was fashionable.

Early chairs were not necessarily built for comfort. Roman men used couches, and only magistrates and women used chairs. Children and cobblers used stools. Chairs were used in the Middle Ages by nobles, but usually in a form that could be easily assembled and disassembled. In the 19th century, ladies' chairs had no arms in order to allow for wide skirts and had lower backs than gentlemen's chairs. Ladies were expected to sit upright, while gentlemen, with arm rests and higher backs on their chairs, were permitted to relax.[19] Leather was early used for stools and chair seats, but cloth upholstery was not used for furniture until the 17th century. It was in common use by 1650.

Industrialization and mechanization of textiles and other processes affected furniture manufacture, but many operations still required significant hand labor. Casters were added to furniture in the U.S. by 1838, enabling the householder or servant to move furniture more easily for cleaning. Metal springs were introduced into furniture in the first half of the 19th century, providing sturdier construction and, where properly used, more comfort.

Among notable modern examples of chairs are a tubular steel rocking chair exhibited by R. Winfield of London in 1851 and the Mies van der Rohe chair manufactured in 1926. By 1929 American manufacturers were making aluminum furniture. The transparent "blow chair" was developed in 1967, the first inflatable chair.

Cushions were used in Egypt from early times and by subsequent civilizations. Nonetheless, the Egyptians used a wooden headrest, not a pillow on their beds. The Chinese and Japanese used wooden and ceramic headrests. The Romans did use soft cushions and the nap, sheared off wool fabric during the fulling process, for pillow stuffings. Our modern pillows are mainly made of foam rubber, first produced in 1929 by Dunlop Latex

Development Laboratories, England, and used for seats in the Shakespeare Memorial Theater in 1932.

Mattresses were initially made of piled straw, pine needles or boughs, leaves, and other organic matter that could be used to cushion a hard surface. Later, such materials were stuffed into sacking and other materials, such as feathers, cotton felt, and horse hair were also used. Unfortunately, while such materials provided comfort, they also provided a haven for vermin. Bedbugs, rats, fleas, and lice remained a problem until the development of improved cleaning products, insecticides, and synthetic materials.[20.] Coiled springs, first used in chair and sofa seats, were used in some mattresses by 1870, but they were expensive and not sufficiently robust. By 1925 Zalmon Simmons had developed and was selling his innerspring Beautyrest mattress. The foam rubber developed by Dunlop in 1929 was also used in mattress construction.

The waterbed was a California invention. Charles Hall, a student at the University of California at Berkeley, developed it in 1965, and it was popularized in 1970 by Michael Zamoro in California. By the following year thermostatically controlled models were available.

Pieces of fabric used as cloaks or wrappings were probably among the earliest items made on looms and were also used as blankets. Weaving techniques are discussed more fully in chapter 5. Our modern electric blankets evolved from the heating pad developed by Dr. Sidney Russell in 1912. Thermega, London, began marketing the electric blanket in 1927, and in 1946 the Simons Company, Petersburg, Virginia, marketed an electric blanket with a thermostatic control. Early electric blankets were unreliable and it was difficult to maintain a constant temperature. With improved materials, better temperature control, and lower cost, they have become a fixture in most U.S. homes.

The Murphy bed, which folds up into a wall, was developed by Lawrence Murphy and patented in 1918. It was originally developed to conserve space, which was then at a premium owing to San Francisco's rapid growth. The beds were popular in the 1920s and 1930s. Because of steel rationing during World War II, production was suspended; it was resumed after the war, but suffered competition from the folding sofa bed. Recently there has been a resurgence of interest in such beds for small guest rooms.

The rolltop desk is an American invention, but it owes some of its inspiration to two 18th-century French desks and to an English desk. The *bureau à cylindre* had a curved, solid wood top that required a large housing chamber to hide the cover when open. The *secretaire à abattant* had a hinged, flat, solid wood cover that opened to reveal a number of drawers and pigeonhole compartments. The English tambour desk, popular in the late 18th century, had a cover of wooden strips glued to a canvas or linen backing and opened to the side, not the top. The American rolltop desk was devised by Abner Cutler in 1850. The desks became popular after being displayed at the American Centennial Exposition in 1876 in Philadelphia and were soon found in many offices. The original desks have now become collectors' items; modern manufacturers still make the desks but in new materials.

Table 8.4. Decoration, Furnishings, and Home Maintenance

Dates	*Event*
B.C.	
5000	Plaited rush floor coverings used (EB).
1483-1411	Earliest known tapestry made (EB).
499-201	Carpets used in royal graves at Pazyryk, Altai Mountains, southern Siberia (EB).
250	Philo of Byzantium suggested use of bronze springs for catapults (DID).
100	Embroidered carpet used at Noin Ula, northern Mongolia (EB).
A.D.	
750	Beds were popular in France and Germany (TH).
>1100	Egypt and Syria produced carpets (DBE).
1255	Eleanor of Castile introduced Spanish carpets to England (EB).
1300-1400	Folding chairs, stools, trestle tables, boxes with compartments, and collapsible beds used (EB).
>1400	Drawers developed for chests (EB).
>1450	Carpetmakers guild organized in Paris (WB).
1482	Carpets recorded in the accounts of the duke of Bedford (1ST).
>1500	Pile carpets made in England (EDE).
1508	Herman Schinkel, Delft, printed and sold wallpaper according to trial testimony (1ST).
1509	First known wallpaper made by Hugo Goes, Steengate, York, used at Master's Lodging, Christ College (1ST).
>1600	The Dutch developed tables with leaves (EB).
1601	King Henry IV set up 200 workmen from Flanders to make tapestries in Gobelin Paris factory (TH).
1620	Pierre Dupont and Simon Laudet started manufactory (for carpets) (DBE).
1648	Mirrors and chandeliers manufactured at Murano near Venice (TH).
1650	Leather uphosltery used for furniture (TH).
1665	Road-coach springs introduced into England (DID).

1680	Earliest surviving example of flocked wallpaper used at Worcester (EB).
18th c	Bookcases with adjustable shelves devised in England (EB).
1700	Commode became a popular piece of furniture (TH).
>1700	*Bureau à cylindre*, French desk with solid, curved wooden cover, used (USAIR).
	Secretaire à abattant, a French desk with hinged flat cover, drawers, and pigeonholes, used (USAIR).
	Gateleg tables developed in England (EB).
1701	Royal Charters granted to weavers in Axminster and Wilton for making carpets (TH).
1720	Wallpaper fashionable (TH).
>1750	Tambour, English desk with a cover of wooden strips glued to canvas or linen backing, popular (USAIR).
1785	Christophe-Phillippe Oberkampf invented the first machine for printing wallpaper (EB).
1800	Jean-Baptiste Reveillon and Joseph Dufour were popular designers of French wallpapers (EB).
1805	Semi-elliptical laminated springs were patented by Obadiah Elliot (DID).
1822	Thomas Hancock made rubber sheets (1ST).
1830	Edwin Budding, Stroud, Glouster, patented an improved lawn mower (DID).
1838	Furniture casters were patented by John A. Philos and Eli Whitney Blake, New Haven, Connecticut, for use on bedsteads (patent 821) (FFF).
1840	First English machine-printed wallpapers produced by Lancashire printer (EB).
1844	Kamptulicon (floor covering) patented by Elijah Galloway, U.S. (DID).
1845	E. B. Bigelow constructed the power loom for manufacturing carpets (TH).
<1850	Springs introduced into furniture construction during the first half of the century (EB).
1850	Rolltop desk devised by Abner Cutler, Buffalo, New York (USAIR).
1851	Tubular steel furniture, rocking chair, exhibited by R. W. Winfield, London (1ST).
1860	Frederick Walton invented Cork linoleum (TH).
1862	William Morris's first wallpaper designs available (EB).
1867	Barbed wire patented by Lucien B. Smith, Kent, Ohio (1ST).
1868	M. Kelly patented Kelly Diamond for barbed wire (1ST).
1870	Horse-drawn mowing machine introduced (DID).

	B. F. Goodrich, Akron, Ohio, manufactured rubber hoses (EOOET).
1873	Joseph F. Glidden, DeKalb, Illinois, designed a commercially successful barbed wire (WB).
1874	Machine for manufacturing barbed wire fencing introduced (EB).
1876	Halcyon Skinner, U.S., perfected the power-driven Axminster carpet loom (WB).
	The rolltop desk shown at American Centennial Exposition, Philadelphia (USAIR).
1893	Steam-driven mowing machine introduced (DID).
1897	First gasoline-powered mower built by Benz Co., Stuttgart, and by Coldwell Lawn Mower Co., Newbury, New York (1ST).
1900	Horse and Pony mower introduced by Ransomes, Sims and Jefferies, Suffolk, manfacturers of agricultural machinery (TT).
1912	Heating pad, forerunner of electric blanket, was developed by Dr. Sidney Russell, U.S. (TT).
1918	Murphy bed developed and patented by Lawrence Murphy (CDIS).
>1920	Asphalt tiles developed in the U.S. (EB).
1925	Zalmon Simmons, U.S., marketed his innerspring Beautyrest mattress (EOOET).
1926	Mies van der Rohe tubular steel chair manufactured by Thonet, Germany (1ST).
1927	Electric blanket marketed by Thermega, London (1ST).
1929	Foam rubber produced at Dunlop Latex Development Laboratories, Fort Dunlop, Birmingham (1ST).
	American manufacturers made aluminum furniture (TH).
>1930	Decorative linoleums marketed (EB).
1932	Foam rubber used for seats in Shakespeare Memorial Theater (1ST).
1933	Vinyl asbestos floor tiles introduced at the Chicago World's Fair (EB).
1937-39	Otto Bayer discovered polyurethanes (TT).
1941	Polyurethane commercially produced by I. G. Farbenindustrie (TT).
1946	Simons Company, Petersburg, Virginia, marketed an electric blanket in the U.S. with thermostatic control (FFF).
>1948	Vinyl floor tiles became generally available (EB).
1949	Tufted carpet first made in the U.S. (TT).

1954	Polypropylene invented by Dr. Natter from petroleum by-products, less costly than nylon, for carpets (TT).
1957	Polypropylene commercially produced (TT).
1958	Electric lawn mower with rotating blade introduced by Wolf (WABI).
>1960	Epoxy floors developed (EB).
1965	Charles Hall showed his water bed at San Francisco Cannery Gallery (CDIS).
1967	Transparent "blow chair" developed in Milan, the first totally inflatable chair (TT).
1968	Washable wallpaper manufactured by Dupont, U.S. (TT).
1970	Michael V. Zamaro, California, marketed a water-filled vinyl mattress, popularizing the waterbed (TIME).

Home Maintenance Tools

Various tools that help us to maintain and protect our homes include lawn mowers, hoses, and fences. Modern householders have sometimes eliminated grass and its care by planting ground cover that does not require cutting or by graveling or cementing over yards. For those who have lawns, grass still needs to be tended and cut. In earlier times, although the wealthy had gardens, grass was most often used as fodder, fuel, or fertilizer, and for thatching roofs; grass cutting had to be done by hand with scythes. Grass mowing for aesthetic purposes has become common only in modern times.

Lawn tennis became a popular sport in the 1870s and made the lawn mower an essential tool. The first lawn mower was patented by Thomas Plucknett in 1805. Edwin Budding patented an improved mower in 1830. Mr. Budding's mower was used at Regent's Park and on country estates. The horse-drawn mower was introduced in the 1860s and used in response to the rising popularity of lawn tennis. Steam-driven mowers were introduced in 1893, and the first gasoline mowers were manufactured by Benz, Germany, and the Coldwell Lawn Mower Company of New York in 1897. The popular British Horse and Pony mower was introduced by Ransomes, Sims & Jefferies Ltd., Ipswich, in 1900. The electric mower with a rotating blade was introduced by the Wolfe Company in 1958.

Gardening and lawn care also make considerable use of the rubber hose. The evolution of rubber technology was described in chapter 5. Once Charles Goodyear had developed his vulcanization process, the mass production of rubber became feasible. Although Thomas Hancock of London had developed a number of rubber products, few commercial applications were found in the United States. In 1870, Benjamin Goodrich founded the B. F. Goodrich Company in Akron, Ohio, to manufacture rubber fire hoses; the company later produced domestic garden hoses as well.

263

In 1871, the company was also producing rubber rollers for mangles and clothes wringers, bottle stoppers, rubber rings for home canning, and gaskets. With municipal pumped water supplies of sufficient pressure and sturdy rubber hoses, the gardener was able to water lawns and flowers without carrying buckets or watering cans.

Fencing established boundaries and in many early cultures mud embankments, thorn, brush, or stone were used depending on availability. Western Ireland is covered with dry stone walls while England has hedgerows. Wire fences became popular in the 19th century as manufacturing technology evolved, and by mid-century barbed wire fencing was used. Lucien Smith, Kent, Ohio, patented barbed wire in 1867. His design was followed rapidly by others, most notably M. Kelly, who patented Kelly's Diamond in 1868. The first commercially successful design is credited to Joseph F. Glidden, De Kalb, Illinois, designed in 1873. A machine for manufacturing barbed wire fencing was introduced in 1874.

We have developed the means to feed, clothe, and shelter ourselves. We have also devised the means to alter our environment and to make it more comfortable and attractive. We can control the temperature of our dwellings regardless of outside temperatures. We are protected from wind, rain, or storm. We no longer have to adapt to the climate, but instead can, at least inside our buildings, adapt the climate to ourselves.

9
Lighting

The image of a torch illuminating the darkness of ignorance is no accident. The warmth and light of a fire promise safety and protection, and our continued attraction to fireplaces is evidence of fire's hold on us. Throughout our history, we have sought a variety of means to keep darkness at bay. Now we can even control the cycle of night and day.

There were many tasks to occupy the housewife and the householder in the dark evening hours. There was flax or wool to card and spin into thread, clothes to mend or sew, or tools and weapons to sharpen or repair. Fire was used for both heating and light, but firelight was not adequate for many tasks; it would require too much wood or fuel to generate enough light and also generate excessive heat. Too, fire was not portable. Lamps and other lighting media were developed to meet these needs. Artificial lighting allows us to work when we want, not just when the sun is shining.

A particularly useful source for lamps and the development of lighting is Leroy Thwing's *Flickering Flames*.[1] Caroline Davidson's *A Woman's Work Is Never Done* provides insights into English usage and developments.[2]

Early Lighting

The earliest lamps appeared about 70,000 B.C. as hollowed stones for holding grease-soaked moss or other absorbent material. A particularly well-formed and decorated sandstone lamp was found in a French cave and dated from roughly 18,000 B.C. Its sleek form indicates that lamps must have been in existence for some time. A crude mine lamp made of chalk, obviously a workman's tool, was dated 10,000 B.C.

Figure 9.1. Lighting

YEAR	ELECTRIC	GAS	LAMPS
1987			
	George Kovacs's lamps (U.S.)		
1975			
	Metal halide lights		
1950			
	Fluorescent light		
	Coiled tungsten filament, frosted bulbs		
1900	Electric Christmas lights	Coin gas meter	
	Edison/Swan incandescent light	Incandescent gas mantle	
		Gas meter	Stearic acid candles
1800		Water gas	Nurse lamp
			Argand lamp
		Coal gas	
1600			
1200			
400			Candles
			Wicks
-1200			
-4400			
-10800			
			Lamps
-23600			
-59200			
-110400			
-212800			
-407600			
			Fire

Oil-burning lamps appeared about 3000 B.C. Oil from olives and other sources including fish and animal fat were used in lamps. By 1000 B.C. wicks were in use. The wick provided a place for the flame to form and drew up oil from the lamp basin by capillary action to feed the flame. The size and decorative features of lamps increased during Greek and Roman times. After the decline of the Roman Empire, lighting in northern Europe relied on more primitive lamps using animal fats, fish oil, or wood in the form of rushlights, splints, and candles.

Rushlights and Splints

Disrupted communications, intermittent warfare, and uncertain sources of alternatives for high quality oil were major factors in the development of lighting during the early Middle Ages. Access to materials suitable for lighting varied by locality. Rushlights were among the most common form of lighting used in northern Europe until the advent of gas, kerosene, and electric lighting. Rushlights are made with rush piths dipped in animal fat and were used in England by 800 A.D. They were dim and sometimes smoky, but were cheap and easy to make with readily available rushes that only required drying and dipping in animal fat. Use of animal fat for lighting had to compete with other uses including tanning, grease for wheels or equipment, and in some places, soap making.

Other materials used for lighting were splints and candle coal. Splints, made from the heart of pine slivers or resinous bog fir, were usually 18 to 24 inches long, and were used in the north where adequate supplies of the right types of wood were available. Splints burned with a bright light for several hours, but occasionally gave off sparks. Candle or cannel coal, a form of coal that burned with a bright light, was used in parts of Wales and northern England.[3]

Candles

Candles were used during Roman times. A date-marked candle from 1 A.D. was found at Vaison near Orange, and candle-making apparatus was found in the ruins of Herculaneum of 79 A.D. During the Middle Ages, beeswax candles were rare and expensive and mainly used by the wealthy or in churches. Sometime before 1740 American whalers discovered spermaceti wax from whales. This wax burned with a bright, steady flame, but was expensive and highly taxed in England. Common candles were made from tallow with linen or cotton wicks making them more expensive than rushlights. Cotton, except when worn to rags, was too scarce and valuable to be used for wicks, while tallow had other uses besides candles.

Snuffing the candle originally meant trimming the wick to make it burn longer and brighter, not putting it out. Snuffing was required because wicks then in use did not burn down with the candle, but extended beyond the wax. Periodically the wick had to be snuffed, or cut back to keep the candle burning and producing the maximum amount of light.[4] Rushlights did not require snuffing. In 1820 Cambacres discovered that braiding the wicks made candles self-snuffing, and in 1825 M. Chevreul and J. Guy-Lussac patented a process for making improved candles with stearic acid. These two discoveries were combined by 1831 when self-snuffing, stearic candles were sold in Paris. These candles burned brighter without the constant attention required for older types of candles. By this time, however, other substances including gas were coming into use for lighting.

Lamps

During the Middle Ages some beautiful metal lamps were designed for palaces and churches, but most homes relied on rushlights or the hearth fire for lighting. Scraped horn was used for lanterns in 890 by King Alfred. By 1415 the London streets were lit by lanterns, and in 1681 they were lighted by suspended oil lanterns.

Da Vinci turned his attention to lighting; in 1490 he designed a water-filled tube as part of a cylindrical glass lamp chimney to enhance the brightness of the flame. Water lenses were used by cobblers and lacemakers to enhance the brightness of the light.[5] The light-enhancing property of glass as well as its decorative appearance contributed to the popularity of pressed glass for oil lamps in the 19th century.

Major advances in lighting occurred in the late 18th and early 19th centuries. The invention by Aime Argand of the Argand lamp (1784) with a hollow wick to reduce flickering, smoke, and smell significantly improved lighting. Lance added a glass chimney to the Argand lamp in 1789 and Count Rumford also made improvements by placing the oil reservoir level with the wick and adding translucent shades in 1790.

The Betty lamp, a form of cruise lamp, was popular with American colonists. The cruise lamp originated in Scotland during the Middle Ages and consisted of an iron pan for holding oil or fat and a trough to hold the wick. Some models had a second pan below the first to catch excess oil dripping from the wick. The Betty lamp had a crescent-shaped arm that fitted into loops of linked chain so the lamp could be raised or lowered.[6]

The 19th century is rich in patents for lamps burning a large variety of fuels including fish oil, lard, and other animal fats. Among these were the Nurse lamp and warmer, patented in 1812, the Delmar Kinnear patent lamp of 1851, and the Davis lard lamp in 1856. The advent of gas and the later electric light replaced these, and they are now only found in the collections of ardent lamp collectors or as emergency lamps in rural places. Even most campers now use gas or electric lanterns.

Lamps required light oil such as whale oil or colza oil from the rape plant. Lamp fuel remained scarce until the discovery of oil in Titusville, Pennsylvania, in 1959. Lamps also required a means of lighting such as a paper or splinter of wood lit from the fireplace or matches.

Matches

The ability to create fire when and where needed was a major achievement. Early people first carried glowing coals and later learned to make fire with flint or a drill. The fire drill looked like a small bow with a pointed stick looped through the bowstring. The bow was moved back and forth causing the stick to turn thus creating friction. Generally enough heat was generated to ignite small kindling.

Matches were not devised until the 10th century. The Chinese invented matches in 970, and Marco Polo mentioned them in his writings. It was some time before matches were invented in Europe.

Sulfur matches were mentioned in 1530 in England and used by Roger Boyle in 1680. By 1780 phosphorous matches were in use in England and France. Match development continued until John Walker invented the sulfur friction match, the first practical match, in 1826. This was followed by Charles Sauria's strike-anywhere match in 1831. By 1832 the manufacture of friction matches was well established in Europe.

Gustaf Pasch, Sweden, improved the safety of matches in 1844 by placing some of the igniting ingredients on the striking surface. Pasch's match was patented by John Lundstrom in 1855. Packaging of matches also progressed. Book matches were devised by Joshua Pusey in 1892 and first produced by the Diamond Match Company in 1896. In the original version, the striking surface was too close to the match head and presented a safety problem. By World War I Diamond had resolved this problem by moving the striking surface.

Matches contained white and yellow phosphorous, substances injurious to the health of both the makers and the users. These particular types of phosphorous, and compounds formed from them, caused burns and were corrosive to skin, eyes, and mucous membranes. A nontoxic phosphorous compound was developed by Georges Lemoine, France, in 1864, but not patented and used until 1898. The Diamond Match Company bought the U.S. patent registered in 1910, but was unable to use it to produce matches because of climatic differences until William Fairburn adapted the formula for U.S. manufacture in 1911. The United States government had placed a heavy tax on white and yellow phosphorous matches to discourage use of these highly toxic chemicals in 1910. Fairburn's discovery provided an alternative for the match industry and saved it from financial ruin.

In 1943, Raymond Cady, Oswego, New York, developed a waterproof match that lighted even after eight hours in water for use in the Pacific during World War II.

Automatic pilot lights, electric heating and cooking, and electric lighting have reduced the need for matches. For many of us who have fireplaces, they are still necessary. With the advent of the lighter and the decline in smoking, matches are no longer as widely used, but they are still popular with the energy conscious, campers, and others needing an easy, reliable means to create fire.

Fuel

While lamp design and variety were improving, the lack of cheap, readily available lamp fuel frustrated lamp builders. A number of experiments were undertaken to find safe, inexpensive substitutes for whale oil. In 1820 Professor Robert Hare described his camphene mixture as a "burning fluid." It was patented in 1830 by Isaiah Jennings, and in 1835 Henry Porter patented Porter's Patent Portable Burning Fluid. The camphene, a mixture of alcohol and turpentine, required care in handling and in use, so some insurers forbade its use.[7]

Kerosene, a derivative of oil and coal, was discovered early in the 19th century but did not become practical until mid-century; it was not widely available until the discovery of oil reserves in America. In 1830 Karl von Reichenbach discovered paraffin (kerosene), and in 1847 James Young produced it from shale oil. Abner Gesner distilled kerosene from coal and sold it as lamp oil in 1846. Gesner patented his process in 1854, the same year John and George Austin invented a kerosene lamp. In 1856 Young introduced the first British kerosene lamp. Kerosene could also be extracted from petroleum; consequently, the discovery of oil in Titusville, Pennsylvania, in 1859 provided a cheaper fuel for lamps. Kerosene had the advantage over gas and electric lights of being easily transportable. Kerosene lamps could be moved from place to place at will. Gas did give a brighter light, but only after the incandescent mantle was developed late in the 19th century.

Table 9.1. Lamps, Early Lighting, and Matches

Dates	Event
B.C.	
500,000	Fire used (HEP).
70,000	Earliest lamps used were hollowed stones for holding fat or grease (DBE).
20,000	Fine sandstone lamp used in French cave (FFH).
10,000	Crude chalk miner's lamp used (FFH).
3000-2501	Oil burning lamps used (TH).
1000	Wicks used in lamps (DBE).
A.D.	
1	Date-marked candle left at Vaison near Orange (DID).
79	Candle-making apparatus was left in the ruins of Herculaneum (DID).
>800	Rushlights used in England (FFH).
890	Scrapped horn lamps used by King Alfred (DID).
970	Thao-Ku recorded Chinese "light-bringing slave," five-inch stick later sold in Hanchow markets (DID).
1270	Marco Polo mentioned Chinese matches (DID).

1415	London streets first lit by lanterns (DID).
>1490	Leonardo da Vinci designed a water-filled tube as part of a cylindrical glass chimney for a lamp (EDE).
1530	Sulphur matches first mentioned in England (DID).
1680	Robert Boyle used sulphur-tipped splints drawn through phosphorus-impregnated paper (DID).
1681	London streets lit by suspended oil lanterns (DID).
1694	Lampposts erected in the City of London by Convex Light Company (1ST).
1712-40	New England whalers discovered spermaceti wax from whales made excellent candles (CD).
1780	Phosphorus matches used in England (DID).
	"Phosphorus candle" of phosphorus-impregnated paper in glass tube introduced to France (DID).
1784	Aime Argand invented the Argand lamp, an iron chimney with hollow wick to reduce flickering, smoke, and smell (DBE).
1789	Lance added glass chimney to Argand lamp (DID).
>1790	Count Rumford further improved the Argand lamp (FFH).
1805	Matches tipped with potassium chlorate and sugar dipped in sulphuric acid to ignite invented by Chanel (DID).
1810	Peltier Instant Light Box matches were introduced from France to England (DID).
1812	William Harvey, Boston, patented the Nurse lamp and warmer (FFH).
	James Mallory, New York, patented a lamp (FFH).
1816	Fransçois Derosne, Paris, credited with making the first phosphorus friction match (EB).
1820	Professor Robert Hare, in the *American Journal of Science and Arts*, described "burning fluid," camphene (FFH).
	Cambacres discovered braided wicks for candles made self-snuffing candles (FFH).
1825	M. Chevreul and J. Guy-Lussac patented improved manufacturing methods for candles using stearic acid (EB).
1826	John Walker devised the sulphur friction match (1ST).
1827	First recorded sale of sulphur friction matches by John Walker (TH).
1828	Samuel Jones, London, produced the Promethean match, a glass bead, wrapped in paper, that contained acid (EB).
1830	Karl von Reichenbach, Blansko, Moravia, discovered paraffin (TH).
	Isaiah Jennings patented "burning fluid," camphene (FFH).

J. F. Kammer, Germany, invented matches that used yellow phosphorus, sulphur, and potassium chlorate (DID).

1831 Self-snuffing, stearic acid candles sold in Paris (CD).

Charles Sauria, France, developed a method for making matches easy to ignite (TH).

1832 Manufacture of friction matches well established in Europe (TH).

1835 Henry Porter patented Porter's Patent Portable Burning Fluid, camphene (FFH).

1836 American patent for the manufacture of white phosphorus matches granted to A. D. Phillips (TH).

1842 First machine-made matches were manufactured (DID).

1844 Gustaf Pasch, Sweden, proposed safer matches by placing some ingredients on striking surface (TH).

1845 Anton Schroetter, Vienna, patented amorphorus phosphorus matches (DID).

1846 Abraham Gesner, U.S., distilled kerosene from coal and sold it as lamp oil (CD).

1847 James Young produced paraffin (kerosene) from shale oil (DID).

1848 Bottger produced the first safety matches (TH).

>1850 Pressed glass lamps were popular in the U.S. (EB).

1851 Delmar Kinnear Patent lamp patented (FFH).

1852 John E. Lundstrom and brother, Jonkoping, Sweden, made matches proposed by Pasch in 1844 (DID).

1854 Gesner patented his process for making kerosene (WB).

John H. and George W. Austen, New York, invented a kerosene lamp (1ST).

1855 John E. Lundstrom, Sweden, patented the safety match (1ST).

1856 James Young, Bathgate, Glasgow, Scotland, introduced the first British kerosene lamp (1ST).

Dr. Gesner, London, produced kerosene by distillation (DID).

Davis patented a lard-burning lamp (FFH).

1859 Moses G. Farmer, Salem, Massachusetts, installed electric lamps in his parlor that were powered by a galvanic battery (1ST).

1860 Wax match-making machine invented by Louis Perrier, Marseilles (DID).

1864 George Lemoine, France, developed phosphorus sequisulfide matches, a nontoxic form of phosphorus (EB).

1892	First book matches patented by Joshua Pusey, Lima, Pennsylvania (1ST).
1896	Book matches manufactured by Diamond Match Company, Barberton, Ohio (1ST).
1898	E. D. Cahen and H. Sevene patented and used Lemoine's discovery of a nontoxic phosphorus compound (EB).
1910	U.S. heavily taxed white and yellow phosphorus matches (EB).
	Diamond Match Co. bought U.S. patent for Lemoine's compound, but was unable to use it (EB).
1911	William Armstrong Fairburn, U.S., adapted French red phosphorus formula for production in the U.S. (EB).
1921	Friedrich Bergius hydrogenated coal to oil (TH).
1934	Persplex commercially produced by Imperial Chemicals Industries for use in airplane windows and light fittings (TT).
1943	Raymond Davis Cady devised a waterproof match for use in the Pacific (EB).

Gas Lighting

Gas lighting experiments began in the 18th century. Coal gas was first obtained by Dr. Stephen Halcs from the distillation of coal in 1727. Coal gas or coke oven gas is produced from the distillation of bituminous coal and was used for both heating and lighting. The hot gases from a coke oven were bubbled through water to remove such elements as tar, ammonia, and light oil. In 1760 George Dixon, Cockfield County, Durham, used coal gas to light a room. There were attempts to use gas to light the interior of buildings in Germany and England, and in 1792 William Murdoch used coal gas successfully to light his entire home.

Once suitable methods had been developed for producing gas, gas companies were formed to supply and market gas. By 1805 the National Gas Light and Heat Company, London, had established a gasworks, and by 1807 gas lights were used in some London streets. By 1821 Fredonia, New York, was using gas for street lighting.

Other improvements included gas meters, new lamps, and bottled gas. The gas meter was devised in 1815 by William Clegg. As noted under cooking and heating, the coin-in-the-slot gas meter developed by R. W. Brownhill, England, in 1887, was also a significant contributor since the gas company was paid in advance for the gas and the poor used only what they could afford. In 1816 the fishtail or batswing gas burner was introduced. Access to gas was broadened with the advent of bottled gas, first marketed in 1825 by the Provincial Portable Gas Company, Manchester, England.

A number of inventors worked on processes for making water gas which was made by forcing steam and air over hot coke or coal to yield a hydrogen and carbon monoxide fuel. Felice Fontana produced water gas in 1780. In 1823 William de Vere and Henry Crane patented a process for illumination with water gas. In 1833 Jobard invented a process for making water gas from resins, and in 1847 Stephen White and H. Paine, Massachusetts, also patented a water gas process.

While glass chimneys on lamps reduced the flickering of the flame, the major improvement in gas lighting was the development of the gas mantle. The platinum mantle was introduced in 1848 and J. A. Hagg introduced the atmospheric gas burner in 1868, but the major advance occurred when Karl von Welsbach invented the incandescent gas mantle in 1885. This type of mantle used a loosely woven fabric soaked with chemicals. When the mantle was installed the fabric burned away leaving a residue of chemicals that glowed brightly when the gas lamp was lit. The new, relatively inexpensive mantle generated from 3 to 10 times the light of other devices.[8]

Gas lighting spread rapidly in the late 19th century but was soon replaced by another newcomer already under development: the electric light.

Table 9.2. Gas Lighting

Dates A.D.	Event
1727	Coal gas first obtained from the distillation of coal by Dr. Stephen Halcs (DID).
1760	George Dixon lit a room in his house at Cockfield County, Durham, using coal gas (DID).
1780	Felice Fontana produced water gas (TH).
1786	Earliest attempts at internal gas lighting, Germany and England (TH).
1792	Illuminating gas used in England for the first time (TH). Murdoch, Scottish engineer, used coal gas to light his home (EB).
1805	Gasworks established by National Gas Light and Heat Company, London (1ST).
1807	Gas street lights in London (TH).
1815	Gas meter devised by William Clegg (1ST).
1816	Fishtail or batswing gas burner introduced (DID).
1817	Gasworks, public utility, Manchester Corporation, established in Manchester, England (1ST).
1821	Fredonia, New York, used natural gas for street lamps (1ST).

1823	William de Vere and Henry S. Crane patented a process for illumination by water gas (DID).
1825	Bottled gas marketed by Provincial Portable Gas Company, Manchester, England (1ST).
1833	Jobard, Belgium, invented a process for making water gas from resins (DID).
1847	Stephen White, Manchester, and H. M. Paine, Worcester, Massachusetts, patented a process for making water gas (DID).
1848	Platinum mantle burner for gas lamps introduced (DID).
1868	J. A. Hagg introduced an atmospheric gas burner (DID).
1885	Karl Auer von Welsbach invented the incandescent gas mantle (TH).
1890	Gas mantle combined with Bunsen's gas burner to create a lamp (DID).
1897	Inverted gas mantle invented by Kent (DID).
1898	Practical use natural gas used for lighting of Heathfield Station, Sussex (1ST).

Electric Lighting

Experiments with electric lighting began somewhat later, but like gas the major developments occurred late in the 19th century, following the industrial revolution and building on many scientific discoveries about the nature of electricity. The first electric lamp was developed by James Bowman Lindsay, Dundee, Scotland, in 1835. Lindsay, having satisfied himself that light could be generated by electricity, abandoned his experiment.

Many of the first efforts with electric lighting were with arc lights for street lighting. The light was created by an electrical discharge between electrodes, and was bright and reliable, but bulky and complicated to install and maintain. Experimental electric arc lamps were installed in le quai Conti, Paris in 1841. Insulated electric cables were manufactured in 1847 at London, and an electric arc lamp was erected on the north tower of Hungerford Bridge two years later in 1849. The city of Lyons installed permanent electric lights in 1857 and London followed suit in 1878.

A number of developments important to the availability of electrical power occurred in the next several years. In 1884 the electric transformer was brought into commercial use for an electric lighting system for the Metropolitan Railway in England. The same year a practical steam turbine was designed to drive a dynamo in the Gateshead Lamp Works. In 1886 the hydroelectric installation at Niagara Falls was begun.

Critical for home lighting was the development of suitable bulbs, because arc lights were too massive and complex for domestic use. The light bulb was invented by Heinrich Goebel in 1854. In 1859 Moses Farmer installed in his parlor electric lamps of his own design powered by a galvanic battery located in his cellar. Edison and Swan independently invented incandescent electric light bulbs in 1878. Edison began commercial manufacture of his bulbs in 1880, the same year New York streets were first lit by electric lights. Newcastle upon Tyne installed street lights in 1881 using Swan's new bulb. Special types of bulbs were also developed. Edward Johnson, an associate of Edison, had the first Christmas tree lit with electric lights, but the first commercially produced tree lights did not appear until 1901.

The filament for the incandescent light was not durable, and considerable effort was expended to find an improved filament. In 1902 osmium was tried, but it was too expensive and thus not suitable despite its efficiency as a light producer. In 1905 tantalum was used, and in 1906 tungsten was incorporated into the design. By 1908 a process for producing a more durable tungsten filament had been developed, and the coiled tungsten element was produced in 1934.

Further improvements were the use of nitrogen in the bulb in 1913 to prolong the life of the filament, the removal of pips, (glass imperfections caused during manufacture) in 1919, thus strengthening the bulb, and the internal frosting of the bulb to provide a diffuse light source in 1925. The Corning Glass Works developed a machine to blow light bulbs in 1929, further mechanizing their production and contributing to the availability of a low-cost, high-quality modern light bulb. The modern light bulb costs one fiftieth of the cost of lighting five generations ago.[9.]

Fluorescent light developments began in 1901 when Peter Copper-Hewitt demonstrated the liquid mercury lamp tube. Edison had applied for a patent on a fluorescent light in 1896, but did not use his design for commercial applications. Albert W. Hall added improvements to the fluorescent light in 1927, and in 1935 General Electric Company showed fluorescent lights publicly in Cincinnati. The first practical application was at a centenary dinner for the U.S. Patent Office in Washington, D.C., in 1936, but General Electric and Westinghouse did not begin commercial production of fluorescent lamps until 1938. White fluorescent lamps were produced in 1951. Continued improvements have been made in both fluorescent and incandescent lighting. The most recent development is metallic halide lamps, introduced in 1964. By 1971 Richard Sapper, Italy, designed a high-fashion halogen lamp that was manufactured by Artemide. George Kovacs in the United States introduced his halogen lamp in 1979.

Table 9.3. Electic Lighting

Dates A.D.	Event
1835	Electric lamp developed by James Bowman Lindsay, Dundee, Scotland (1ST).
1841	Experimental electric arc lamps installed in le quai Conti, Paris (1ST).
1847	Insulated electric cables manufactured by Gutta Percha Company, London (1ST).
1849	Electric arc lamp erected in the north tower of Hungerford Bridge (1ST).
1854	Heinrich Goebel invented the first form of the electric light bulb (TH).
1857	Lyons, France, introduced permanent electric street lights (1ST).
1878	Electric street lighting was installed in London (TH). Swan and Edison independently developed the incandescent electric light bulb (1ST).
1880	First commercially produced incandescent electric light bulbs manufactured at Menlo Park, New Jersey (1ST). New York streets first lit by electric lights (TH).
1881	Incandescent electric street lamps were installed in Newcastle upon Tyne (1ST).
1882	First Christmas tree lit with electric light bulbs installed by Edward C. Johnson, New York City (1ST).
1884	Electric transformer used in Metropolitan Railway, England, electric lighting system (1ST). Practical steam turbine designed by Charles Parsons, Gateshead, to drive the dynamo in local lamp works (1ST).
1886	Hydroelectric installations begun at Niagara Falls (TH).
1896	Edison applied for a patent on the fluorescent light (FD).
1901	Peter Cooper-Hewitt, U.S., demonstrated the forerunner of the fluorescent lamp tube using liquid mercury (TT). First Christmas tree lights sold commercially by the Edison General Electric Co., Harrison, New Jersey (1ST).
1902	First electric light using an osmium metal filament, more efficient, but too expensive, developed (TT).
1905	Electric light built with tantalum metal filaments marketed by Siemens and Halske, Berlin (TT).
1906	Tungsten filament used in commercially produced electric light bulb (TT).

1907   Strip Lighting, Moore Tubing, used to illuminate Savoy
        Hotel court (1ST).
1908   Coolidge, U.S. scientist, developed a prototype of a durable
        tungsten filament with his new process (TT).
1909   Benjamin D. Chamberlin, Washington, D.C., first filed a
        patent on a machine to make glass light bulbs (patent
        491812) (FFF).
1913   Nitrogen-filled electric bulb developed by Irving Langmuir,
        General Electric, that improved the life of the tungsten
        element (TT).
1914   Corning Glass Works, Corning, New York, began using
        Chamberlin's machine further modified by others to
        make glass light bulbs (FFF).
1916   Gas-filled tungsten filament incadescent bulb introduced
        with thirteen times more light than carbon bulbs
        (ACTB).
1919   Pips, glass imperfections left by manufacturing process,
        eliminated thereby strengthening bulb (TT).
1925   Internally frosted glass bulbs introduced by British
        manufacturers for improved diffuse light source (TT).
        Benjamin D. Chamberlin, Washington, D.C., received U.S.
        patent 1551935 on a glass bulb manufacturing machine
        (see also 1909, 1914) (FFF).
1927   Albert W. Hall added improvements to fluorescent lights
        (TH).
1929   Machine blown electric bulb developed by Corning Glass
        Works (WB).
1935   Fluorescent lighting first shown publicly in Cincinnati by
        General Electric Company (1ST).
1936   First practical application of fluorescent lighting used in
        Washington, D.C. (1ST).
1938   General Electric and Westinghouse began commercial
        production of fluorescent lamps (1ST).
1951   White fluorescent lighting tubes produced using Peter
        Ranby's discovery of halophosphate (TT).
1964   Metal halide lamp introduced (MHEST).
1971   Richard Sapper, Italy, manufactured halogen lamps
        designed by Artemide (VIS).
1979   George Kovacs, U.S., introduced his first halogen lamps
        (VIS).

From fire and flame, which required a fuel supply and care to ensure
continued burning, we have progressed to instant light available by pushing a

switch, a safe, odorless lighting medium available at any time. Gone was the need for the storage of smelly, inflammable fuel and the task of cleaning and polishing lamp chimneys. We now have the option of undertaking activities at our convenience instead of waiting for the light of day. Light somehow seems to give comfort and promise safety and warmth. The old tie has changed, but it has not been fully broken. While we no longer need light to keep animals at bay, the lighted window remains a sign of welcome.

10
Tools

One factor that distinguishes humans from apes is humans' extensive use of tools. Tools have been indispensable to humans and their development. Many tools are used, but only those few that are particularly important to daily life are discussed. This chapter includes a variety of different technologies and devices not treated elsewhere. Power sources are essential to metal production and almost everything we do. Metal working and the development of power sources are included because they are central to so many of our daily needs. Obviously metal technology is important to food production, food preparation, housing, and lighting. Clocks are important as time measuring devices, but they are also incorporated in many other devices and appliances, as are computers and microprocessors. Paper and writing instruments have long provided basic means of communication and the transmission of knowledge from one generation or cultural group to another as well as shaping the course of modern communications technologies.

As in chapter 8, the sources used for basic tools and technologies have been diverse and it is difficult to cite only a few as representative. Particularly useful, however, was Gimpel's *The Medieval Machine*,[1] the *Encyclopaedia Britannica*,[2] and the *World Book Encyclopedia*.[3] Specific articles are cited in the Notes and in the Sources.

Power Sources

Humans and animals were the early sources of power. Some of the profound changes wrought by the use of horse power, dating to the 9th century, were discussed in chapter 2. Once the rigid horse collar was available, the horse became the main source of power and transport for those without access to water power, reliable winds, or bodies of waters including oceans, seas, rivers, and lakes allowing access and transport to other locations. Water was used to a limited extent by the Romans, but not truly exploited until the Middle Ages. Windmills can be traced from the 7th century. Steam, as a power source, is a recent invention that came into general use in the latter part of the 18th century. Steam helped reduce the geographic restrictions inherent in wind and water power. Power became humankind's servant with the advent of electric power and its widespread use in the 20th century.

Figure 10.1. Technologies Overview

YEAR			
1987			
	Fuel cells		
			Optical disc
			Apple II
1975			
			Personal computer
		Digital clock	Minicomputer
1950	Atomic power		
	Solar heating experiments	Timex	ENIAC computer
	Rural electricity		
1900	Stainless steel		Punched card
	Aluminum process		
	Oil	Electric clock	Charles Babbage
1800	Gas		Jacquard loom
	Steam	Wrist watch	
	Sheffield steel	Pendulum clock	
1600			
		Traveling clock	
		Alarm clock	
1200	Steel (blast furnace)		
	Rigid horse collar		
	Windmills		
400			
	Water mills	Hour glass	
	Iron	Water clocks	
-1200	Solar architecture		
		Sundial	
	Bronze		
-4400			
-10800			
-23600			
-59200			
-110400			
-212800			
-407600			
	Fire		

282

Water and Wind

The Romans used water mills mainly for grinding grain into flour. Strabo of Greece mentioned a water mill at Cabria in 63 B.C., and Vitruvius, a Roman architect, described the mill machinery in 11 B.C. Water mills were used extensively in the Middle Ages and were important in the early industrial development of New England. Water and horse power were the main sources of power until the invention of the steam engine in the 18th century.

Vitruvius also described the post-type windmill. The horizontal windmill evolved around 600 A.D. in Iran and Persia. The first mention of a windmill is a reference to a Persian windmill builder, Abu Lu'lu'a, in 644. By 900 the horizontal mill was in use in Arab countries and by 1000 it was being used in the Low Countries. A post-type windmill in Normandy was mentioned by Leopold de Lisle in 1180, and one was in operation at Weedley, York, by 1185. Mills of various types, including wind, water, and tidal mills, were used for grinding grain, processing and fulling of cloth, metalworking, beer making, and pumping water.[4]

Windmills became more popular in the United States for use on the prairies as Americans moved westward. Winds were readily available, and the invention in 1854 of a self-governing windmill by Daniel Halladay of Connecticut made unattended operation practical. His device turned itself out of the wind when the breeze became too strong. By the 1880s windmills had become essential machines on U.S. farms. One large Texas ranch had 335 mills pumping water for cattle. With the advent and spread of electric power, many windmills were shut down.[5]

Wind power is again being used as an energy source encouraged by governmental support for renewable energy sources. "Wind farms" have been established at suitable sites in California to generate electricity. One such farm supplies power to 50,000 homes. Current costs are 12 to 15 cents per kilowatt-hour higher than the five cents for conventionally generated electricity, but competitive with the more expensive nuclear power plants. Improved and less costly equipment could further reduce costs. Estimates are that 8 percent of California's electricity needs by the year 2000 could be supplied by wind.[6]

Wind and water power are available only in those areas having sufficient and regular supplies of the motive forces. Farm animals supplied much of the power elsewhere until the 18th century when steam power began to evolve as a major new source of power.[7]

Steam

The use of steam power began when Thomas Savery patented a steam engine in 1698. It was manufactured in 1702 for draining water from mines.

Thomas Newcomen, Tipton, Staffordshire, also built a steam engine in 1712 and continued to operate it for thirty years.

Much of the credit for making the steam engine a versatile tool belongs to James Watt and the condenser he invented in 1764. This made the engine more efficient since the main cylinder no longer had to be cooled for condensing the steam. Watt perfected his engine in 1775. By 1782 he had invented a double-acting rotary steam engine. Watt's engine was quickly adapted in the textile industry and was a major factor in the industrial revolution.

To make steam requires fire and a fuel source. Wood and coal were the two main fuels until the late 19th century when diesel oil and gasoline became available.

Fuels

Wood had been the major fuel for much of human history. Clearing for agriculture, coupled with heavy reliance on wood for fuel, led to the depletion of forests. By the 16th century the British were forced to use other sources, particularly coal. By the mid-19th century oil had been discovered in adequate quantities for commercial use and was used as a lighting fuel. Its heavy use as a source of power has occurred mainly in the 20th century. Its major use has been as a fuel to power vehicles and ships. A more detailed discussion was given in chapter 8.

Electricity

A number of scientists, including Benjamin Franklin, experimented with electricity. In the 19th century interest in electricity was related to its use for lighting. By the end of the century it was being applied to cooking and heating as well. The development of the electric motor facilitated the use of electricity for other applications, including those requiring mechanical power. Electric motors are used to drive many devices including household appliances such as clothes washers, dryers, can openers, refrigerator compressors, and exhaust fans. Thomas Davenport, Rutland, Vermont, patented an electric motor in 1837, and the first practical storage battery was devised by R. L. G. Plante in 1859.

Significant in creating and distributing electrical power were the electrical transformer, the dynamo, and the electric motor. The transformer allows the power from a high density source to be reduced to a level suitable for household use; Michael Faraday built the first electric transformer in 1831. The dynamo is a generator of electricity and was demonstrated by Hypolite Pixii, Paris, in 1832 while the rotating dynamo was demonstrated by Joseph Saxon in 1833 at Cambridge.

Street lighting was one of the first applications of electric power. By 1857 Lyons had permanent electric street lighting. The invention of the incandescent light by Edison and Swan added further impetus to the development of electric power.

By 1881 an electric generating station providing power for both domestic and public use was in operation at Godalming, Surrey. Edison designed the first hydroelectric plant in 1882, and by 1896 the Niagara Falls plant had opened. The Tennessee Valley Authority was established in 1933, and the Hoover Dam was completed in 1936.

It is hard now to understand the profound effect inexpensive, readily available electric power had on society and in particular on household technology. Farms and rural users had been slow to benefit from the new appliances and from machines available in the urban areas that required electric power to operate. The rural electrification programs changed that and made the same resources more readily available to all.

Municipal and individual water supplies could be pumped by electric power: water no longer had to be hauled long distances or pumped manually, and hot water supplies could be available in each home. With constant water supplies, septic and sewage systems became practical. Electric sewing machines and washing machines reduced effort and electric refrigerators and freezers made storage and preservation easier. Milking machines and other agricultural equipment could be used. Wood or coal burning fireplaces or stoves are no longer the major means of heating for most homes. Even for those not using electricity for heating, it powers the fans in most heating and air conditioning systems. It also powers other tools and appliances including clocks and computers. The entire fabric of life has changed dramatically.[8]

Atomic Power

In addition to hydroelectric and coal-burning power plants, other energy sources were also being developed. Otto Horn of the Berlin Chemical Institute discovered nuclear fission in 1939. Wartime research led to the development of the atomic bomb, but also to the use of nuclear power for commercial, educational, medicinal, and domestic uses. By 1951 electric power had been produced from atomic energy at Arcon, Iowa, and in 1954 the Russians opened the first atomic power station at Oblinsk. They also used volcanic steam to generate electricity, in 1973.

Accidents at Three Mile Island, a nuclear power plant near Harrisburg, Pennsylvania, and, more recently, at Chernoybl in the Ukraine cast a pall over the use of nuclear energy for electrical power generation. These accidents, coupled with cost overruns on construction projects, reports of inadequate safety measures, and careless operation, have increased public awareness of the potential dangers of nuclear power if improperly handled, and the consumer lobbies--demonstrating against nuclear energy--have become more forceful in recent years. Efforts to develop alternative energy sources, particularly renewable sources such as wind or solar power, have waxed and waned. The United States is still heavily dependent on oil and coal for power generation.

Fuel and Photovoltaic Cells

Power for the space shuttle comes from fuel cells that provide electric power and also generate water for drinking, rehydrating foods, and washing. A fuel cell generates electricity by a chemical reaction between a fuel and an oxidizer, consequently a battery is also a fuel cell, but, more commonly, the term is used to describe devices using gaseous agents, usually oxygen and hydrogen or carbon monoxide. Fuel cells convert energy efficiently with little or no energy loss in the form of heat. Unfortunately, they are also expensive to produce and operate, in part because of the high energy required to produce the gaseous raw materials needed. A San Francisco building was equipped with fuel cells to generate power and hot water in 1985.

Solar energy for heating and water supplies was discussed in chapters 7 and 8. Most solar or photovoltaic collectors convert only about 11 percent of the sunlight they receive into electrical energy. Arco Solar, an Atlantic Richfield Corporation subsidiary, developed thin silicon cells in 1985, reducing the cost significantly over older types of photovoltaic cells. Most cells generate electricity at a cost of fifty cents per kilowatt-hour, approximately ten times the cost of conventional power. Arco's new technology is expected to reduce fuel cell cost to a competitive level by 1990. The thin film silicon could be applied to building windows, providing both power and some insulation. In the future our windows could become an effective power source.[9]

Table 10.1. Power Sources

Dates	Event
B.C.	
3000	Wheel used (WB).
800-700	Etruscans used hand cranks (TH).
63	Strabo recorded the existence of a water mill at Cabria (MM).
22-11	Vertical (post) windmill described by Vitruvius, a Roman (DBE).
A.D.	
>300	Ausonius described a water-driven sawmill on River Moselle used to cut marble (DID).

>600	Horizontal windmill was used (DID).
644	Earliest reference to windmills was to Abu Lu'lu'a, Persian windmill builder (EDE).
700	Waterwheels used in Europe (TH).
800	Rigid horse collar adopted in Europe increased the load a horse could pull (MM).
900-1000	Horizontal windmill used in Low Countries (DID).
1010	Water-driven hammer used at Schmidmuhler (MM).
1170	Earliest recorded tide mill used in Britain, Woodbridge, Suffolk (DID).
1180	Post-type windmill in Normandy mentioned by Leopold de Lisle (DBE).
1185	Windmill used at Weedley, York (DBE).
1230	Villard de Honnecourt sketched a water-driven sawmill (MM).
1328	Sawmill invented (see 300, water-driven sawmill described by Ausonius); metal technology improved blade life (TH).
1592	Windmills used in Holland to drive mechanical saws (TH).
1633	Wind sawmill used at the Strand, London (TH).
1690	Denis Papin, France, devised a pump with a piston raised by steam (TH).
1698	Thomas Savery, London, patented the first working steam engine for raising water (CD).
1702	Steam engine for draining mines manufactured by Thomas Savery (1ST).
1707	Denis Papin invented the high pressure boiler (TH).
1712	First practical steam engine erected by Thomas Newcomen, Tipton, Staffordshire (used for thirty years) (DBE).
1764	James Watt invented the steam condenser, the first step to efficient steam engine (TH).
1774	John Wilkinson's boring mill facilitated the manufacture of steam engine cylinders (TH).
1775	James Watt perfected the steam engine (TH).
	Pierre-Simon Girard invented the water turbine (TH).
1831	Electric transformer built by Michael Faraday, London (1ST).
1832	Dynamo demonstrated by Hypolite Pixii, Paris (1ST).
1833	Rotating dynamo was demonstrated at Cambridge by Joseph Saxon (1ST).
1837	Electric motor patented by Thomas Davenport, Rutland Vermont (1ST).

1845   William M'Naught developed the compound steam engine (TH).

1854   Daniel Halladay, Connecticut, developed a self-governing windmill (CDIS).

1857   Lubricating oil manufactured by Price's Patent Candle Company, Battersea, London (1ST).

1859   R. L. G. Plante developed the first practical electric storage battery (TH).

1868   Filter pump invented by Robert Wilhelm Bunsen (DID).

1881   First electric power station providing current for public and domestic use operated at Godalming, England (1ST).

1882   Edison designed the first hydroelectric plant (TH).

1886   Kieselguhr filter was patented by Heddle and Stewart (U.K.) and Weischmann (U.S.) (DID).

1888   Nikola Tesla created the electric motor (TH).

1896   Niagara Falls hydroelectric plant opened (TH).

1908   Middle Eastern oil first struck in commercial quantities (1ST).

1911   William Armstrong Fairburn, U.S., adapted French red phosphorus formula for production in the U.S. (EB).

1923   Chicago Crawford Ave. Power Station used excess steam to drive turbines to regenerate heat (TT).

1929   First ocean thermal power plant operated at Mantanzas Bay, Cuba (TT).

1933   Tennessee Valley Authority established, major electrification project (TT).

1936   Hoover Dam completed (TH).

1939   Nuclear fission discovery published by Otto Horn, Chemical Institute, Berlin (1ST).

1951   Electric power produced by atomic energy at Arcon, Iowa (TH).

1954   First atomic power station at Oblinsk, U.S.S.R., began producing electrical current for industrial use (1ST).

1957   New dry cell battery developed by P. R. Malloy Co., lasted several times longer than previous dry cell (TT).

1981   California "wind farms" established in Altamont Pass (TR).

1985   NASA investigated use of photosynthetic bacteria for converting solar energy to hydrogen for shuttle (USA).

        Arco developed thin-film silicon for photovoltaic cells to make fuel cells competitive with current technology (FORT).

Metals

Metalworking technologies can be traced to at least 4000 B.C. The first metallic tools were made of gold, copper, or meteoritic iron. Metals at or near the earth's surface and soft enough to be worked easily were used first. Later, as people discovered the usefulness of metals, they also learned how to extract them from deeper in the earth.

The first smelting furnaces dating from 2300 B.C. were found at Tuyn, Ur. By 2000 B.C. the Hittites used bellows with their smelting furnaces, creating higher temperatures and thus broadening the types of ore that could be worked. Theodorus of Samos described ore smelting and casting in the 5th century B.C. Sometime after 300 B.C. the Chinese were using a double-acting piston bellows for continuous blast in their smelting furnaces.

During the Middle Ages water power was applied to metal technologies as well as to other areas. Water power was used to break up the ore for processing and for forging blooms of iron (intermediate masses of metal from the forge or the puddling process). Gimpel indicates a water-powered hammer may have been used for these purposes as early as the 12th century.[10] By 1323 water power was used to drive the bellows to increase the furnace heat for processing iron. By 1380 the first European blast furnace had evolved. Steel and cast iron required high furnace temperatures.

Mechanical power reduced the manpower required to process iron and increased the processing heat, thereby providing more, higher quality iron. With abundant iron produced at lower cost, it could be used instead of wood for common tools increasing their durability and effectiveness.[11]

Developments in the 18th century included a method for converting cast iron into malleable iron, a puddling process for wrought iron, and the application of steam to rolling mills, all of which contributed to increased metal production at lower cost. Puddling involves stirring the molten iron with oxidizing agents to turn pig iron into malleable iron, which can then be worked into a variety of products. Unlike malleable iron, cast iron is brittle and difficult to work, consequently, it usually must be cast into the desired shapes. The improved processes provided a ready supply of lower cost cast iron and of malleable iron, which were used to provide ovens, boilers, stoves, and later water pipe.

Steel processing dominated the 19th century with the development of large open-hearth furnaces. Aluminum was also manufactured in 1855 in Paris, but was not widely used until after 1886 when the electrolytic process was developed independently by Charles M. Hill in the United States and by Paul Heroult in France. This process provided a more cost effective way to mass produce aluminum. As was noted in chapter 4, aluminum was used for the manufacture of cooking utensils in 1890, but did not come into common use until after the turn of the century.

The early 20th century saw the development of stainless steel and its application to cutlery. It was first developed in 1904 by Leon Guillet, France, but not really used until cast by Harry Brearley, U.K., in 1913. It was used for cutlery in 1915 and by 1921 International Silver, Meriden, Connecticut, was marketing table knives with stainless steel blades. Metal technologies and new alloys continue to evolve aided by the availability of cheap power sources.

Table 10.2. Metals and Metal Working

Dates	Event
B.C.	
3500	Bronze used (WB).
2300	Tuyn smelting furnances used at Ur (DID).
2000	Hittite smelting furnaces used bellows (DID).
1500-1000	Iron used in Syria and Palestine (TH).
1000-900	Iron used in Greece (TH).
900-800	Iron and steel used (TH).
700-600	Iron soldering used (TH).
600-500	Ore smelting, casting, water level, lock and key, carpenters square, and lathe described by Theodorus, Samos (TH).
>300	Double-acting piston bellows for continuous blast used in Chinese smelting furnances (DID).
A.D.	
1323	Water-driven bellows used for a furnace (MM).
1380	First European blast furnace in operation (MM).
1640	Coke made from coal (TH).
1740	Steel smelting using "crucible" process improved by Benjamin Huntsman (TH).
1754	First iron-rolling mill established at Fareham, Hampshire, England (TH).
1758	Blast furnace installed by John Wilkinson, Bilston, Staffordshire, England (1ST).
1762	Cast iron converted into malleable iron at Carron Ironworks, Stirlingshire, Scotland (TH).
1784	Henry Cort introduced the first puddling process for manufacturing wrought iron (TH).
1790	First steam-powered rolling mills built in England (TH).
1795	Joseph Bramah invented the hydraulic press (TH).
1800	Richard Trevithick constructed the light press steam engine (TH).
1810	Sarah (Sister Tabitha) Babbitt devised a circular saw (HDI).

1836	Electroplating patented in Great Britain by George and Henry Elkington (1ST).
1855	Austrian engineer Franz Koller developed tungsten steel (TH).
	Aluminum manufactured by Henry Deville, Salindres Foundry, Paris (1ST).
1856	Sir Henry Bessemer introduced the converter into his process for making steel (TH).
	Sir William Siemens made ductile steel for boiler plating (TH).
1863	Open-hearth steel furnace was developed by Matin brothers in France based on Siemens process (TH).
1880	Carnegie developed the first large steel furnace (TH).
1886	Charles M. Hall, U.S., and Paul Heroult, France, independently discovered the electrolytic process for aluminum (FHTC).
1904	Leon Guillet, French scientist, made the first stainless steel, but failed to capitalize on it (TT).
1913	Stainless steel first cast by Harry Brearley, Sheffield, England (1ST).
1916	Stainless steel introduced to the U.S. (ACTB).
1971	Water used for cutting, pioneered by McCartney Manufacturing Company (TT).

Writing

Civilization cannot exist without communication. We use many means of communication in our homes and daily lives including the telephone, radio, television, books, newspapers, and magazines. Product directions, appliance care and use instructions, and food preparation and storage directions are a few examples of information manufacturers and retailers want and need to communicate to consumers and to repair personnel.

Although the telephone, radio, and television represent significant achievements, none of these has been included here. The means of communication such as paper, pens, pencils, and typewriters, all of which may be found in the home, have been included, but the development of writing itself and of the printing press has not. Paper is important as a low-cost medium for printed information and equally for product packaging, cleaning and hygiene products such as paper towels and toilet paper, decorating aids such as wallpaper, and even disposable clothing.

Writing and reading are the right and left hands of written communication processes. Writing is the encoding of information into symbols, while reading is the decoding of those symbols into meaningful information. Reading is essential to identify the products we use and how they should be used. There is a growing concern within the United States over adult literacy problems. Some manufacturers have resorted to package graphics to overcome product identification and use problems by semiliterate or non-English-reading customers. Even television, primarily a visual and oral medium, carries some print information in the form of slogans, product names, telephone numbers or addresses, or advertising. The use of computers relies heavily on print information in the form of instructions and messages. While Apple's Macintosh and the Xerox Star computers pioneered the use of symbols, there is still much print information required in order to employ these machines as more than fascinating toys.

The development of writing was vital to communication. Its importance in the household often goes unrecognized. How could householders make lists for shopping and laundry without the ability to read and write, or the tools with which to write? How can consumers correctly select and use appliances or household products if they cannot read the directions for use? This overview traces the history of papermaking, the development of the pen, the evolution of the pencil, and, last, the invention and development of the typewriter.

Paper

Writing requires something to write on and to write with. Papyrus was used in Egypt to make writing materials around 2500 B.C. Paper manufacture began in China by 105 A.D. and spread from there westward to Samarkand by 751. The Arabs began making paper at Baghdad by 793 and brought paper manufacture to Spain in 1150. The first English paper mill was established at Dutford in 1590.

Papermaking was a slow and tedious process from breaking the rags down to fibers to the creation of a single sheet of paper at a time. Until the late 18th century, when cloth manufacture was mechanized, all cloth was handmade and fabric was carefully conserved. Clothing was mended and patched until no longer wearable. Rags had many uses, including rugs, cleaning, and stuffing for pillows, and as candle wicks, so rags for papermaking were in short supply.[12]

Mechanization of papermaking began with a machine developed by the Dutch in 1750 to reduce rags to fibers and was furthered in 1798 when Nicolas Louis Robert of France developed a moving screen belt for continuous paper manufacture, making paper production more efficient and wallpaper more economical. Henry and Sealy Fourdrinier improved Robert's machine and patented the improved model in 1807. A cylinder paper machine was patented by John Dickinson, an English papermaker, in 1809.

In 1800 Moritz Illig, Germany, discovered that paper could be sized in a vat with rosin and alum, reducing both the time and the cost of paper sizing. Sizing was required to provide a smooth surface for printing or writing. Illig published his method in 1807, but it was not widely used until twenty-five years later.

In 1840 Friedrich Gottlob Keller invented wood pulp paper. His process, using ground wood pulp, reduced the reliance on rags and made modern, inexpensive paper possible. In 1852 soda pulp was made from wood in England, and in 1867 Tilghman patented a sulfite pulp in the U.S. In 1884 Carl F. Dahl, Germany, invented kraft pulp for making paper bags. Some modern householders recycle paper grocery bags for garbage and trash or even as wrapping paper for packages.

One consequence of the sulfite pulp process was the rapid deterioration of such paper because of the high residual acid content. This is why newspapers and some paperback books do not last and why libraries now have serious problems with brittle books that in time crumble to dust.

Pens

Writing instruments were developed early. Hollow reeds were used in Egypt and Rome, and by 50 B.C. the Romans were using sharpened goose quills. St. Isidore of Seville claimed quills were first used for pens in 635, which may indicate either a decline in use after the fall of Rome or the slow spread of innovations to Spain. In any event, quills were in common use in Europe by 1250. The first fountain pens were reported in 1656 by Dutch travelers from Paris. In 1663 Samuel Pepys used a "silver reservoir pen" presented to him by William Coventry.

Pen developments continued throughout the 18th century. The "pen without end" was invented by Jean F. Coulon de Thevenot, France, in 1740 and Johann Jansen, Aachen, introduced steel pens in 1748. Harrison, Birmingham, made cylindrical steel pens cut like quills in 1780. Bryan Donkin invented the two-piece steel pen nib in 1803, while Joseph Wise invented the metal nibbed "perpetual pen."

A series of further patents related to fountain pens were granted, including one to D. Hyde, Reading, Pennsylvania, for a continuous-flow fountain pen in 1830; William Baddeley invented a plunger-filled fountain pen in 1833, but the first practical fountain pen was developed in 1884 by Lewis E. Waterman. George Parker invented the lever-fill pen in 1904, and by the 1920s a disposable cartridge pen was available, followed by the stainless steel pen in 1926. M. Perraud of Jiff-Waterman developed an ink-cartridge in 1935.

The ballpoint pen was conceived of in the 19th century, but a practical model was not developed until shortly before World War II. The first ballpoint was patented by John L. Loud, U.S., in 1888 and, although manufactured, was not successful. Ladislas Biro invented the first successful ballpoint pen in 1938 and patented it in 1943. Eterpen Company, Buenos Aires, manufactured the first successful pens based on Biro's patents. Biro had some difficulty in the U.S. because of Loud's earlier patent, but outside the U.S. ball-point pens are known as *biros*. Today the ballpoint pen is the most common pen; fountain pens are used by only a few.

Felt marker pens and fiber pens were developed in the 1950s. Sydney Rosenthal, Richmond Hill, New York, applied for a patent on the Magic Marker pen in 1953. It was the first felt-tipped marker pen and originally intended for the art supply market, but was soon adapted for a variety of purposes. The first fiber-tipped pen, Pentel, appeared in 1960 marketed by the Japan Stationery Company, Tokyo.

Pencils

Pencils did not evolve as early as pens, and were first described by Konrad von Gesner in his treatise on fossils in 1565. By 1584 they were being made from graphite dug at Borrowdale, Cumberland. Casper Faber of Stein near Nuremberg, Germany, developed and sold the first Faber pencils in 1767, and Nicolas J. Conte, St. Cenari, Normandy, invented a pencil of pulverized graphite and potter's clay in 1795. The mechanical pencil was invented in 1915.

Tradition has it that Magalhaens, Portugal, invented the eraser after 1750; erasers were mentioned by Joseph Priestly in 1770. India rubber erasers were sold in 1770 in London, but the pencil with the attached eraser was not patented until 1858 by Hyman Lipman, Philadelphia.

Typewriters

The first recorded attempt to invent a typewriter is a British patent of 1714 for "an artificial Machine or method for the Impressing or Transcribing of Letters Singly or Progressively one after another, as in Writing, whereby all Writing Whatever may be Engrossed in Paper or Parchment so Neat and Exact as not to be distinguished from Print."[13] Another attempt was recorded in France in 1784 for a machine embossing characters for the blind. In the United States, a "typographer" was patented in 1829, and the Thurber Patent Printer was patented in 1849, but the first practical typewriter was patented in 1868 by American Christopher Latham Sholes. Sholes was inspired in 1867 by an account of a British machine in *Scientific American*. A contract was signed in 1873 with E. Remington and Sons, gunsmiths of Ilion, New York, and the first machine was marketed in 1874. The typewriter ribbon was patented in 1886, and a manifold copy ribbon was patented in 1888.

The first typewriter would only produce capital letters. A shift key was added in 1878. Also available at this time was a typewriter with a double set of keys, one for capitals and one for lowercase letters. With the development of touch typing, the shift key model with the more compact key set dominated.

A number of enhancements were added, including the development of visible writing, allowing the typist to see what was being typed, in 1883, and the front-stroke machine developed in 1890 from work by John N. Williams, U.S. Jacob Wortman's Visible Writing machine was patented in 1893 and sold to John Underwood, who established the Underwood Typewriter Company in 1895. The electric typewriter was invented by Thomas Edison in 1872 and later developed into the ticker-tape machine. The electric office typewriter was pioneered by James Smathers in 1920, and the IBM Selectric with a spherical typing element was introduced in 1961. Electronic memory typewriters were introduced in 1973.

Portable typewriters were developed in the late 19th century, but were heavy and slow typing. George Blickensderfer developed the first U.S. portable in 1892, but the first successful portable machine was marketed in 1909. By 1950 home use portables were being marketed, and in 1956 Smith-Corona announced their electric portable. By 1981 Brothers had introduced the low-cost dot matrix typewriter.

The tools of writing are inexpensive and make it possible to communicate with others, to make notes and lists, and to augment memory. A personal letter is still one of the cheapest forms of communication. Computers and word processing may further change this process.

Table 10.3. Writing Tools

Dates	Event
B.C.	
3000	Writing used in the Tigris-Euphrates Valley (WB).
2500-2000	Papyrus used in Egypt (TH).
300	Greeks and Egyptians used hollow calamus reed pens; they filled the core with ink and squeezed the reed to release ink (WB).
50	Sharpened goose quills used in Rome for pens (WB).
A.D.	
105	Paper manufacture began in China (EB).
635	Quills first used for pens according to St. Isidore of Seville (they had been in use in Rome, 50 B.C.) (DID).
751	Paper manufacture reached Samarkand, Central Asia (EB).

793	First paper made in Baghdad (EB).
1150	Paper manufactured by the Arabs in Spain (TH).
1238	Water-powered paper mill built in Xativa, Valencia, Spain (MM).
1250	Goose quill used for writing (TH).
1550	Sealing wax used for first time (TH).
1565	Konrad von Gesner, Zurich, described pencils in his treatise on fossils (1ST).
1584	Pencils made from graphite dug at Borrowdale, Cumberland (1ST).
1590	First English paper mill built at Dutford, England (TH).
1615	Envelope used in Geneva (WABI).
1656	First fountain pen reported from Paris by two Dutch travelers (1ST).
1657	Fountain pens manufactured in Paris (TH).
1663	Samuel Pepys used "silver reservoir pen" presented to him by William Coventry (1ST).
1748	Steel pens introduced by Johann Jansen, Aachen, Germany (1ST).
1750	Machine developed for reducing rags to fiber for papermaking, Holland (EB).
>1750	Magalhaens, Portugal, invented an eraser (WABI).
1767	Casper Faber of Stern, Germany, developed and marketed the first Faber pencils (AMERW).
1770	India rubber eraser sold in London (1ST).
	J. Priestly mentioned an eraser (WABI).
1780	Cylindrical steel pens cut like quills made by Harrison, Birmingham (DID).
1784	Embossing machine for the blind devised (EB).
1786	Ezekiel Reed, American inventor, built a nail-making machine (TH).
1795	Nicolas Jacques Conte, St. Cenari, France, invented a pencil of pulverized graphite and potters' clay (DID).
1798	Nicolas-Louis Robert, France, constructed a moving screen belt for continuous papermaking (EB).
1800	Moritz Frederick Illig, Germany, discovered paper could be sized in vats using rosin and alum (EB).
1803	Bryan Donkin invented the first two-piece steel pen nib (DID).
	Joseph Wise invented the metal-nibbed "perpetual pen" (DID).
1807	Illig published his method for vat sizing of paper (EB).

	Henry and Seely Fourdrinier developed an improved version of Robert's papermaking machine (EB).
1809	John Dickinson, English papermaker, patented the cylinder paper machine (EB).
	First two British patents were granted referring to the fountain pen (DID).
1819	John Schaffer invented the Penographic pen with quill-shaped brass nib (DID).
1820	Joseph Gisllot, Birmingham, invented a new steel pen nib (DID).
1823	Hawkins and Mordan patented the ever-pointed pencil (DID).
	Hawkins and Mordan invented pen nibs of horn and tortoise shell (DID).
1828	John Mitchell, Birmingham, introduced machine-made metal pens (EB).
1829	Typographer patented in the U.S. (EB).
1830	James Perry patented his steel slit pen (TH).
	D. Hyde, Reading, Pennsylvania, patented a continuous flow fountain pen (PA).
1833	William Baddeley invented a plunger-filled fountain pen (DID).
1840	Friedrich Gottlob Keller first used ground wood pulp for paper in Germany (EB).
1841	Maquet, France, began industrial manufacture of envelopes in Paris (WABI).
1843	Charles Thurber, Norwick, Connecticut, obtained U.S. patent 3228 on his Thurber Patent Printer, a typewriter (FFF).
1844	Wood-pulp paper process invented by Friedrich Gottlob Keller (TH).
1849	Machine for manufacturing evelopes patented by Jesse K. Park and Cornelius S. Watson, New York City (FFF).
1851	Richard H. Pease, Albany, New York, produced Christmas cards (FFF).
1852	Soda pulp for paper made from wood in England (EB).
1858	Pencil with attached eraser patented by Hyman Lipman, Phildelphia (1ST).
1863	Johann Faber, Stein, Germany, patented the screw-top ever-pointed pencil case (DID).
1867	U.S. patent issued to Tilghman for sulphite pulping (EB).
	Scientific American described a British typewriter (EB).

1868	Luther Sholes patented the first practical U.S. typewriter (EB).
>1870	Wealthy Americans exchanged Christmas cards (FORB).
1872	Thomas Edison developed the first electric typewriter (EB).
1874	Louis Prang, Roxbury, Massachusetts, sold Christmas cards in the United Kingdom (FFF).
1875	Louis Prang, Roxbury, Massachusetts, sold Christmas cards in the United States (FFF).
1883	"Visible writing" typewriter developed (EB).
1884	Lewis E. Waterman developed a practical fountain pen (PA).
	Carl F. Dahl invented kraft pulp in Danzig (EB).
1886	George K. Anderson, Memphis, Tennessee, patented the typewriter ribbon (patent 349026) (FFF).
1888	John L. Loud, U.S., patented and manufactured a ballpoint pen, but the pen was not successful (DID).
	Dr. William Champion Deming, New York City, devised the "hatching cradle," an incubator for infants (FFF).
1890	Front stroke typewriter developed by John N. Williams (EB).
1892	First U.S. portable typewriter patented by George Blickensderfer, Stamford, Connecticut (patent 472,692) (FFF).
1893	Visible writing typewriter developed by Herman L. Wagner, Brooklyn, New York (patent 497560), later sold to John T. Underwood (FFF).
1895	Underwood Typewriter Company founded in New York (FFF).
1904	George Parker invented the lever-fill fountain pen (TT).
1906	Jack Sapirstein established American Greeting Co. (FORB).
1909	First successful portable typewriter introduced (EB).
1910	Joyce Hall established Hallmark, greeting card manufacturer of Kansas City (FORB).
1915	Mechanical pencil invented (TT).
>1920	Disposable cartridge pens available (WB).
	Electric office typewriter marketed (EB).
1924	Spiral binding, Spirex, invented by Staale (TT).
1926	Stainless steel pens introduced (EB).
1935	M. Perraud, Jif-Waterman, invented the ink cartridge (WABI).
1937	Scotch tape developed by Richard Drew, 3M Company (TT).

1938	Ladislas Biro invented a ballpoint pen, the first successful such pen (TH).
1943	Ladislas Biro patented his ballpoint pen (1ST).
1945	First successful commercially produced ballpoint pens produced by Eterpen Company, Buenos Aires (1ST).
1950	Home use portable typewriters available (RB).
1953	Sidney Rosenthal applied for a patent on Magic Marker, a felt-tipped marker pen (FORB).
1956	Smith-Corona announced an electric portable typewriter (EB).
1960	First fiber-tipped pen, Pentel, marketed by Japan Stationery Company, Tokyo (1ST).
	IBM Selectric typewriter introduced (EB).
1973	Electronic memory typewriter introduced (EB).
1981	Brother low-cost dot matrix typewriter introduced.

Clocks

Timekeeping devices have intrigued inventors, who have devised a variety of ways to mark time, including water clocks, candles, and hourglasses. The Egyptians were using sundials by 1400 B.C., but they also used water clocks since sundials were only useful during the day. The Greeks and Romans used water clocks and sand hourglasses, which appeared during the 1st century A.D. Calibrated candles were in use by 870 A.D.

Time measurement during the Middle Ages was related to prayer and religious observances. Early clocks divided the daylight into equal periods, but since a winter day was shorter than a summer day, winter day hours were shorter than summer day hours. Water clocks were reset every day.[14]

The Chinese developed the first mechanical clocks, but it was the Europeans who refined and developed them. The Chinese invented the first clock with an escapement in 724 A.D. and the first mechanical clock in 1088. The first European mechanical clock was made by the Austin canons, 1283, and set up in Dunstable Priory, Bedfordshire. The first equal hours clock (each hour, regardless of the length of daylight, was of equal length) was developed for Saint Gothard Church, Milan, in 1335. The first alarm clock appeared in Wurzburg, Germany, between 1350 and 1380 and was used to help the monks maintain religious observances. The first modern alarm clock was made in 1787 by Levi Hutchins of Concord, New Hampshire.

Other improvements include traveling clocks and pendulums. By 1585 Bartholomew Newsam had constructed the first English traveling and standing clocks, and by 1631 the London Clock Makers Company was incorporated. Pendulum clocks were made by 1656, and the famous Black Forest clocks were being made by 1738.

Clocks came in all sizes. The forerunner of the watch, a "pocket clock," was mentioned in a 1462 letter to the Marchese di Manta. Invention of the watch is credited to Peter Henlein around 1504 or 1509. Wrist watches were itemized in the accounts of Jacquet-Droz & Leschot, watchmakers, Geneva, in 1790. More recently, the self-winding watch was patented in 1923. The low-cost Timex was introduced in 1946; Timex made the watch an easily purchased item for many people. By 1957 Hamilton had introduced the electric watch, followed in 1960 by Bulova's electronic watch, Accutron. With the advent of the digital watch it is now often cheaper to buy a new watch than to replace the batteries, which is exactly what advertisements recommended for the Swatch watch, marketed in 1982.[15]

Electricity was applied to clocks in the 19th century and by the 20th century fully electric models were being marketed. By 1839 Carl August Steinhall had invented a battery-powered electric clock, and electric clocks were battery-driven until the development of electric power generating and distribution systems, which became available along with electric lighting in the late 19th century. The application of synchronous electric motors to clocks in 1918 made current driven electric models practical and they soon became broadly available. The first electric clock with a cathode ray display and with no moving parts appeared in 1965.

Clocks and timers have been important for their use in other machines and appliances, particularly in the stove, the programmed washer, and the microwave. Timers allow us to turn lights and heating systems on and off, to start our morning coffee, to wake up to music, or to cook our food while we are working or away from home.

Timers and clocks help us to mark the passage of time and to regulate various processes. We cannot control time, but we can use these devices to help us use it more effectively.

Table 10.4. Clocks

Dates	Event
B.C.	
700-600	Assyrian water clock used (TH).
200-150	Gears and ox-driven water wheel used (TH).
A.D.	
724	Chinese invented the first clock with an escapement (EB).
870	Calibrated candles used to measure time (TH).
1088	Water-driven mechanical clock devised in Peking (1ST).
1283	First European mechanical clock made by the Austin canons and set up in Dunstable Priory, Bedfordshire (1ST).

1335	First recorded equal hours clock at Saint Gothard, Milan (MM).
1350-80	Earliest alarm clock at Warzburg, Germany, used to help monks track religious observances (RMRB).
1462	"Pocket clock" mentioned in a letter to the Marchese di Manta (1ST).
1504-9	Peter Henlien devised the watch (EB).
1585	Bartholomew Newsam constructed the first English traveling and standing clocks (TH).
1631	London Clock Makers' Company incorporated (TH).
1656	Christian Huygens made the first pendulum clock (1ST).
1704	Jeweled watch movement patented by Facio de Duillier and Peter Debaufre (1ST).
1738	First clocks made in the Black Forest district (TH).
1787	Levi Hutchins, Concord, New Hampshire, developed the modern alarm clock (RMRB).
1790	Wristwatch itemized in the accounts of Swiss watchmakers Jacquet-Droz and Leschot, Geneva (1ST).
1839	Carl August Steinheil built the first electric clock (TH).
1840	Sir Charles Wheatstone devised an electric clock (DID).
1847	Alexander Bain, Edinburgh, invented an electric clock (DID).
1923	Self-winding watch patented in Switzerland by John Harwood (1ST).
1928	Self-winding watch manufactured by Harwood Self-Winding Watch Company, London (1ST).
1946	Timex watch introduced (TT).
1957	Hamilton Watch Company, Lancaster, Pennsylvania, introduced the electric watch (FFF).
1960	Bulova Watch Company introduced the Accutron, an electronic watch using a precision tuning fork (FFF).
1965	First electric clock with no moving parts using a cathode ray display introduced (TT).
1982	Disposable fashion watch, Swatch, introduced (WABI).

Computers

Tools enable humans to overcome their limitations of strength, endurance, or skill. One tool, the computer, helps with some mental limitations. Jacquard's punched card loom had some elements of a computer in its control of the weaving of complex patterns, but credit for the idea of the computer is generally accorded to Charles Babbage. He designed one in 1835 but never succeeded in building his "analytical engine." The first analog computer was built by Vannevar Bush in 1930. The first digital computer, the Mark I, was completed in 1944 by Howard Aiken, Harvard. The first digital computer built with vacuum tubes was the ENIAC built at the University of Pennsylvania in 1946. In the 1970s Victor Anatasoff successfully contested the patents developed from ENIAC, but by then the patents had expired.

Computers were being commercially manufactured by 1951, and by 1954 the first business computer was in use. The transistor, developed by Bell Laboratories in 1948, led to smaller, more efficient machines, requiring less energy and cooling, and ultimately to printed circuits, microchips, and the computer on a chip crucial to the complex functions required for the incorporation of computers into home appliances.

Digital Equipment Corporation developed the minicomputer in 1962. The first model, LINC (*L*aboratory *In*strument *C*omputer), was really a desk unit with many of the features of the modern personal computer. In 1969 Alan Kay and Ed Cheadle, U.S., designed the FLEX, another desk top machine with graphics capability but, like the LINC, expensive. These machines were followed in 1971 by John Blankenbaker's Kenbak I, a personal computer, but only a few were sold. Troung Trong Tri of France invented the MICRAL, a French microcomputer, in 1973 but, like Blankenbaker, he went bankrupt. IBM also developed an early personal computer, the SCAMP, a portable in 1973, which never went beyond the prototype stage.

The first mass-produced personal computer, the Altair, was sold in 1975, and the first Apple computer, Apple I, was shipped in 1976. The Apple II was introduced in 1977 and improved versions are still sold. By 1977 Commodore and Tandy were offering the PET and the TRS 80 Model 1, respectively. The Osborne 1 and the IBM PC were offered in 1981. Osborne was the first personal computer offered with software and all necessary features including two disc drives, an RS 232 serial interface and an IEEE 408 parallel interface for peripherals, and an integral seven-inch monitor. The Osborne was also the first mass-marketed "portable" computer. Unfortunately, while the computer was popular and sold quickly, the firm ran into financial and managerial difficulties and by 1984 had declared bankruptcy. By 1985 it had emerged from Chapter 11, but it was no longer a major force in the personal computer market.

IBM's entry spelled difficulties for a number of smaller manufacturers and the shake-out began in earnest in 1984. Tandy saw reduced profits. Texas Instruments abandoned its home computer, the TI-99. Coleco did the same with its home computer, the Adam, in 1985. Profits for almost all companies, except IBM, declined. IBM introduced the XT in 1983 and the AT and (for home use) the PC Jr in 1984. PC-DOS replaced CPM as the operating system of choice. The AT put the power of an IBM 360 main frame computer on the desk top. IBM announced the withdrawal of the PC Jr in 1985, causing some to question the viability of home computers.

Apple, after trouble with the Apple III and the initial slow acceptance of the Lisa, built on concepts first seen in the earlier Xerox Star, introduced the Macintosh in 1984. The Fat Mac followed in late 1984, and the "open" Mac with IBM compatibility and open board slots for greater adaptability was introduced in 1987, the same year Compaq introduced its 386-based personal computer and IBM announced its PS/2 series, which replaced the PC, XT, and AT models.

The Macintosh provided easy access to superior graphic capabilities and incorporated a "mouse," reducing reliance on keyboard entry for all commands. The ability to manipulate individual bits of a screen display provided a sophisticated drawing capability. Coupled with a word-processing facility, it made the production of forms, graphic aids, newsletters, invitations, and a host of printed items truly child's play. All that was lacking was standard software packages and color.

By 1985 Lotus had announced JAZZ, its version of its popular spreadsheet package 1-2-3 for Mac, and other manufacturers were also offering new or modified software. Apple had also announced a superior laser printer, but, unfortunately, it was priced beyond the reach of most home users. Atari and Commodore announced color machines similar to the Macintosh, but less expensive in 1985. It is too early to tell whether they will succeed.

AT&T also entered the personal computer market in 1984. Its initial offering was competitive with the IBM PC. AT&T's strategy is still evolving, but early signs indicate a potential coupling of telephones and computers.

Computers are being incorporated in stoves, microwaves, refrigerators, washers, and even smaller appliances. Home security systems and heat control systems also use microcomputers, and fast-food restaurants are also using them.[16]

With bar code readers, a running inventory of all the foods in the home cupboards or refrigerators could be maintained. The computer could track the number of items, and in conjunction with the week's menu, could print a grocery list. With access to prices in various stores, it could even determine where to buy and transmit the order to the store.

Such a system is not here yet, but most of the equipment and even the software is available. With a little ingenuity it could be built. The system would have to make the entry and recording of information automatic. Householders are unlikely to be willing to key in product codes. Too, it would require stores willing to deliver merchandise. Technology makes it possible.[17]

Table 10.5. Computers

Dates	*Event*
A.D.	
1835	Charles Babbage designed his analytical engine (EB).
1855	First practical programmed computer built by George Scheutz, Stockholm, and exhibited in Paris (1ST).
1889	Punched card system created by Herman Hollerith (TH).
1930	Vannevar Bush built the first analog computer (EB).
1942	First electronic brain or automatic computer developed in the U.S. (TH).
1944	Mark I computer completed by Howard Aiken, Harvard (EB).
1946	Electronic brain (computer) built at University of Pennsylvania (TH).
1948	Transistor developed by John Bardeen and Walter Brattain, Bell Laboratories, Murray Hill, New Jersey (TH).
1951	First commercial application of transistors (1ST). First commercially manufactured computer, UNIVAC I, installed at U.S. Census Bureau, Philadelphia (1ST).
1954	First business use computer, Lyons Electronic Office, LEO, used at J. Lyons & Company, Cadby Hall, London (1ST).
1962	Digital Computer Corporation produced LINC, Laboratory Instrument Computer, the first desktop computer (PCW).
1969	Alan Kay and Ed Cheadle developed FLEX, an early, expensive graphics desktop computer (PCW).
1973	Micral, French microcomputer, marketed (WJ). IBM engineers developed the first portable computer, the SCAMP, but it was not produced commercially (PCW).
1975	Altair, first mass-produced personal computer, offered for sale (CDIS).
1976	Apple I microcomputer developed by Apple Computer Co. (CDIS).
1977	Commodore offered PET; Tandy offered Radio Shack TRS-80 microcomputers (CDIS).
1981	Osborne I and IBM PC microcomputer marketed (CDIS).
1983	IBM XT and Apple LISA microcomputers marketed (CDIS).
1984	Apple Macintosh, Fat Mac, IBM AT, AT&T PC, and Epson PX-8 personal computers debuted (CDIS).

1985   IBM announced the end of the IBM PC Jr. while Apple did
        the same for Lisa personal computers (WJ).
1987   Compaq announced the 80386-based computer (WJ).
        Apple introduced the Open Mac with open board slots for
        greater flexibility and IBM compatability (WJ).

11
Health and Children

Important to our happiness and well-being has been our health and our families. This chapter highlights some of the developments in certain aspects of health care and the care and raising of children. Popular books that were useful in providing background information include *The World Almanac Book of Inventions*,[1] and the *Encyclopaedia Britannica*.[2]

Health Care

Home remedies for simple ailments have been used throughout history. Some home remedies used in Appalachia are described in *The Foxfire Book*.[3] For much of human history, the housewife has treated minor ailments and provided first aid. She had to be knowledgeable in the use of herbs and home remedies. A description of the origin of a few of these common modern medications and aids found in many households follows.

Household Medication

One of the older home treatments, Epsom salts (hydrated magnesium sulfate), is still a standby in many modern medicine cabinets. Epsom Salts are used as a relaxing bath aid for treating bruises and sprains and taken as a drink for laxative purposes. Mineral springs at Epsom, Surrey, were discovered in 1618, and Epsom salts were first prepared from the springs by Nehemiah Grew in 1695.

A number of tonics and sedatives were marketed in the late 19th century. Several soft drinks were originally marketed as tonics. Hires Root Beer (1870) was marketed as Hires Herb Tea, a drug extract, and both Moxie (1895) and Coca-Cola (1886) were first sold as nerve tonics. Miles Nervine, a sedative, was introduced in 1882 by Miles Laboratory, which also later introduced Alka-Seltzer.

Effective headache powders and medication are mainly a 20th-century phenomenon. While home remedies and some opiates were used earlier, side effects and dependency concerned doctors and users. Willow bark tea was an early remedy for headache and was later found to contain salicylic aldehyde, from which salicylic acid and eventually acetylsalicylic acid, the active ingredient in aspirin, were derived. In 1829 a French chemist, Henri Leroux, extracted salicyn from willow bark, and Charles Gerhardt, University of Montpellier in France, first derived acetylsalicylic acid in 1853, but made no use of it. A Bayer chemist, Felix Hoffman, developed aspirin in 1893, and aspirin in powder form was introduced commercially in Germany by Bayer in 1899. By 1915 Bayer was marketing aspirin in tablet form. Disprin, soluble aspirin, was introduced by Rickett and Colman in 1939, while Roche Laboratories introduced the tranquilizer Valium in 1963.

Aspirin did have some side effects for some users, and is known to irritate the stomach lining. Efforts to develop substitute pain killers continued. McNeil Consumer Products began selling Tylenol, acetaminophen, in 1955. The company was purchased in 1959 by Johnson and Johnson, and by 1960 the FDA had approved nonprescription sale of Tylenol, but it was 1975 before Johnson and Johnson sold it generally.

In 1982 packages of Tylenol capsules were removed from store shelves after several were found to have been contaminated with cyanide. Several people died, and the resultant public outcry led to tamper-proof packaging for Tylenol as well as other nonprescription drugs. A reoccurrence of tampering in 1986 caused Johnson and Johnson to withdraw capsules and to introduce the caplet, an elongated tablet. The tamperings highlighted the vulnerability of consumers and manufacturers and induced many manufacturers to investigate new types of packaging. While improvements have been made, no completely effective solution has been found.[4]

Two health care products use petroleum jelly. Robert Cheesebrough was inspired to develop Vaseline Petroleum Jelly when he observed oil-rig workers using oil residue to treat cuts and burns. He extracted a white jellylike substance from petroleum and marketed it as a healing balm. It has been used for treating diaper rash, burns and as a skin softener. It was used with cucumbers as a skin cream by women around the turn of the century. Vick's Vaporub, originally Vick's Magic Croup Salve, was introduced in 1905 and used to treat chest congestion, colds, and croup. Vick's Vaporub consists of menthol, camphor, turpentine spirits, and eucalyptus oil in a petroleum jelly base and is applied externally or used in steam vaporizers.

Two antacids, Alka-Seltzer and milk of magnesia, are kept in many medicine cabinets. Alka-Seltzer was introduced in 1931 and further popularized during the 1960s by the advertisement line "What shape is your tummy in?" and marketed as a relief aid for overeating. It is an effervescent tablet containing bicarbonate of soda, phosphate, aspirin, and citric acid. Phillips' milk of magnesia was introduced in 1873 by Charles Phillips of Glenbrook, Connecticut. The basic ingredients of milk of magnesia were extracted from dolomite and sea water in 1936. Milk of magnesia was also marketed as a laxative.

Another laxative, Ex-Lax, was first marketed in 1905. Originally named "Bo-Bo" by Max Kiss, its developer, it consisted of phenolphthalein combined with chocolate and was considered by many an improvement over other, less pleasant tasting laxatives.

Feminine personal hygiene products were first introduced commercially in 1921 when Kimberly-Clark introduced the Kotex sanitary napkin. Shortages of cotton and wool during World War I led to the development in 1918 of wood cellulose, celucotton, for bandages and dressings. Nurses adapted these dressings as sanitary napkins, and Kimberly-Clark, the maker of the dressings, began marketing them commercially in 1921. Earle C. Haas, a physician of Denver, Colorado, adapted surgical tampons for the same purpose and patented the tampon in 1936, marketing them commercially that year.

Tampons, however, were not new. Tamponlike devices consisting of rolled lint and wood covered with absorbent materials were described by Hippocrates, and the Egyptians, Babylonians, and Assyrians used two different types of tampons--the affluent used softened papyrus and the poor used softened water reeds. Soft, high-grade wool, specially carded, combed, and rolled, was used by Byzantine women. Roman women used greased or waxed wool tampons. Active women in countries where clothing is light usually chose tampons, while those engaged in more sedentary activities and in countries where clothing is more voluminous used external pads.[5]

Toxic shock syndrome, associated with tampon use, came to public attention in 1980. It had first been identified by James Todd, a U.S. pediatrician in a 1978 *Lancet* article on an infection in children aged 8 to 17 years. It was not linked to tampon use until 1980. According to Edward H. Kass, the polyester foam and polyacrylate rayon used in certain tampons, primarily Rely, absorb magnesium from the body and promote the growth of bacteria toxin, thereby causing toxic shock. Rely was withdrawn from the market in September 1980. Other manufacturers made changes to their products and added a warning to their packaging.[6]

The first adhesive bandage was developed in 1882 by Paul Beiersdorf, a German pharmacist. He patented a manufacturing method for a plaster-coated bandage, called Hansaplast. In 1920 Earle Dickson, a cotton buyer for Johnson and Johnson, devised a temporary dressing for his wife using surgical tape, bandage, and crinoline. To use the dressing, the crinoline was peeled off. Dickson's invention was marketed by Johnson and Johnson as the Band-aid, the first self-adhesive bandage, in 1921.

Q-Tips were first devised in 1926 by Leo Gerstenzay as a cleaning aid for infants. He was inspired in 1922 by watching his wife use a cotton wad on a toothpick for cleaning their baby daughter. He called his invention Baby Gays, but later renamed it Q-Tips.

Thermometers are found in many home medicine kits. Galileo devised the first thermometer in 1593, but it was not accurate. An improved thermometer using alcohol in a sealed glass tube was devised by Ferdinand II, duke of Tuscany, in 1641. The modern mercury thermometer was devised by Gabriel Daniel Fahrenheit of Germany in 1714. He also invented the Fahrenheit temperature scale. Anders Celsius, Uppsala, Sweden, devised the Celsius or centigrade temperature scale in 1742.

The thermometer was first described for medical purposes by the Italian physician Santorio in 1626 using a water thermometer, but it was not particularly useful because of its inaccuracy. Jean Rey, a French physician, independently developed a liquid thermometer in 1632. The modern clinical thermometer was devised by English physician Allbutt in 1867.

Care outside the Home

Medical care outside the home includes a broad range of activities, services, and products. This section, however, is limited to eyeglasses, hearing aids, teeth, and cosmetic surgery.

Early man faced a dimming old age. Eyeglasses were not invented until the latter part of the 13th century, and bifocals did not appear until the 18th century. Spectacles were first mentioned in 1289 by Sandro di Popozo in his *Traite de la amille* as having been recently invented. The *Dictionary of Inventions* attributes the invention of eyeglasses to Italian monk Nicolas Bullet in 1285. The *World Almanac Book of Inventions* attributes it to Salvino degli Armati of Florence who developed a means for magnifying objects in 1280. The *Encyclopaedia Britannica* cites Alessandro di Spina. Marco Polo indicated that glasses were in use in China by 1275.

Regardless of their source, eyeglasses were welcomed and further improvements were made. The first concave lenses for nearsighted people were in use by 1517 when Raphael painted Pope Leo X with such a pair. Eyeglass frames fitting over nose and ears were invented in 1746 by French optician Thomin.

Bifocals were invented in the 18th century and improved in the late 19th and early 20th centuries. Bifocals were first suggested in 1716, but their invention is attributed to Benjamin Franklin in 1784 according to the *Encyclopaedia Britannica*. They were improved in 1884 when cemented bifocals were invented, in 1889 when Borsch developed an improved means of welding the bifocal lenses together, in 1908 with fused bifocals, and in 1910 when Bertram and Emerson developed one-piece bifocals. Cemented lenses are fastened together with glue, whereas in fused lenses, the lenses are fastened together under heat and pressure, causing the material at the join to melt and bond.

Developments affecting eyeglasses in the 20th century include lightweight, unbreakable organic glass, the French Varilux lens in 1959, and the Omnifocal lens developed in 1962. The latter two lenses eliminated the discontinuity in bifocal lenses caused by joining lenses of two different focal lengths. Dr. Stookey of Corning developed polychromatic glasses in 1964 that darkened or lightened depending on the ambient light.

Contact lenses were developed by A. E. Fick in 1887, and methylmethacrylate was used in such lenses in 1938. Small lenses covering only the cornea appeared in 1950, followed by the bifocal contact lens in 1958 and the new permeable plastic soft contact lens in 1965.

In 1986 two new lenses were developed for bifocal wearers, the Alges and the Hydrocurve II. The Alges lens, developed by University Optical Products of Florida, is designed in a bull's-eye pattern with the reading lens in the center of the lens and the distance lens on the periphery. The Barnes-Hind Hydrocurve lens has aspheric curves with a flatish center for distance and stronger magnification at the periphery. Both are soft lenses. Cosmetic lenses are now available in fashion colors; the wearer can be blue-eyed one day and brown-eyed the next.

Hearing aids were developed somewhat later than eyeglasses. Ear trumpets using the principle of bone conduction were in use by the 17th century. In the late 18th century, the audiphone or dentiphone, a cardboard or celluloid device shaped like a fan, was invented. The edge was held in the mouth and the fan was bent toward the sound.

The development of modern electronic hearing aids began with the work of Alexander Graham Bell; his invention of the telephone was an outcome of his experiments to help deaf children. The first patent for a hearing aid was registered in 1880 by Francis D. Clarke and M. G. Foster using bone conduction. The first electronic hearing aid was developed in 1901 by Miller Reese Hutchinson, but was large and inconvenient to use. It was marketed by Acousticon. In 1923, the Marconi Company launched the Otophone, which used an electronic valve and weighed sixteen pounds. A smaller model, weighing about four pounds, was developed in the 1930s. A. Edwin Stevens produced a small, practical, battery-operated unit, the 2 1/2 pound Amplivox, in 1935. By 1952 Sonotone Corporation was manufacturing and selling a 3 1/2 ounce transistorized hearing aid. The transistorized hearing aid led to the incorporation of hearing devices in eyeglass frames. By 1985 a digital hearing aid had been patented that used two microphone pickups and tailored its output to the wearer's particular hearing loss.

Dental prostheses have a long history, with crowns and fillings dating at least to Etruscan times. An interesting source on dental history is *Dentistry: An Illustrated History* by Melvin E. Ring.[7] Gold crowns were first used by the Etruscans around 600 B.C. The modern crown was devised by Pierre Fauchard in the 18th century and popularized in the U.S. by W. H. Dwinelle in 1856. Teeth have been filled with many types of materials, but gold has long been favored. Fillings were done by the Etruscans in gold, and solid gold leaf fillings were discussed by Arculanus in 1484. Amalgam fillings of mercury and silver were used in 1819 by Bell and by 1850 cast metal fillings were used. Gutta-percha was used for fillings in 1847. By 1854 gold leaf was replaced by spongial gold and then with adhesive gold in 1855. Vulcanite was used in 1915 for molding teeth fillings.

The making of artificial teeth was described in 1598 by Jacques Guillemeau, France, using wax, powdered coral, and pearls. Guillemeau is erroneously sometimes listed as having invented this method, but it was in use for some time when he published his description. Alexis Duchâteau, a French pharmacist, attempted to make dentures from a mineral paste in 1770, and in 1778 Dubois des Chemant of Paris produced durable dentures using impressions from plaster casts.

Unfortunately for George Washington, progress was slow; he used a pair of ivory false teeth that discolored readily and also picked up and retained food odors and taste. To reduce the effects of the latter, Washington soaked his dentures in port, but this further aggravated the staining. Porcelain teeth, less porous than ivory and consequently less retentive of odors and stains, were produced in the U.S. in 1817. By 1906 cast gold, embossed steel, and porcelain were all used for dentures. Synthetic resins were first used in Germany in 1935, and by 1979 Dr. Sozio of the U.S. had developed a solid ceramic that did not require reinforced metallic support.

Modern cosmetic surgery can be traced to the work of Dr. Roe, U.S., who described cosmetic rhinoplasty (modifying the shape of the nose for cosmetic reasons) in 1887 and in another paper published in 1891. Dr. Suzanne Noel began performing face lifts in 1925. Improved surgical techniques were developed using the superficial muscula-aponevrotic system of the face demonstrated by French surgeon Vladimir Mitz in 1973.

Breast reduction and enlargement techniques were also developed. Hyppolyte Morestin, France, performed breast surgery to remove a tumor in 1903 and described a method for mammary hypertrophy in 1905 and again in

1907. Contributions to mammary reduction were made by H. Biesenberger in 1928 and by E. Schwarzmann in 1930. In 1962 sacs of silicone rubber gel replaced sponge pads for breast implants.

Table 11.1. Health Aids

Dates	Event
B.C.	
600	Etruscans developed gold crowns and bridges for teeth (EB).
A.D.	
1268	Roger Bacon proposed spectacles (DID).
1275	Marco Polo reported that the Chinese wore eyeglasses (WB).
1280	Salvine degli Armati, Florentine physicist, devised a means for magnifying objects (WABI).
1285	Spectacles invented by the Italian monk Nicolas Bullet (DID).
1289	Sandro di Popozo's *Traite de conduite de la famille* mentioned spectacles (1ST).
1352	Hugh of Provence painted by Tommasco da Moderna wearing eyeglasses (EB).
1484	Arculanus discussed the use of solid gold leaf fillings (WABI).
1517	Raphael portrait of Pope Leo X depicted concave eyeglass lens for nearsightedness (EB).
1593	Galileo devised the thermoscope, the first thermometer (EB).
1598	Jacques Guillemeau, France, described use of wax, powdered coral, and pearls for artificial teeth (used primarily as a filling) (WBW).
c. 1600	Ear trumpet used to enhance hearing (WB).
1618	Mineral spring discovered at Epsom, Surrey (DID).
1625	Italian physician Santorio used a water thermometer for medical purposes (WABI).
1632	Jean Rey, French physician, independently devised a liquid thermometer (EB).
1641	Ferdinand II, Grand Duke of Tuscany, developed the sealed glass alcohol thermometer (EB).
1695	Epsom salts first prepared from Epsom springs by Nehemiah Grew (DID).
1714	Gabriel Daniel Fahrenheit devised the mercury Fahrenheit thermometer (EB).

1716	Bifocals first suggested (EDB).
c. 1728	Pierre Fauchard, France, author of *Le chirurgien dentiste*, devised the dental crown (WABI).
1742	Anders Celsius devised the Celsius or centigrade temperature scale (EB).
1746	Modern type of eyeglass frame invented by French optician, Thomin (WABI).
1770	A. Duchateau, a French pharmacist, attempted to make dentures from mineral paste (WABI).
>1775	Audiophone or dentiphone hearing aid used (WB).
1778	Dubois des Chement, Paris, produced durable dentures using plaster impressions (WABI).
1784	Benjamin Franklin invented bifocals (EB).
1790	John Greenwood invented the dental drill (HS).
1817	Porcelain teeth produced in the U.S. (WABI).
1819	Amalgam fillings of mercury and silver were used for dental fillings (WABI).
1829	Henri Leroux, French chemist, extracted salicyn from willow bark, a first step toward aspirin (WABI).
1847	Gutta-percha used as a dental filling (WABI).
1850	Cast metal fillings used for teeth (WABI).
1853	Charles Gerhardt, University of Montpellier, France, first derived acetylsalicylic acid (aspirin) (WABI).
1854	Spongial gold replaced gold leaf filling for teeth (WABI).
1855	Adhesive gold used by Robert Arthur, U.S., in dental fillings (WABI).
1856	W. H. Dwinelle popularized the dental crown in the U.S. (WABI).
1859	Robert Cheesebrough discovered oil-rig workers used oil residue to treat cuts and burns (Vaseline) (CDIS).
1867	English physician Thomas Allbutt devised the modern glass medical thermometer (WABI).
1872	Alexander Graham Bell began experiments to help deaf that led to the development of the telephone (WB).
1873	Charles Phillip, Glenbrook, Connecticut, devised and marketed Phillips' milk of magnesia (EOOET).
1879	Richard S. Rhodes, Riverpark, Illinois, patented the Audiophone in the U.S. (patent 219828) (FFF).
1880	First hearing aid patent was registered to Francis D. Clarke and M. G. Foster (DBE).
1882	Miles Nervine, a sedative, marketed (WABI).
	Paul Beirsdorf, Hamburg, Germany, devised an adhesive bandage, Hansaplast (WABI).

1884	Cemented bifocals devised (EB).
1887	First contact lens developed by A. E. Fick (DID).
	Dr. John Roe, U.S., performed cosmetic rhinoplasty (WABI).
1889	Borsch developed an improved means of welding bifocal lenses together (WABI).
1893	Felix Hoffman, Bayer chemist, developed aspirin (WABI).
1899	Aspirin in powder form introduced commercially by Bayer AG, Leverkusen, Germany (1ST).
	J. L. Borsch invented a more economical procedure for welding the two lenses of bifocal glasses (WABI).
1901	Electronic hearing aid first devised by Miller R. Hutchinson, but was bulky and difficult to use (WABI).
1905	Vick's Magic Croup Salve, later renamed VapoRub, introduced by Vick Chemical, North Carolina (TT).
	Max Kiss, New York, introduced Ex-Lax, a chocolate-based laxative (EOOET).
	Hyppolyte Morestin, France, performed mammary hypertrophy (WABI).
1906	Gold, porcelain, and embossed steel used for dentures in the U.S. (EB).
1908	Fused bifocals introduced (EB).
1910	One piece bifocals developed (DID).
	Bentron and Emerson of the Carl Zeiss Company developed one piece bifocals (WABI).
1915	Volcanite used for molding teeth fillings (TT).
	Aspirin tablets first retailed by Bayer (1ST).
1918	Wood cellulose, Celucotton, used as a substitute for cotton for bandages and dressings (TT).
1920	Earle Dickson, cotton buyer for Johnson and Johnson, devised a temporary dressing, later marketed as the Band-Aid (CDIS).
1921	Johnson and Johnson introduced the Band-Aid, first stick-on bandage (TT).
	Kimberly-Clark marketed Kotex sanitary napkins (WABI).
1922	Leo Gerstenzang began to develop the idea for Q-Tips (CDIS).
1923	Marconi Co. launched an electronic valve operated hearing aid, the Otophone, which weighed 16 pounds (DBE).
1924	Cosmetic mammoplasty available (TT).
1925	Dr. Suzanne Noel began performing cosmetic face lifts (WABI).

1926	Leo Gerstenzang began marketing Baby Gays, later renamed Q-Tips, for cleaning baby's ears, nose, etc. (CDIS).
1928	H. Biesenberger, Vienna, contributed to techniques for mammary reduction plasty (WABI).
	Dr. George Nicola Papanicolaou, U.S., devised the Pap smear test for cervical cancer, but the test was not fully accepted until 1941 (HS).
1930	E. Schwarzmann, Vienna, contributed to techniques for mammary reduction plasty (WABI).
>1930	Vacuum-tube hearing aid that weighed four pounds developed (WB).
1931	Miles Laboratories, Indiana, introduced Alka-Seltzer (TT).
1935	A. Edwin Stevens produced the Amplivox, the first practical hearing aid at two and one-half pounds (DBE).
	Synthetic resins used for false teeth in Germany (EB).
1936	British scientists extracted the basic ingredients of milk of magnesia from dolomite and sea water (TT).
	Commercial tampons available (EBM).
1937	Earl Hass patented tampon and founded Tampax Co. (patent filed and tampons sold in 1936) (WABI).
1938	Contact lens of methylmethacrylate developed (DID).
1939	Disprin, soluble aspirin, introduced by Rickett and Colman (TT).
1950	Small contact lens covering only cornea developed (DID).
1952	Harold Ridley, English physicist, invented a polymethylmethacrylate lens for implantation in the eye (WABI).
	Sonotone Corp. manufactured and sold a three and one-half ounce transistorized hearing aid (EDB).
1953	Transistorized hearing aid introduced (WB).
1955	McNeil Consumer Products first marketed Tylenol pain reliever (NYT).
	Lightweight, unbreakable organic lenses for glasses made from a thermo-hardening material (WABI).
1958	First bifocal contact lens developed (TT).
1959	Johnson and Johnson acquired McNeil and rights to Tylenol pain reliever (NYT).
	French Society of Opticians invented the Varilux lens for replacing bifocals and trifocals (WABI).
1960	U.S. Food and Drug Administration approved Tylenol pain reliever for sale without a prescription (NYT).

1962	Sacs of silicone rubber gel replaced sponge pads for breast implants (TT).
	Omnifocal lens eliminating bifocal line developed (TT).
1963	Roche Labs introduced Valium (TT).
1964	Photochromatic glasses invented by Dr. Stookey, Corning Glass, New York, darken on exposure to sunlight (TT).
1965	New permeable plastic for soft-contact lens developed (TT).
1973	French surgeon Valdimir Mitz demonstrated superificial musculo-aponevrotic system for face lifting (WABI).
1975	Johnson and Johnson sold Tylenol pain reliever to the general public (NYT).
1976	Weinstein, U.S., developed the disposable oral thermometer (WABI).
1978	Toxic shock syndrome associated with tampon use identified (EBM).
1979	Dr. Sozio, U.S., developed a solid ceramic for dentures (WABI).
1980	Procter & Gamble withdrew Rely tampon from the market (EBM).
1982	Tylenol poisonings case led to tamper-proof packaging for nonprescription drugs (NYT).
1985	Valium patent expired opening the way to generic manufacturers (CDIS).
	Dr. Edward Kass identified magnesium absorbing fibers as increasing risk of toxic shock (WJ).
	Digital hearing aid patented by A. Maynard Engebretson and Gerald R. Popelka, St. Louis (IEEE).
1986	Soft bifocal contact lens, Alges lens, developed by University Optical Products, Florida (CDIS).
	Soft bifocal contact lens, Hydrocurve II, developed by Barnes-Hind of Revlon Vision Care (CDIS).

Children

For much of human history, children represented not just more mouths to feed, but more hands to share the work. With decreased infant mortality and changing life-styles, children have become nonproductive members of the family and a matter of choice--a luxury, not a means to ensure sufficient labor for a family farm, craft or income to support the family unit, and parental support for parents in their old age.

Child Care

Before pasteurization was developed, children fed with milk from nonhuman sources had high mortality rates. Attempts were made to feed nursing children from artificial sources, as is evidenced by ancient feeding vessels found at Cyprus, 900 B.C., and in Italy, 800 B.C. Pieces of rag or sponge were used to regulate the flow of liquid food. Rubber nipples were not introduced until 1835. The Turtle glass nursing bottles were introduced in 1864 and the HUB vented nursing bottle in 1887. Glass bottles were seen as a significant improvement since the inside of the bottle could be seen and encouraged greater cleanliness.[8]

Figure 11. Children

YEAR

YEAR		
1987		
	Superabsorbent diapers	
	Reusable tab diapers	
1975		
	Prenatal blood transfusion	Aprica stroller
	Disposable diapers	Disposable bottle
1950		
	Canned baby food	
1900	Morphine used in childbirth	Folding push chair
	Infant food supplements	
	Glass nursing bottles	Baby carriage
1800	Rubber nipple	
	First verifiable cesarean delivery	
1600		
	First cesarean delivery (?)	
1200		
400		
	Pottery infant feeding vessels	
-1200		

Disposable bottles are a recent phenomenon. Hospitals adopted the Beneflex feeding system of disposable bottles in 1963. Playtex had introduced bottles with disposable liners for home use before 1965. The collapsible plastic liners eliminated the need for sterilizing bottles. Bottle feeding meant the father as well as the mother could share in feeding the child.

Food supplements in liquid form for children were being marketed in the latter part of the 19th century. Mellin's Food and Murdock's Liquid Food were advertised heavily for both infants and invalids (1875). Canned strained baby food was offered by Clapp in 1922, and Gerber began offering strained baby food in 1927.

The 1970s was a time for questioning a number of practices including bottle feeding of babies. Women in both developed and lesser-developed countries were encouraged to breast feed. In lesser-developed countries, sanitation, inadequate supplies of safe water, dilution of infant foods, and improper use all contributed to a growing problem of infant health. Mothers' milk also contains protective antibodies not present in commercial infant foods. Nestlé was blamed for much of the problem, but it was not clear that the infant foods themselves were at fault, but rather improper use and unsanitary conditions. In 1981 Nestlé and other baby food manufacturers agreed to modify marketing practices in Third World countries.[9]

In developed countries nursing practices depend on personal preference and life-style. Some working women still prefer to bottle feed because of the convenience and relative freedom it offers and the ease of sharing feeding responsibilities with fathers.

Two other innovations of significance for child care were the baby carriage or stroller and disposable diapers. The modern baby carriage was devised by Charles Burton, New York, in 1848; the first regular manufacture in London began in 1850. A folding metal push chair was developed in 1904, and a folding baby carriage appeared in 1906. Kassai of Japan began manufacture of the Aprica stroller in 1970 and introduced the Aprica stroller to the U.S. market in 1980.

The first mass-marketed disposable diapers, Pampers, were test marketed by Procter & Gamble in 1961 in Peoria, Illinois, and marketed nationally in 1968. In 1983 Kimberly-Clark introduced reusable tabs followed in 1985 by Procter & Gamble's superabsorbent Ultra Pampers able to absorb 700-800 times their weight in liquid. Uni-charm of Japan introduced the first superabsorbent diapers to the Japanese market in 1982, three years ahead of Procter & Gamble. The same materials are also used in some sanitary napkins and in aids for incontinent adults.

Use of disposable diapers has had a number of effects. The use of commercial diaper services has declined, and it is difficult to find good quality cloth diapers. While disposable diapers have eliminated washing and drying of diapers, they have also created an added burden for waste disposal, but most parents would agree they have made life easier.[10]

Procter & Gamble patented a disposable bib with a gravitationally operated pocket for catching food spills and began limited marketing of it as Dribbles in 1986.

Childbirth

One of the few written sources on gynecology to survive is the work of Soranus of Ephesus (100 A.D.). His work was used by Islamic and European doctors until well into the 17th century. He described contraception and abortion, the diseases of women, and childbirth. Only limited progress was made in these areas until the 17th century.

Other surviving books on childbirth were written by two midwives. Louise Boursier, France, was the first to do so in 1609. Her descendant, Angélique le Boursier du Coudray, published a book in 1759 that lasted through six editions. Coudray also used a plaster model of the torso to demonstrate the progress of childbirth.

Cesarean childbirth--delivery via surgical removal through the abdomen instead of naturally via the birth canal--was associated with the Roman Julii family, whose cognomen Caesar, meaning "to cut," was linked to the birth of a child in this fashion. Normally, it was only performed on dying women in advanced pregnancy, and usually resulted in the mother's and sometimes also the child's death. Jacob Nufer, a pig farmer, claimed to have performed the first successful cesarean operation--mother and child surviving--on his wife in 1500, but there are some doubts about the truth of his story. François Rousset, personal physician to the duke of Savoy, in 1581 published a report of 15 cesarean sections performed over 80 years. One set dealt with a woman who was reported to have survived six successful cesarean operations. The first fully documented cesarean was performed by Jeremiah Trautman in 1610 with the mother surviving for 25 days. The first verifiable cesarean in the British Isles was performed by Mary Donally, Charlemont, Ireland, in 1738, after the mother had been in labor for twelve days.

Childbirth is not without pain or danger. In 1847 the first child was born under anesthesia, and Ignaz Semmelweiss, Hungarian physician, discovered the connection between childbed puerperal fever and septicemia. While others may have postulated a similar link, credit is usually accorded to Semmelweiss. Unfortunately for women, it took some time before Semmelweiss's antiseptic methods were widely accepted. His work and the application of carbolic acid as an antiseptic by Joseph Lister in 1865 eventually saved many lives.

In 1902 Von Steinbuchel, Graz, introduced the use of a drug, scopolaminemorphine, to ease pain during childbirth. British and U.S. physicians adopted it in 1917. By the 1970s, women and their doctors had changed from a heavy reliance on anesthesia during childbirth to little or none. It became a matter of personal choice for each woman assisted by the advice of her doctor.

Treatment of the fetus has also become possible. Testing for certain conditions before birth allows the mother to abort if she chooses or the doctor to treat some conditions in the womb.[11] The first prenatal blood transfusion was performed in 1963 in Auckland, N.Z.

Until modern times the odds were heavily against survival for premature infants. While incubators were used to hatch eggs, it was some time before a similar apparatus was devised for infants. The incubator was reinvented in 1588 by Giovanni Battista della Porta based on ones used in ancient Egypt. In 1666 Cornelius Drebbel, Holland, added a thermostat to the incubator, and John Champion, London, patented his in 1770. The incubator was repatented in 1846. Dr. William Deming devised his "hatching cradle," an incubator for infants, in 1888. Dr. Alexander Lion, Nice, also developed an infant incubator in 1891.

Table 11.2. Children and Childbirth

Dates	Event
B.C.	
2700	*I Ching* advised use of quicksilver as an oral contraceptive (WA).
1850	Petri papyrus described recipes to prevent conception (EB).
1100	Chinese philosophers recommended "thinking of other things" at the moment of ejaculation (WA).
900	Ancient feeding pots used small piece of sponge or rag to nurse infants in Cyprus (SA).
800-700	Feeding vessels for infants used in Italy (SA).
A.D.	
>100	Soranus of Ephesus, Greek gynecologist, described contraceptive methods then in use (EB).
500	Aetius, Greek physician, recommended the use of half a pomegranate to catch semen for contraception (WA).
1500	Jacob Nufer claimed to have performed the first successful cesarean on his wife (1ST).
1581	Francoise Rousett, France, published his report of cesarean surgery performed over 80 years (EMH).
1588	Giovanni Battista della Porta designed an incubator after those used in Ancient Egypt (DID).
1609	Louise Boursier, French midwife, published a book on obstetrics (WL).
1610	Jeremiah Trautman, Witterberg, performed the first fully documented cesarean surgery (EMH).

1666	Cornelius Drebbel of Alkmaar, Holland, invented an incubator (Athenor) fitted with a thermostat (DID).
1701	Reference by William Burnaby in his play, *The Ladies' Visiting Day*, to contraceptive sheath (1ST).
1733	First baby carriage built at Chatsworth by William Ken for the third duke of Devonshire (1ST).
1738	First verifiable cesarean operation in the British Isles performed by Mary Donally at Charlemont, Ireland (1ST).
1759	Angelique le Boursier du Coudray, descendant of Louise, published a book on midwifery (WL).
1770	John Champion of London patented an incubator (DID).
1785	First human artificial insemination conducted by M. Thouret on his wife (1ST).
1826	Early birth control manual, Richard Carlile's *Every Woman's Book*, London, was published (1ST).
1835	Rubber nipples introduced for infant nursing bottles (HG).
1838	First vaginal cap or pessary developed by Dr. Friedrich Wilde, Berlin (1ST).
1846	Incubator repatented (DID).
1847	Hungarian physician I. Semmelwis discovered connection between childbed puerperal fever and septicemia (TH).
	First child born alive under anaesthesia was Wilhelmina Carstairs (1ST).
1848	Charles Burton, New York, devised the first modern baby carriage (RMRB).
1850	First regular manufacture of perambulators began in London (1ST).
1864	Turtle nursing bottle introduced for infant feeding (ECTL).
1865	Dr. Joseph Lister used carbolic acid as an antiseptic in an operation (DID).
1866	Dr. James M. Sims, Women's Hospital, New York City, performed human impregnation by artificial insemination (FFF).
>1870	Rubber condoms sold (EOOET).
1875	Mellin's Food and Murdock's Liquid Food sold as infant or invalid supplement (HG).
1881	Dr. Aletta Jacobs, Amsterdam, offered birth control advice; she recommended the use of the diaphragm (WL).
1884	First artificial insemination with donor, other than the husband, performed by Professor Pancoast, Philadelphia, Pennsylvania (1ST).

1887	HUB vented nursing bottle for infant feeding patented (ECTL).
1888	Dr. William Champion Deming, New York City, devised the "hatching cradle," an incubator for infants (FFF).
1891	Baby incubator devised by Dr. Alexander Lion, Nice (1ST).
1902	Von Steinbuchel, Graz, introduced Scopolamine morphine for childbirth (TT).
1904	Folding metal push chair developed, a boon to parents and servants (TT).
1906	E. Baumann, Paris, devised the folding baby carriage (WABI).
1916	Margaret Sanger opened a birth-control clinic, Brooklyn, New York (1ST).
1917	British and U.S. physicians used Scopolamine morphine to ease childbirth (TT).
1920	U.S.S.R. legalized abortion on demand (1ST).
	Karl Spire and Arthur Stoll extracted the alkaloid drug ergotamine from ergot for migraine and abortion use (TT).
1921	Dr. Maire Stopes established Mothers' Clinic that dispensed birth-control information in London (1ST).
1922	Canned baby food manufactured by Harold H. Clapp, Rochester, New York (1ST).
1928	Rhythm birth control method devised by Dr. Harmann Knaus, Berlin (1ST).
1930	State supported birth control clinics were established in Bangalore and Mysore by Mysore government (1ST).
>1930	Thin latex rubber condoms introduced (EOOET).
1935	Iceland introduced legalized abortion on medico-social grounds (1ST).
1939	Artificial impregnation of a rabbit performed by Dr. Gregory Pincus, Clark University, Worcester, Massachusetts (FFF).
1952	Contraceptive pill of phosphorated hespidin produced (TH).
1954	First clinical tests made of oral contraceptive pill (1ST).
1960	First commercially produced oral contraceptive pill, Enovid, marketed by G. D. Searle, Skokie, Illinois (1ST).
<1961	Disposable diapers marketed; Pampers was the first successful mass-marketed brand (PG).
1963	Hospitals adopted Beneflex disposable bottles for feeding infants (TIME).

First prenatal blood transfusion performed by Professor George Green, Auckland, New Zealand, on Mrs. E. McLeod (1ST).

1967   Royal assent given to the British Abortion Act (1ST).

1968   British Abortion Act became effective (1ST).

1970   Kassai, Japan, began manufacture of baby strollers (BUSWK).

Pampers disposable diapers marketed nationally (WJ).

1973   U.S. Supreme Court reaffirmed a woman's right to abortion on request (TIME).

1977   Warner Lambert offered home pregnancy test (FORT).

1978   Louise Brown, first test-tube baby, was born (TT).

1980   Kassai introduced its Aprica baby stroller to the U.S. market (BW).

1982   Uni-Charm, Japan, introduced superabsorbent disposable diapers in Japan (BUSWK).

1983   Kimberly-Clark introduced disposable diapers with reusable tabs (WJ).

1984   GIFT, gamete intrafallopian transfer, developed for artificial insemination (SCI8).

Procter & Gamble patented a disposable bib with gravitationally operated pocket, to catch baby's dribbles (WJ).

1985   Improved monoclonal home pregnancy test introduced (FORT).

Monoclonal Antibodies, Mountain View, California, offered OvuStick for determining peak fertility (FORT).

Procter & Gamble developed new superabsorbant Pampers; they absorb 700-800 times their own weight in fluid (WJ).

1986   Procter & Gamble introduced Dribbles, a disposable bib for infants (WJ).

Dr. Beatrice Courzinet and Dr. Gilber Schaison, France, tested a "morning after pill" for contraception (TIME).

Reproduction

Because of the economic value of children in agrarian societies and the high infant mortality rate throughout much of human history, a premium was placed on the ability of women to bear healthy children. High infant mortality rates required high fertility rates to offset losses. Women who were unable to conceive were at a disadvantage and in many societies could be set aside by the husband. Methods to improve the ability to conceive were chancy and unlikely to succeed. It was not until fairly modern times that help was available.

Fertility

The first artificial insemination of a human was performed by M. Thouret on his wife in 1785. A donor other than the husband was used in 1884 by Professor Pancoast, Philadelphia. The first test-tube baby--fertilization done in vitro, outside the mother's body--Louise Brown, was born in 1978. A new process has been announced that allows fertilization to take place in the fallopian tubes. GIFT, gamete intrafallopian transfer, was developed by Ricardo Asch and co-workers at the University of Texas (1984). The first baby conceived by this method was born in April 1985.

Other means of treating infertility include treatment with drugs such as clomiphene to induce ovulation, surgery to correct structural flaws or remove scarred tissue, or drugs to treat other conditions inhibiting normal conception. Both partners may be treated depending on where the problems are identified. Fertility drugs and techniques have resulted in a number of multiple births and have raised serious ethical questions, particularly in cases where none of the infants survived. Other ethical issues have been raised concerning donors of sperm or eggs and surrogate motherhood.

Home pregnancy tests have recently undergone a number of improvements. Warner Lambert first introduced a home pregnancy test in 1977. In 1985 monoclonal techniques (using the cloning of a single cell) were applied and introduced by Monoclonal Antibodies of Mountain View, California, to provide a simpler and more reliable pregnancy test. Also developed by Monoclonal and others were tests to determine the time of ovulation for either encouraging or discouraging conception: Monoclonal was marketing Ovusticks while Tambrands had introduced First Response.

Contraception

Contraception has been practiced for much of recorded history. The *I Ching*, 2700 B.C., advised the use of quicksilver as an oral contraceptive, and Chinese prostitutes of that period drank lead for the same purpose. Contraceptive methods are described from 1850 B.C. in Egypt through the Petri papyrus. The Egyptians favored a mixture of crocodile dung and honey in the vagina. Indian women used feathers while Persian women used sponges. By 1100 B.C. Chinese philosophers were recommending that women should control their passions and think of other things at the time of the male's ejaculation.[12]

Soranus of Ephesus, 2d century A.D., described contraceptive methods then in use. His work was a major source for Islamic and European doctors until the 17th century. By the 6th century, Aetius, another Greek physician, recommended inserting half a pomegranate into the vagina to catch semen, a forerunner of 19th-century pessaries.

William Burnaby, in his play *Ladies' Visiting Day* of 1701, referred to a contraceptive sheath. Sheaths made of animal skin had been used from early times, but mainly as protection from venereal disease. The word *condom* was adopted from that of the earl of Condom, personal physician to Charles II, who recommended the use of sheep's gut to Charles to reduce the likelihood of contracting syphilis during intercourse. Rubber condoms were marketed in the 1870s,and the modern latex models were introduced in the 1930s. It has been recommended in the 1980s for protection against acquired immune deficiency syndrome (AIDS).

A variety of works during the 18th and 19th centuries dealt with conception and contraception. The audiences for these varied. Richard Carlile's *Every Woman's Handbook*, published in London in 1826, was aimed at women rather than doctors.

Methods of birth control were still primitive, but evolving, in the 19th century. The vaginal cap or pessary was developed by Dr. Friedrich Wilde, Berlin, in 1838. At least twenty-six pessaries were patented in the United States between 1850 and 1885.

A number of women doctors were concerned about the effects of frequent births on women's health. Dr. Aletta Jacobs, Amsterdam, provided birth-control advice to women and in 1881 recommended the use of the diaphragm, a flexible disk that covered the cervix. Margaret Sanger opened her birth-control clinic in Brooklyn, New York, in 1916 followed by Dr. Maire Stopes's Mothers' Clinic of London, established in 1921. Sanger's clinic was closed by the police soon after opening, but the publicity and associated trial helped to change attitudes and enabled doctors to prescribe birth-control methods in certain circumstances without violating the law.[13] State-supported clinics were established in Bangalore and Mysore by the Mysore Government in 1930.

The interuterine device (IUD), was originally used by Arab camel drivers to prevent pregnancies in female camels. It was not until 1928 that Ernst Frafenberg, a German physician, devised the first scientific metal coil for use by women. In the 1980s some IUDs were linked to diseases causing infertility; the Dalkon Shield was the best known of these.[14]

The rhythm method of birth control was devised by Dr. Harmann Knaus, Berlin, in 1928. A more certain method, the phosphorated hespidin contraceptive pill, was produced in 1952, and the first clinical tests were made of an oral contraceptive pill in 1954. Commercial production of Enovid was begun by G. D. Searle in 1960 at Skokie, Illinois.

In addition to preventing conception, another method of birth control involved terminating an unwanted pregnancy. In some societies this was considered normal, but in others it was regarded as a form of murder. In the

19th century pills to induce abortion included Madame Kostell's French Monthly Pills, Dr. Peter's French Renovating Pills, Dr. Monroe's French Periodical Pills, Dr. Merveau's Portuguese Female Pills, and Madame Drunette's Lunar Pills.[15]

Dr. Beatrice Courzinet and Dr. Gilbert Schaison, Hospital de Bicentre, Paris, completed tests in 1986 of a morning-after pill for terminating unwanted or unsafe pregnancies developed by Etienne-Emile Baulieu, a French biochemist. The drug, RU 486, blocks the action of progesterone and triggers menstruation. So far no undesirable side effects have been detected, but it has not yet been approved by the USFDA for use in the United States.[16]

Abortion Legislation

The U.S.S.R. legalized abortion on demand in 1920 followed in 1935 by Iceland legalizing it on medico-social grounds. The British Abortion Act became effective in 1968. The U.S. Supreme Court decision in Roe v. Wade (1973) confirmed women's right to choose abortion. Subsequently the Pro-Life movement and the Reagan Administration have urged a reversal and federal funds can no longer be used for abortions.[17]

As elsewhere, technology has offered more choices in what to use and in more effective products. Standards of health and hygiene are rising. More children now survive, and methods of feeding and care are better.

While the list of technologies required to support the modern household has not been exhausted, those mentioned highlight some of the impact of technology on the home and standards of living. Technology provides the fabrics worn, the foods eaten, the means used to prepare those foods, the tools used in their preparation, the water drunk, the heat for homes, and protection from the weather. The final chapter explores these effects further and draws together the various issues raised in this work.

12
Choices

In Western society technology has freed us from the need to plan each hour, each day, each month. It has given us choices in how we live our daily lives; that has been the major theme of this book. We can now decide how, when, and where to do particular tasks. We can focus on the things that are important to us, and these can be quite different for each person. We can sew our clothes if we choose, or make pasta, but we do not have to. For most of us, life is no longer devoted to purely survival related activities. We have control over our discretionary time and we have more of it.

A number of things have given us that time and control over it, but perhaps foremost in modern times is electric power. We are able to trade our labor for money, and to use that money for many things, including our servant, electricity. Our furnaces all have electric fans and electric control systems, and our homes are lit by electricity. Our food may be cooked by it. That food is certainly preserved by it, since most refrigerators run on electricity. Our vacuum cleaners, irons, washing machines, microwaves, coffee pots, can openers, even knives use electricity. It also powers our communication systems: telephones, television, radio. We can adjust our living environment through heating and air conditioning. We can light our homes whenever we want and at almost any level of brightness. Daylight and darkness are less restrictive of our activities than ever before. We can live underwater, on mountains, in the tropics, in the Arctic regions, in outer space, and even on the moon. Without electricity most of our choices and control of our lives would disappear.

Western society lives with greater comfort and ease than ever before in human history. Even our poor have more than many kings of older times, and more than the poor of other, less-developed nations. That does not make their deprivation any less real in comparison to other members of their society.

Our population is aging; older people are staying healthy longer and, with improvements in domestic technology, have the ability to remain independent longer. It is abundantly clear that our present system of social welfare and social security cannot continue. With the baby boomers joining the ranks of the retiring in twenty or so years, there will be too few workers to support the system. The retirement age will have to rise. The senior citizens will continue to work full- or part-time. As a consequence we will see more career changes and a growing need for continuing education. Our universities will be faced with more older students with different attitudes and goals. It will not be a repeat of the veterans flooding campuses after World War II. There will be similarities, but also differences. Most of these new students will have been trained earlier and will now be looking for ways to move into new careers.

Demands for living accommodations and facilities will also change. Families with two wage earners are already seeking ways to reduce their work loads; they are sharing more with each other and with children. This will increase the emphasis on labor saving devices and reinforce the throwaway mentality: if something breaks, replace it, don't fix it.

Homes will also be smaller, requiring less care. They will incorporate improved security and sensor systems controlling more functions automatically. They will be energy conserving, mixing a variety of energy sources, and allowing easy switching between sources.

The diffusion and adoption of technology and technological advances will be uneven. Certain groups and certain societies will benefit before others. The ability to afford technology and technology's products is one limiting aspect. Education, attitudes, and knowledge are also important.

Much of what we know and use within the home depends on what tools our mothers used, what our friends use, and what we see advertised. This is one of the reasons new products are generally slow to catch on. The microwave and the food processor are two examples. New tools require us to learn new habits.

Sometimes it is difficult to separate cause and effect. Because potters made more dishes, and they became available at low cost, women had more dishes to do. Because the sewing machine enabled women to make clothes faster, they made more clothes. Because women made more clothes, they had more laundry to do. Because hot water, soap, and washing machines were available, women washed more. Which is cause and which is effect?

People had few garments and few dishes, not because that was all they needed or wanted, but because that was all they could afford or acquire. It required considerable labor to clothe a family, as was discussed in chapter 5. Glazed pottery dishes washed after each use with hot water and soap are more sanitary than wooden dishes only rinsed or cleaned with sand; thus promoting healthier families and fewer sick members to nurse. Properly cared for, pottery dishes will also last longer. Timing is also critical, however, and modern cleaning products and convenient hot water supplies are primarily a 20th-century phenomenon. Conditions in Appalachia as described in the Foxfire series remained unchanged for many years, but electrical power, the automobile, and television have rapidly brought this region into the mainstream of American society.[1]

Much of the increased hours of housework by women who do not work outside the home, the conditions encouraging their housekeeping habits, and the technologies eliminating some tasks altogether are relatively recent phenomenon. The impact of personal choice is also significant. Each householder has to make choices of what tasks to do. Our society has equated dirt, smell, and disease. Our homes, clothing, and dishes must be spotless or we may be subject to the specter of germs and illness. Ruth Cowan in *More Work for Mother* questioned how much cleanliness is necessary.[2] Household standards differ. Working women and men are devoting less time to household chores, are taking more shortcuts, and are no longer doing certain tasks. Making beds and dusting are among the first casualties; ironing is a close second.[3]

The advent of abundant pure water supplies and the means to have hot water available for washing and cooking changed lives dramatically. Even in the 20th century such changes are occurring in developing countries. The recent availability of a fresh water aqueduct in Colombia has changed life there.[4] Formerly fresh water supplies required, for one family, a daily two-hour trip to the river and back. The aqueduct has provided fresh water in the home and saved time to devote to other tasks. The impact on family health and the decrease in childhood illness are also important. Pure water available in the home has saved time and improved health and general cleanliness. Water no longer has to be carried long distances, consequently more of it is available, and available whenever it is wanted.

Easily accessible supplies of hot water have made cleaning easier, although new detergents will wash well in cold water, reducing the need for hot water. A hot shower is still the preferred way for personal cleaning, although a cold shower has its uses and adherents. Even the Romans provided a means of heating their bath water.

Waste disposal has also been changed. Our general health has been improved by modern waste removal systems. We are less likely to have our water supplies contaminated by sewage or human waste matter. It is now a community, rather than an individual, responsibility.

Survival living demanded that everyone contribute: men, women, and children. With the change to urban societies, people move beyond survival needs. Men and women work at other tasks to gain the means to live. Cloth is no longer made at home, but purchased, and now entire garments are purchased. Food is not grown at home, but bought. Extended storage may still be a problem, but preservation techniques were well known from prehistoric times, and technology has kept abreast of contemporary life-styles and needs.

Major problems in urban societies are sources of food, fuel, clothing, water, and a means of waste disposal. Transportation and storage of food were limits for much of human history. The development of steam and gasoline engines changed such limits significantly. We have still not fully solved water supply and waste disposal problems, but they are substantially better than in medieval times. Food storage problems have almost been eliminated. With rapid transport from growing areas, and with a variety of preservation methods, we do not need to store foods for half a year in order to survive winter. We can have fresh fruits and vegetables all year long.

Table 12 provides an overview of how significant some of these changes have been for Western society. We have eliminated many tasks; certainly grinding grain for flour or spinning thread and weaving are not done by most families. A study recorded the hours devoted to housework in 1930 as fifty hours per week. For 1970 it had risen to fifty-five hours per week but if a wife worked, only twenty-six hours per week were required.[5] This may confirm the old adage "work expands to fill the time," but it also says much about choice and how we spend our time. Working wives must simplify and concentrate on the essentials. Technology allows them to manage what would otherwise be two full-time jobs.[6]

Notable about the items in table 12 is that we now buy most products, rather than make them individually. The two services in the table, laundry and cooking, are still largely done at home or at least by the householder, but the time required is significantly less and, more importantly, laundry at least can be done when convenient, at night, early morning, or whenever time is available or when at work or play.

We buy summer clothes in the winter and winter clothes in the summer, which is probably a carry over from our earlier need to make clothes in relation to the seasons. Sheep were sheared in the late spring, the wool processed and made into cloth and then into garments in the fall. Cotton and flax were gathered and processed in the fall and made into cloth and clothing in the winter months.

During the past fifty years, we have changed from ice boxes to refrigerators, from defrosting refrigerators to frost-free freezers and refrigerators, from stoves to microwaves, from canned vegetables to frozen vegetables, from drying clothes outside to using a dryer, from washing cloth diapers to using disposable diapers, from using handkerchiefs to using paper tissues, from girdles to panty hose, from ironing to wash-and-wear clothes, from peeling potatoes to pouring them out of a box.

Table 12. Changes in Task Time

TASK	BY HAND	MECHANIZED	NOW
Cooking		3 times faster (Microwave)	Buy, 1 hour
Grinding Flour		40 times (Mill)	Buy, .25 hour
Spinning Thread		6 times to 1000 times	Buy, .25 hour
Cloth	1 hour /yard (setup 5 hours)	8 times	Buy, .5 hour
Garment	14 hour	1 hour (Sewing machine)	Buy, 1.5 hour +
Laundry	1-4 days	1 day	Buy, 2 hours
Bread	24 hour	2 hour	Buy, <.1 hour
Gelatin	24 hour	1/2 hour	Buy <.1 hour
Lighting	24 hour		100 times, .25 hour

Technology enables those of us fortunate enough to live in Western society to do more, encourages us to want more, and provides the means to achieve a higher standard of living and better health care. Technology is neutral. It is how we use it and what we use it for that makes it good or bad.

We have gained, but also lost. Few of us could make a garment from scratch. Some of us only know the simplest of sewing tasks. Without patterns, modern thread, and a sewing machine, we would not know where to start.

For most of us, the trade has been worthwhile. We have gained the opportunity to choose how we want to spend our time. We can join clubs, go to school, pursue hobbies, read, watch television, visit friends; the choice is ours. The future promises even more leisure time. It will be a major challenge for our society, as we live longer, how we remain an interested, motivated, learning society or whether we become a passive consumer society. Some of us will choose one path, some the other. The technologies we use in our daily life provide us with the opportunity to choose. The question for each of us is, what do we do with the time we now have?

Notes

Preface

1. Caroline Davidson, *A Woman's Work Is Never Done: A History of House Work in the British Isles 1650-1950* (London: Chatto & Windus, 1982).

2. Ruth Schwartz Cowan, *More Work for Mother: The Ironies of Household Technology from the Open Hearth to the Microwave* (New York: Basic Books, 1983).

3. Ann Oakley, *Woman's Work: The Housewife, Past and Present* (New York: Vintage Books, 1976).

Chapter 1

1. E. F. Carter, *Dictionary of Inventions and Discoveries* (New York: Crane, Russak, 1976).

2. Patrick Robinson, *The Book of Firsts* (New York: Bramhall House, 1982).

3. Charles Panati, *The Browser's Book of Beginnings* (Boston: Houghton Mifflin, 1984).

4. Valerie-Ann Giscard d'Estaing, *The World Almanac Book of Inventions* (New York: World Almanac, 1985).

5. *Encyclopaedia Britannica* (Chicago: Encyclopaedia Britannica, 1980).

6. *World Book Encyclopedia* (Chicago: Field Enterprises, 1975).

7. Kenneth B. Taylor, "The Economic Impact of the Emerging Global Information Economy on Lesser Developed Nations," in *The Global Economy: Today, Tomorrow, and the Transition,* ed. Howard F. Didsbury, Jr. (Bethesda: World Future Society, 1985), 147-62.

8. H. Michael Lafferty, "Lighting up Their Lives: REA Turned Farms on 50 Years Ago," *Columbus Dispatch,* 26 May 1986, F1.

9. *Encyclopaedia Britannica,* 5:937.

10. Marek Zvelebil, "Postglacial Foraging in the Forests of Europe," *Scientific American* 254 (May 1986):110, 112.

11. *World Book Encyclopedia,* 4:810.

12. Davidson, *A Woman's Work*, 77.

Chapter 2

1. Panati, *The Browser's Book*, 113.

2. L. Sprague de Camp, *The Ancient Engineers* (New York: Ballentine, 1963), 7-9.

3. Reay Tannahill, *Food in History* (New York: Stein and Day, 1973).

4. Panati, *The Browser's Book.*

5. *Encyclopaedia Britannica,* 5:425.

6. Wolfgang Ziegler, *Dumont Guide: Ireland,* trans. Russell Stockman (New York: Stewart, Talbori & Chang, 1984), 242, 290.

7. Myma Davis, *The Potato Book* (New York: William Morrow, 1972), 10.

8. *World Book Encyclopedia,* 25:602.

9. Barbara Friedlander, *The Secrets of the Seed: Vegetables, Fruits and Nuts* (New York: Grosset & Dunlap, 1974), 153-65.

10. Barbara Karoff, "In Praise of Peanuts," *USAIR* 8 (January 1986):14, 16.

11. Peter Wynne, *Apples* (New York: Hawthorn, 1975), 3-13.

12. Noel D. Vietmeyer, "Lesser Known Plants of Potential Use in Agriculture and Forestry," *Science* 232 (13 June 1986):1382.

13. Friedlander, *Secrets of the Seed*, 98-99.

14. Patricia A. Taylor, "Time for Tomatoes," *USAIR* 8 (July 1986):12.

15. Ibid., 14.

16. Regina Schrambling, "All about the Avocado," *USAIR* 8 (March 1986):68.

17. Joel Schapira et al., *The Book of Coffee and Tea: A Guide to the Appreciation of Fine Coffees, Teas, and Herbal Beverages* (New York: St. Martin's, 1975), 5.

18. Ibid., 6-10.

19. Robinson, *The Book of Firsts*, 74.

20. Schapira, *The Book of Coffee*, 6-10.

21. Ibid., 148-78.

22. Peter J. Reynolds, *Farming in the Iron Age* (Cambridge: Cambridge University Press, 1976), 16.

23. *Encyclopaedia Britannica,* 1:326.

24. *World Book Encyclopedia,* 10:132-33.

25. *Encyclopaedia Britannica,* 16:193.

26. Ibid., 5:937.

27. Ibid., 17:169.

28. George Edwin Fussell, *Farming Techniques from Prehistoric to Modern Times* (New York: Pergamon, 1966).

29. Michael Partridge, *Farm Tools through the Ages* (New York: Promontory Press, 1973).

30. Percy W. Blandford, *Old Farm Tools and Machinery: An Illustrated History* (Ft. Lauderdale, Fla.: Gale Research, 1976).

31. Margaret W. Rossiter, *The Emergence of Agricultural Science: Justus Liebeg and the Americans, 1840-1880* (New Haven: Yale University Press, 1975).

32. *Encyclopaedia Britannica,* 5:970.

33. Jean Gimpel, *The Medieval Machine: The Industrial Revolution of the Middle Ages* (New York: Penguin, 1983), 32-39.

34. *Encyclopaedia Britannica,* 5:970-71

35. Gimpel, *The Medieval Machine,* 39-43.

36. *Encyclopaedia Britannica,* 1:338.

37. Sonia L. Nazario, "Florida in Winter Is No Vacation Paradise for a Cane Cutter: Sugar Growers Import Labor from West Indies to Do a 'Near Impossible' Job." *Wall Street Journal,* 3 January 1985, 1.

38. Melvin Kranzberg, and William H. Davenport, *Technology and Culture: An Anthology* (New York: Schocken Books, 1972).

39. George J. Church, "Too Much of Good Thing," *Time* 128 (8 September 1986):22; *Encyclopaedia Britannica,* 1:315.

40. Gimpel, *The Medieval Machine,* 9.

41. Ibid., 19-20.

42. Ibid., 14.

43. Ibid., 1-38.

44. Reynolds, *Farming in the Iron Age*, 11.

45. Robinson, *The Book of Firsts*, 10.

46. Marjorie Sun, "Engineering Crops to Resist Weed Killers," *Science* 231 (21 March 1986): 1360-61; "Breeding the 'Pesticide' Right into the Crop," *Business Week*, 28 April 1986, 71.

47. *Encyclopaedia Britannica,* 1:341.

48. Ibid.

49. Ibid.

50. Zvelebil, "Postglacial Foraging," 106.

51. "Fish on the Farm," *Forbes* 137 (7 April 1986): Matt Moffett, "Looking to Lure New Diners, Promoters of Catfish Work to Spruce up Its Image," *Wall Street Journal*, 2 June 1986, 17.; Lester R. Brown, "Fish Farming," *Futurist* 19 (October 1985):18-25.

Chapter 3

1. N. P. Spanos, and J. Gottlieb, "Ergotism and the Salem Witch Trials," *Science* 194 (24 December 1976):1390.

2. Faye Kinder, Nancy R. Green, and Natholyn Harris, *Meal Management*, 6th ed. (New York: Macmillan, 1984).

3. Caroline Davidson, *A Woman's Work Is Never Done: A History of House Work in the British Isles 1650-1950* (London: Chatto & Windus, 1982).

4. Reynolds, *Farming in the Iron Age*, 31.

5. Jacques McNish, "They Make Pickles the Old-Fashioned Way, and It's Disgusting," *Wall Street Journal*, 15 April 1985, 28.

6. *Encyclopaedia Britannica,* 1:426.

7. Ibid., 5:425.

8. Carl W. Hall, and T. I. Hendrick, "Drying of Milk and Milk Products," 2d ed. (Westport, Conn.: AVI, 1971), 3-7.

9. Norman W. Desrosier, "The Technology of Food Preservation," 3d ed. (Westport, Conn.: AVI, 1970), 160.

10. James Trager, *The Enriched, Fortified, Concentrated, Country Fresh, Lip-smacking, Finger-Licking, International, Oral, and Unexpurgated Food Book* (New York: Grossman, 1980), 407.

11. *Encyclopaedia Britannica,* 7:492.

12. Trish Hall, "New Packaging May Soon Lead to Food that Tastes Better and Is More Convenient," *Wall Street Journal,* 21 April 1986, 21.

13. Anthony Ramirez, "In Hot Pursuit of High-Tech Food," *Fortune* 12 (23 December 1985):94.

14. Robert Lindsay, "Canners Suffer as U.S. Chooses Fresh Produce," *Columbus Dispatch,* 26 October 1986, I13.

15. *Encyclopaedia Britannica,* 15:563.

16. Giscard d'Estaing, *The World Almanac Book,* 283.

17. *Encyclpoaedia Britannica,* 15:563.

18. Giscard d'Estaing, *The World Almanac Book,* 154.

19. "Whatever Happened to Irradiated Foods?" *IEEE Spectrum* 21 (November 1984):28.

20. Ibid.

21. Ibid.

22. *Encyclopaedia Britannica,* 11:246.

23. Ibid., 2:597.

24. *World Book Encyclopedia,* 4:641.

25. Davidson, *Women's Work,* 77.

26. *World Book Encyclopedia,* 2:479.

27. Betsy Morris, "How Much Will People Pay to Save a Few Minutes of Cooking? Plenty," *Wall Street Journal,* 25 July 1985, 19.

28. "Procter and Gamble Says New Cookie Is Selling in Six More Regions," *Wall Street Journal,* 21 March 1984, 4.

29. Mark Schacter, "As Long as They Really Schmeck Who Cares if They're Patented," *Wall Street Journal,* 11 February 1986, 35.

30. Russell Shaw, "Bring on the Brie," *Sky* 14 (September 1985):77.

31. Ronald Alsop, "Dannon Co. Stirs into Action as Yoghurt Competition Grows," *Wall Street Journal,* 12 December 1985, 35.

32. "... And New Foods Tap Yoghurt Boom," *Wall Street Journal*, 8 May 1986, 29.

33. *Encyclopaedia Britannica*, 5:245.

34. Cowan, *More Work for Mother*, 78.

35. *Encyclopaedia Britannica*, 11:751.

36. Patricia Van Benthuysen, "No Matter How You Slice It, Spam Is Here to Stay," *USAIR* 9 (September 1987):86.

37. Steve Mufson, "Don't Thank Us, Pepsi; We Just Talked with a Few Bean Buyers," *Wall Street Journal*, 17 July 1985, 33.

38. Amy Dunkin, Scott Screden, and Resa W. King, "Natural Soda: From Health Food Fad to Supermarket Staple," *Business Week*, 14 January 1985, 73.

39. "Udder Excitement," *Time* 128 (29 September 1986):59.

40. Elizabeth M. Whelan, "Sweetener Wars Taking Aim at Aspartame," *Across the Board* 22 (October 1985):52.

41. Trish Hall, "Food Packaging May Be Improved, but Tampering Can't Be Prevented," *Wall Street Journal*, 16 July 1986, 21. Felix Kessler, "Tremors from the Tylenol Scare Hit Food Companies," *Fortune* 113 (31 March 1986):59, 62.

42. Kinder et al., *Meal Management*, 25-26.

43. Ibid., 25.

44. Dorothy Davis, *A History of Shopping* (London: Routledge & Kegan Paul, 1966).

45. "American Supermarket Miracle: New Formats and Technologies Revolutionize the Way We Shop," *Business Week*, 4 May 1987, 133.

46. Kinder et al., *Meal Management*, 26.

47. *Encyclopaedia Britannica*, 15:778.

48. "More People Eating Out," *Futurist* 20 (May-June 1986):60.

Chapter 4

1. *Encyclopaedia Britannica*, 5:728-36.

2. Molley Harrison, *Kitchen in History* (Reading, Berkshire: Osprey, 1972).

3. Earl Lifshey, *The Housewares Story: A History of the American Housewares Industry* (Chicago: Nat'l. Housewares Mfr. Assoc., 1973).

4. Linda Campbell Franklin, *From Hearth to Cookstove: An American Domestic History of Gadgets and Utensils Made or Used in America from 1700-1930* (Florence, Ala.: House of Collectibles, 1976).

5. *Encyclopaedia Britannica*, 8:710; Davidson, *Woman's Work*, 44-57.

6. Ibid., 71.

7. Ibid., 57-63.

8. Ibid., 63.

9. Robinson, *The Book of Firsts*, 141.

10. Davidson, *Woman's Work*, 66-67.

11. *Timetable of Technology* (New York: Hearst Books, 1982):61.

12. Davidson, *Woman's Work*, 133.

13. Ibid.

14. Gertrude Harris, *Pots and Pans* (San Francisco: 101 Productions, 1971), 33.

15. *World Book Encyclopedia*, 1:380.

16. Harvey Green, *The Light of the Home* (New York: Pantheon Books, 1983), 67.

17. *Encyclopaedia Britannica*, 5:385-86.

18. Cowan, *More Work for Mother*, 44.

19. Loris S. Russell, *Handy Things to Have around the House* (Toronto: McGraw-Hill Ryerson, 1979).

20. Linda Campbell Franklin, *300 Years of Kitchen Collectibles* (Florence, Ala.: Books Americana, 1984), 165-73.

21. Franklin, *From Hearth to Cookstove*, 13; Lifshey, *The Housewares Story*, 244.

22. Schapira, *The Book of Coffee* 120.

23. Giscard d'Estaing, *The World Almanac Book*, 151.

Chapter 5

1. *Encyclopaedia Britannica,* 5:1016.

2. Ibid., 18:183.

3. *World Book Encyclopedia,* 19:173.

4. *Encyclopaedia Britannica,* 5:1016.

5. Marjorie P. K. Weiser and Jean S. Arbeiter, *Womanlist* (New York: Athenaeum, 1981): 106.

6. F. R. Cowell, *Life in Ancient Rome* (New York: Wideview/Perigee, 1961), 70.

7. Ernst Samhaber, *Merchants Make History: How Trade Influenced the Course of History throughout the World,* trans. E. Osers (New York: John Day, 1963), 80-83.

8. *1987 Britannica Book of the Year* (Chicago: Encyclopaedia Britannica, 1987), 259.

9. Dorothy D. Prisco and Harold W. Moore, *Fashion Merchandise Information: Textiles and Non-textiles* (New York: John Wiley, 1986), 9.

10. *Encyclopaedia Britannica,* 18:171.

11. Peter Hunter Blair, *Roman Britain and Early England: 55 BC-AD 871* (New York: W. W. Norton, 1963), 131.

12. *Encyclopaedia Britannica,* 18:178.

13. Gimpel, *The Medieval Machine,* 99-107.

14. *Encyclopaedia Britannica,* 18:173.

15. Ibid., 9:425; Umberto Eco and G. B. Zorsoli, *The Picture History of Invention: From Plough to Polaris* (New York: Macmillan, 1963), 172.

16. *Foxfire 2,* ed. Eliot Wigginton (Garden City, N.Y.: Anchor Books, 1973), 184-90.

17. Carter, *Dictionary of Inventions and Discoveries.*

18. R. J. Forbes, *Studies in Ancient Technology,* 2d. ed., vol. 3. (Leiden: Brill, 1964), 187.

19. Gimpel *The Medieval Machine,* 15-16.

20. *Encyclopaedia Britannica,* 5:1016.

21. Ibid., 3:44.

22. *Foxfire 2*, 207-12.

23. *Encyclopaedia Britannica,* 5:1099.

24. Ibid., 7:259.

25. Ibid., 18:186.

26. Ibid., 18:171.

27. *World Book Encyclopedia,* 14:104.

28. Green, *The Light of the Home*, 80-81.

29. Ibid., 82.

30. Carl Kohler, *A History of Costume*, ed. Emma von Sichart, trans. Alexander K. Dallas (New York: Dover, 1963).

31. Charles Panati, *Extraordinary Origins of Everyday Things* (New York: Harper & Row, 1987), 340.

32. Cecil Saint-Laurent, *Great Book of Lingerie* (New York: Vendome Pressa, 1986).

33. *Encyclopaedia Britannica,* II:93; *World Book Encyclopedia,* 2:333.

34. Betty Thornley, "Eve in Action: Current Modes" *Collier's*, 6 July 1929, 28.

35. *Oxford English Dictionary Supplement* (Oxford: Clarendon Press, 1972).

36. Green, *The Light of the Home*, 120-27.

37. *Encyclopaedia Britannica,* 9:160.

38. Green, *The Light of the Home*.

39. Ibid.

40. *Encyclopaedia Britannica*, 4:581.

41. Richard B. Manchester, *Mammonth Book of Fascinating Information* (New York: A & W. Visual Library, 1980), 254.

42. *World Book Encyclopedia*, 2:146.

43. *Encyclopaedia Britannica*, 5:1024.

44. Lifshey, *The Housewares Story*, 356.

45. Green, *The Light of the Home*, 104.

46. Davis, *A History of Shopping*.

47. Kimberley Carpenter, "Catalog Showrooms Revamp to Keep Identity," *Across the Board* 13 (November 1986):45-53.

48. Albert Haas, Jr., "How to Sell Almost Anything by Direct Mail," *Across the Board* 13 (November 1986):45-53.

49. "Shopping Malls 30 Years Old," *Columbus Dispatch*, 12 October 1986, 12G.

Chapter 6

1. Manchester, *Mammoth Book of Fascinating Information*, 380.

2. Davidson, *Woman's Work*, 115-17.

3. Forbes, *Studies in Ancient Technology*, 3:187.

4. Davidson, *Woman's Work*, 141-44.

5. Ibid.

6. American Home Laundry Manufacturers' Association, *Conferences* (Chicago: AHLMA, 1946-).

7. Russell, *Handy Things*.

8. Lifshey, *The Housewares Story*.

9. Robinson, *The Book of Firsts*, 372.

10. Davidson, *Woman's Work*, 144-45.

11. Ibid., 125.

12. *Foxfire Book*, ed. Eliot Wigginton (Garden City, N.Y.: Anchor Books, 1972); 156.

13. *Encyclopaedia Britannica*, 6:284.

14. Davidson, *Woman's Work*, 141-42.

15. Ibid., 145-49.

16. Ibid., 160-63.

17. Green, *The Light of the Home*, 73-74

18. Panati, *Extraordinary Origins*, 210.

19. Cowan, *More Work for Mother*, 218-219.

20. Milee Woods, "Mother Always Told You Not to Take Long Showers," *Columbus Dispatch*, 2 November 1986, 51.

Chapter 7

1. *Encyclopaedia Britannica*, 14:575.

2. Davidson, *Woman's Work*, 141-42.

3. Gimpel, *The Medieval Machine*.

4. Lawrence Wright, *Clean and Decent: The Fascinating History of Bathrooms & the Water Closet and of Sundry Habits, Fashions & Accessories of the Toilet Principally in Great Britain, France, and America* (London: Routledge & Kegan Paul, 1960).

5. Ibid., 6.

6. *Encyclopaedia Britannica*, 19:649.

7. de Camp, *The Ancient Engineers*, 214-15.

8. Ibid., 206-14.

9. Gimpel, *The Medieval Machine*, 1-27.

10. Terry Reynolds, "The Medieval Roots of the Industrial Revolution," *Scientific American* 251 (July 1984):122-30.

11. Roger Lowenstein, "Aqueduct Carries Hope to Colombia Town," *Wall Street Journal*, 19 February 1985; 32.

12. Gimpel, *The Medieval Machine*, 5-6; *Studies in Ancient Technology*, vol. 1, 2d. ed. (Leiden: Brill, 1964), 183.

13. "Will Pumping Cure Denver Quakes?" *Business Week*, 3 August 1968, 54.

14. "Notes from All Over, *Readers Digest,* March 1985, 126.

15. Wright, *Clean and Decent*.

16. Ibid., 26.

17. Robinson, *The Book of Firsts*, 136.

18. Ibid., 137.

19. Wright, *Clean and Decent*.

20. Ken Butti and John Perlin, *A Golden Thread: 2500 Years of Solar Architecture and Technology* (New York: Van Nostrand Reinhold, 1980), 249.

Chapter 8

1. Colin Norman, "No Panacea for the Firewood Crisis," *Science*, 9 November 1984, 676.

2. Davidson, *Woman's Work*.

3. Butti and Perlin, *A Golden Thread*.

4. Davidson, *Woman's Work*, 77.

5. Norman, "No Panacea."

6. Butti and Perlin, *A Golden Thread*, 3, 5.

7. Ibid., 27.

8. Ibid., 19.

9. *Encyclopaedia Britannica*, 8:710.

10. Giscard d'Estaing, *The World Almanac Book*, 166.

11. *Encyclopaedia Britannica*.

12. Giscard d'Estaing, *The World Almanac Book*.

13. *Encyclopaedia Britannica*, 8:711.

14. "Myths and Methods in Ice Age Art," *Science*, 21 November 1986, 937.

15. Ellen Benoit, "Raise High the (Precut) Roof Beam," *Forbes* 137 (10 February 1986) 120-26.

16. *Encyclopaedia Britannica*, 7:407.

17. F. R. Cowell, *Life in Ancient Rome* (New York: Perigee Books, 1961), 24.

18. *Encyclopaedia Britannica*, 7:796.

19. Green, *The Light of the Home*, 97-98.

20. Davidson, *Woman's Work*, 130.

Chapter 9

1. Leroy Thwing, *Flickering Flames: A History of Domestic Lighting through the Ages* (Rutland, Vt.: Charles E. Tuttle, 1958), 6.

2. Davidson, *Woman's Work.*

3. Ibid., 107.

4. Ibid., 105.

5. Thwing, *Flickering Flames*, 119.

6. *World Book Encyclopedia*, 12:52.

7. Thwing, *Flickering Flames*, 59-62.

8. Davidson, *Woman's Work*, 112.

9. R. J. Forbes, *Studies in Ancient Technology*, 2d. ed., vol. 6. (Leiden: Brill, 1966), 125.

Chapter 10

1. Gimpel, *The Medieval Machine.*

2. *Encyclopaedia Britannica.*

3. *World Book Encyclopedia.*

4. Reynolds, "Medieval Roots," 122-30.

5. Michael B. Lafferty, "Windmill Aided Nation's Growth," *Columbus Dispatch*, 6 July 1986, H2.

6. Robert D. Kahn, "Harvesting the Wind," *Technology Review*, November-December 1984, 58.

7. *Encyclopaedia Britannica*, 4:970-73.

8. Michael B. Lafferty, "Lighting Up Their Lives: REA Turned Farms on 50 Years Ago." *Columbus Dispatch*, 26 May 1986, F1.

9. Brian Dumaine, "Arco's Shortcut to Cheaper Sun Power," *Fortune* 112 (4 February 1985):74.

10. Gimpel, *The Medieval Machine*, 67.

11. Ibid., 63-67.

12. *Encyclopaedia Britannica*, 13:966.

13. Ibid., 18:809.

14. Gimpel, *The Medieval Machine*, 168.

15. Giscard d'Estaing, *The World Almanac Book*, 279.

16. Regina Schrambling, "On the Menu," *American Way* 18 (9 July 1986):101-3.

17. Betsy Morris, "Homes Get Dirtier as Women Seek Jobs and Men Volunteer for the Easy Chores," *Wall Street Journal*, 12 February 1985, 31.

Chapter 11

1. Giscard d'Estaing, *The World Almanac Book*.

2. *Encyclopaedia Britannica*.

3. *Foxfire Book*, 230-48.

4. Hall, "Food Packaging," 21; Kessler, "Tremors from the Tylenol Scare," 59, 62.

5. *1982 Medical and Health Annual* (Chicago: Encyclopaedia Britannica, 1983), 231.

6. Ibid., 233-34; David Stipp, "Toxic Shock Linked to Certain Fibers Used in Tampons," *Wall Street Journal*, 6 May 1985, 10.

7. Melvin E. Ring. *Dentistry: An Illustrated History.* New York: Harry N. Abrams, 1985.

8. "Feeding Baby through the Ages," *Today's Health* 42 (April 1964):38-41.

9. *1982 Medical and Health Annual*, 327.

10. Jack Gillis and Mary Ellen R. Fise, *Childwise Catalog: A Consumer Guide to the Safest and Best Products for Your Children Newborn through Age Five* (New York: Pocket Books, 1986), 30-31.

11. "Fetal Surgery outside the Womb," *Reader's Digest*, February 1987, 120.

12. *"Good Housekeeping" Woman's Almanac* (New York: Newspaper Enterprise Assoc., 1977), 29.

13. *Encyclopaedia Britannica,* 2:1066.

14. *Science and the Future, 1987* (Chicago: Encyclopaedia Britannica, 1986), 423.

15. Weiser and Arbeiter, *Womanlist*, 72-73.

16. "The Morning-after Pill," *Time* 128 (26 December 1986):64.

17. Robert Zinti, "New Heat on an Old Issue," *Time* 125 (4 February 1985):17.

Chapter 12

1. *Foxfire: Books 1 through 5*, ed. Elliot Wigginton (Garden City, N.Y.: Anchor Books, 1972-79).

2. Cowan, *More Work for Mother*, 218-19.

3. Morris, "Homes Get Dirtier," 31.

4. Lowenstein, "Aqueduct Carries Hope," 32.

5. Joann Vanek, "Household Technology and Social Status: Rising Living Standards and Status and Residence Difference in Housework," *Technology and Culture* 19 (January 1978):374 (table 3).

6. Charles A. Thrall, "The Comparative Use of Modern Household Technology," *Technology and Culture* 23 (April 1982):193.

B.C. Chronology

4400	Earliest evidence of use of the horizontal loom, depicted on pottery dish at al-Balari, Egypt (EB).
4200	Tall-al-Asmor, Iraq, seals indicated trade in wool from Central Asia (EB).
4000	Flax used by lake dwellers in Switzerland (WB).
	Rice cultivation tools used in China (EB).
	Grapes grown in the Caspian and Black Sea coastal regions (BBB).
	Beer brewed in the Near East (BBB).
	Figs grown in Syria (BBB).
	Wine made in Near East (BBB).
	Sumerians made cheese in Mesopotamia (RM).
3999-3001	Buttons found in Mohenjo-daro (?2500-1700) (DBE).
3500-3000	Linen woven in Middle Egypt (TH).
	Plowing, raking, and manuring techniques used in Egypt (TH).
	Potter's wheel was used in Mesopotamia (TH).
3500	Olive trees cultivated in Syria and Crete (BBB).
	Cakes made from coarse ground grain used in Swiss lake dwellings (EB).
	Bronze used (WB).
3250	Evidence in Sumeria of vases made on potter's wheel (DI).
3200-3000	Cereals grown in Magdeburg-Cologne-Liege area (MM).
	Sumerians made soap from alkalies and fats (DBE).
	Pear used as a medicine in Sumer (BBB).
3200-2780	Laborers depicted in Egyptian tomb decorations eating onions (EB).
3000-2501	Oil-burning lamps used (TH).
<3000	Sugarcane grown in India (BBB).
3000	Egyptians glazed pottery (WB).
	Cotton used in India (EB).
	Asparagus cultivated in Egypt (RM).
	Writing used in the Tigris-Euphrates Valley (WB).
	Soybeans grown in China (BBB).
	Potato cultivated in Peru (BBB).
	Sugar-making techniques employed in India (EB).
	Wheel used (WB).
	Peas grown in the Near East (BBB).
	Irrigation used in the Middle East (WB).
	Radish cultivated in China (BBB).
2999-2001	Buttons used in northern England and Scotland (DBE).
	Safety pin used (but without protected end) (DBE).

2900	Evidence of milk drinking depicted in frieze at Ur, southern Iraq (BBB).
2800	Latrines with crude drains used in Neolithic stone huts on the Orkney Islands (EB).
2737	Tea drunk in China, recorded in the diary of Emperor Shen Nung (BBB).
2700	*I Ching* advised use of quicksilver as an oral contraceptive (WA).
	Rhubarb grown in the eastern Mediterranean lands and in Asia Minor (BBB).
	Horse domesticated in the Ukraine (EB).
2640	Legend reported silk developed by Chinese Empress Hsi-ling Shih (WB).
2600	Leavened bread made in Egypt (BBB).
2500-2000	Papyrus used in Egypt (TH).
	Domesticated chickens kept in Babylon (TH).
	Potter's wheel and kiln used in Mesopotania (TH).
2500-1700	Bathrooms used in private homes in India (EB).
2500	Harrappas at Mohenjo-daro used brick-lined wells and sanitary drainage system of brick (EB).
	Cotton textile traces left in Mohenjo-daro, India; spindle whorls too light for wool also left (DBE).
	Egyptian records mentioned wine making (BBB).
	540 threads/inch linen made in Egypt (WB).
2400	Cantaloupe grown in Iran (BBB).
	Cotton products used in Mexico (EB).
2300	Tuyn smelting furnaces used at Ur (DID).
	Bit and bridle introduced (EB).
2200	Orange cultivated in China (BBB).
2100	Pear seeds buried in Chinese tombs (BBB).
	Sumerian physicians prescribed beer for illness (BBB).
2000	Hittite smelting furnaces used bellows (DID).
	Egyptian and Chaldean cheese molds used (RM).
	Weighted tumbler locks used on Egyptian mummy cases (DID).
	Watermelon grown in Central Africa (BBB).
	"Cold house" timber-lined ice pit used in Ur (FR).
	Woven papyrus sandals worn in Egypt (EB).
	Pre-Incan Peru cotton grave cloth used (EB).
	Chinese milked animals (BBB).
1850	Petri Papyrus described recipes to prevent conception (EB).
1800	Plums grown in Babylon (BBB).

1766-1122	Silk twill damask fragments left in China that required advanced weaving techniques (EB).
1700-1450	Covered drains for waste water and clay pipes used for water in Knossos, Crete (EB).
	Pedestrian walks built in the center of cobblestone streets in Knossos, Crete (EB).
1600	Soapstone pots used in Egypt (WB).
1500-1027	Chinese used the potter's wheel (WB).
1500-1000	Iron used in Syria and Palestine (TH).
1500	Coal used in funeral pyres in Wales (EB).
	Spring-type, pivoted scissors used in Europe (EB).
	Glass vessels made in Egypt and Mesopotamia (EB).
1483-1411	Earliest known tapestry made (EB).
1400	Fine white stoneware made in China (EB).
1350	Assyrian king depicted in frieze shaded by a long handled parasol (RM).
<1200	Hindu Vedas mention use of butter as a food (EB).
1200	Earliest record of fused enameled ware (DID).
	Ramses II, Egypt, ordered apple orchards planted along the Nile (BBB).
1100	Chinese philosophers recommend "thinking of other things" at the moment of ejaculation (WA).
1000	Palestinians salted and dried fish (FR).
	Hops used by the Egyptians to flavor beer (BBB).
	Chinese cut and stored ice (WB).
	Wicks used in lamps (DBE).
10th c	Peach cultivated in China (USAIR).
1000-900	Water supply in Jerusalem carried in subterranean tunnels (TH).
	Iron used in Greece (TH).
900	Ancient feeding pots used small piece of sponge or rag to nurse infants in Cyprus (SA).
900-800	Iron and steel used (TH).
800	Chinese developed arrack from the distillation of beer (BBB).
800-700	Sledges with heavy rollers used (TH).
	Feeding vessels for infants used in Italy (SA).
	Horse-drawn chariots used (TH).
	Etruscans used hand cranks (TH).
800-200	First iron utensils used (TH).
750	Assyrian and Babylonian carvings depicted cooking over brazier (WB).

717-707	Assyrian palace at Dur Sharrukin had jugs of water at each latrine for flushing (RM).
700-600	Assyrian water clock used (TH).
	Iron soldering used (TH).
700	King Sennacherib, Assyria, planted cotton plants (DBE).
7th c	Assyrians, Mesopotamia, imported spices from India and Persia (RM).
699-601	Traces of silk fabric left in Hallstatt sites, Switzerland (DBE).
647	Spinach grown in Iran (BBB).
600-500	Athens used piped water (TH).
	Ore smelting, casting, water level, lock and key, carpenters square, and lathe described by Theodorus, Samos (TH).
600	Part of Cloaca Maxima, Rome's main sewer, was vaulted (EB).
	Etruscans developed gold crowns and bridges for teeth (EB).
	Phoenicians prepared soap from goat tallow and wood ashes and used it as a salve (DBE).
	Chinese used heating stoves (EB).
580	Olive tree introduced to Italy (FR).
500	Artichoke cultivated in central Mediterranean (BBB).
	Carrot grown in Afghanistan (BBB).
500-450	Dams used in India (TH).
	Viticulture developed in Italy and Gaul (TH).
500-400	Wooden barrels used to store wine (TH).
5th c	Confucius wrote of the peach (BBB).
499-201	Carpets used in royal graves at Pazyryk, Altai Mountains, southern Siberia (EB).
401-399	Chestnuts eaten by Xenophon's army in Asia Minor (SSVFN).
400-300	Nailed horseshoes left at Maiden Newton, Dorset (DID).
4th c	Greeks used a "goffering" iron, a cylindrical heated iron bar, to produce pleats in linen garments (EOOET).
	Fabric printing done in India (EB).
>390	Theophrastus described vinegar (used with lead for pigments) (RM).
371	Aristotle and his pupil Theophrastus mentioned coal use (EBHFB).
350	Archestratus's Greek cookbook, *Hedypatheia*, written (EB).
	Aristotle recorded an experiment for removing salt from sea water (DID).

Lacedaemonians, Greece, centrally heated floor of Great
Temple, Ephesus, with flues laid under floor (EB).

328   Alexander the Great found the Indians mining five kinds of
salt (BBB).

327   Alexander's troops found the banana in the Indus Valley
(SSVFN).

326   Alexander found rice and sugarcane grown in India (EB).

325   Sugar first mentioned in western India by Nearchus (DID).

312   Aqua Appia Aqueduct of Rome built (EB).

305   Fourteen Roman aqueducts existed, consisting of 359 miles
with fifty stone arches (EB).

300   Cotton imported into Europe, but was costly (TH).

Greeks and Egyptians used hollow calamus reed pens; they
filled the core with ink and squeezed the reed to release
ink (WB).

Sanskrit law imposed fines for sale of adulterated grains
(MMGT).

>300   Double-acting piston bellows for continuous blast used in
Chinese smelting furnances (DID).

3d c   Bronze Egyptian scissors first used (RM).

Peach reached Persia (BBB).

Peaches cultivated in Persia.

250   Philo of Byzantium suggested use of bronze springs for
catapults (DID).

220   Mandarin oranges grown in China (BBB).

200-150   Gears and ox-driven water wheel used (TH).

200   Chinese used natural gas to evaporate salt from brine (EB).

Pressure aqueduct used at Pergamon with a reservoir at
1,200 feet and cistern at 391 feet (EB).

Asparagus grown in the eastern Mediterranean (BBB).

2d c   Marcus Portius Cato (the Elder) in *De Agricultura* described
grafting of apple stock (PW).

Athenaeus, Greek, wrote *Deipnosophistal*, Greek cookbook
(EB).

Chinese officials responsible for the purity of foods
described in Chinese classics (MMGT).

170   Bread first made by bakers of ancient Rome for public sale
(DID).

150   Lime mortar made by Romans at Pozzuoli (DID).

140-86   Cucumber introduced from India to China (EB).

100   Silk road to China established (EB).

Embroidered carpet used at Noin Ula, northern Mongolia
(EB).

80	Gaius Sergius Orata devised Roman central heating, the hypocaust (AE).
75	Pompey, Roman general, returned to Rome from China carrying silk (EB).
63	Strabo recorded the existence of a water mill at Cabria (MM).
50	Sharpened goose quills used in Rome for pens (WB).
	Potter's wheel used in southern England (DID).
49	Julius Caesar used Aristotle's process of desalting water to supply drinking water to legions (DID).
40	Crank-handled winnowing fan used in China (DID).
30	Blowpipe for glassware used in Sicily (EB).
27	First large Roman public bath built by Agrippa (RM).
27-14	Glass used for windows in Rome (EB).
25	Collapsible umbrella buried in Korean tomb (RM).
22-11	Vertical (post) windmill described by Vitruvius, a Roman (DBE).

A.D. Chronology

Dates	Event
1st c	Romans imported peaches (BBB).
	Vitruvius's *De Architectura,* ten volumes on building materials, hydraulics, and town planning, published (EB).
	Pompeii used a drainage system for waste water, rain, and public baths (EB).
	Romans imported peaches from Persia (BBB).
1	Carruca, heavy plow, used in the foothills of the Italian Alps (DBE).
	Date-marked candle left at Vaison near Orange (DID).
14-37	Apicius, Roman epicure, wrote a cookbook (EB).
50	Athenaeos, Attalia, *Book on the Purification of Water*, discussed filtration and natural desalting (RF).
	Window glass manufactured in Rome (WB).
74	Public conveniences in Rome used running water (RF).
77	Pliny mentioned bronze needles (DID).
	Pliny described the purification of water (RF).
	Pliny indicated Romans preferred German to Gallic soap (prepared with natural soda and fats) (DID).
	Pliny in *Historia Naturalis* described thirty-six varieties of apples (PW).
79	Candy-maker's tools left in ruins of Herculaneum (BBB).
	Candle-making apparatus was left in the ruins of Herculaneum (DID).
	Needles were left in the ruins at Pompeii and Herculaneum (EB).
2d c	Chinese scholar Yang Fu extolled the banana in his *Record of Strange Things* (BW).
100	Pivoted steel scissors were used in Rome (RM).
	Romans in Britain used coal for heating (EB).
>100	Soranus of Ephesus, Greek gynecologist, described contraceptive methods then in use (EB).
105	Paper manufacture began in China (EB).
180	Rotary ventilating fan used in China (DID).
200	Pastry crimper was used in Roman Gaul (EL).
	Chinese added treadle to spinning wheel and used it for silk reeling (EB).
>200	Galen mentioned the use of soap for cleaning (DBE).
286	Sugar first mentioned in China (DID).

290	Chinese used trip hammer mechanism for hulling rice (MM).
4th c	Romans use muscinium for mucus (handkerchief) (RM). House bellows described by Decimus Magnus Ausonius (DID).
300	A print block probably used for textiles was included in a burial at Akhmin in Upper Egypt (EB).
>300	Ausonius described water-driven sawmill on River Moselle used to cut marble (DID).
350	Tea cultivated in China (BCT).
5th c	Seville tailors used pivoted scissors (RM).
400	Clergy wore stockings as a sign of purity (EB).
>400	St. Jerome mentioned the jointed flail (DID).
>450	Turkish tribesmen barter for Mongolian tea (USAIR).
500	Aetius, Greek physcian, recommended the use of half a pomegranate to catch semen for contraception (WA).
552	Europe silk industry was established (TH).
7th c	Arabs introduced banana cultivation into northern Egypt (BBB).
>600	Horizontal windmill used (DID).
618-907	Early form of porcelain made in China (EB).
630	Cotton was introduced to Arab countries (TH).
635	Quills first used for pens according to St. Isidore of Seville (they had been in use in Rome, 50 BC) (DID).
644	Earliest reference to windmills was to Abu Lu'lu'a, Persian windmill builder (EDE).
658-749	Japanese Buddhist priest Gyoki planted tea gardens in temples he built (BCT).
680	Crepe (fabric) first made, Bologna, Italy (DID).
8th c	Arabs learned how to raise silk worms (EB). Cotton grown on Greek mainland (EB). Moors introduced rice to Spain (WB).
700	Waterwheels used in Europe (TH).
>700	Arabs introduced orange to Spain (BBB). Three-field crop rotation used (MM). Soap used in most of southern Europe (EDE).
710	Sugar grown in Egypt (TH).
724	Chinese invented the first clock with an escapement (EB).
729	Emperor Shomu, Japan, inspired one hundred Buddhist priests to plant tea with his gift of powdered tea (BCT).
750	Hops used as beerwort for first time in Bavaria (TH). Beds were popular in France and Germany (TH).
751	Paper manufacture reached Samarkand, Central Asia (EB).

756	Arabs began systematic irrigation of Spanish river valleys (RF).
767	Pope Paul I gave Pepin the Short a jeweled, short-handled umbrella (RM).
>768	Charlemagne introduced linen manufacture to France (WB).
780	Chinese taxed tea (BCT).
793	First paper made in Baghdad (EB).
<800	A crown made of sugarcane described in the *Athavaveda* a sacred Indian text (EB).
800	Japanese began tea cultivation (EB).
	Rigid horse collar adopted in Europe; it increased the load a horse could pull (MM).
800-900	Horseshoes used in Yenesei Region, Siberia (DID).
>800	Rushlights used in England (FFH).
9th c	Moors built waterworks in Cordova, Spain (EB).
	Brie cheese developed in France and served at court of Charlemagne (SKY).
805	Dengyo Dashi, Japanese Buddhist student, returned from China with tea seed and planted it (USAIR).
>827	Arabs introduced silk manufacture to Sicily (EB).
870	Calibrated candles used to measure time (TH).
886-911	Horseshoes mentioned by Leo IV of Byzantium (DID).
890	Scrapped horn lamps used by King Alfred (DID).
>890	Horse for plowing in Norway (MM).
10th c	Moors introduced distillation to Europe (EB).
>900	Horizontal windmill used in Arab countries (DID).
900-1000	Horizontal windmill used in Low Countries (DID).
942	Linens and woolens were manufactured in Flanders (TH).
960	Mariolles cheese produced at Abbey of Thierache, France (RM).
970	Thao-Ku recorded Chinese "light-bringing slave," five-inch stick later sold in Hanchow markets (DID).
987	Watermill used for the manufacture of beer at Saint-Sauveur, Montreuil-sur-mer (MM).
>990	Heavy wheeled plow used (MM).
996	Venice imported cane sugar from Egypt (TH).
	First use of coffee as a beverage recorded by Avicenna, Arabian philosopher and physician (1ST).
11th c	Lemons grown in Burma (BBB).
	Italians produced brandy from wine (EB).
	Naples, Italy, was the birthplace of pieca, forerunner of the pizza (RD).
1002	Arabs introduced the bitter orange to Sicily (BBB).

1010	First useful application in the West of a fulling mill on River Serchio in Tuscany (DID).
	Water-driven hammer used at Schmidmuhler (MM).
1049	Coal discovered near Liege (HFB).
>1057	Venetian daughter of Byzantine emperor, Constantine X Ducas, used a small gold fork (PA78).
1086	Fulling mill in Normandy used for cloth finishing (MM).
1088	Water-driven mechanical clock devised in Peking (1ST).
1095	Records of the priory of Saint Sauveur-en-Rue, Forez, mentioned coal (MM).
12th c	Hopi Indians used coal for heating and cooking (EB).
	Italians introduced silk culture into Europe (EB).
	Venice was the major cotton manfacturing city processing Mediterrean cotton (EB).
	English glassmakers first made crystal glass (WABI).
1100	Fork introduced by the wife of Italian nobleman, Doge Domenico Silvio of Venice (WB).
>1100	Florence was known for fine woolens (EB).
	Egypt and Syria produced carpets (DBE).
1138	Tanning mill belonging to Norte-Dame de Paris mentioned (MM).
1150	Cistercians, Italy, used refuse water to fertilize meadow (RF).
	Paper manufactured by the Arabs in Spain (TH).
	Moors introduced water distillation into Europe (DID).
>1150	Drawers became a male undergarment (WC).
1167	Water supply and sanitation system of the cathedral priory of Canterbury used piped water (MM).
1170	Earliest recorded tide mill used in Britain, Woodbridge, Suffolk (DID).
1180	Post-type windmill in Normandy mentioned by Leopold de Lisle (DBE).
	Glass windows used in English private homes (TH).
1185	Windmill used at Weedley, York (DBE).
1197	Broadcloth first made in England (DID).
1200	Arabs made decoction of coffee from the dried hulls of the coffee beans (BCT).
>1200	Latif, an Arab, described soap (DID).
1202	Assize of Bread, which prohibited peas or beans in bread, was proclaimed (MMGT).
1212	Tiles replaced thatched roofs in London (TH).
1215-94	Huou, China, wrote *The Important Things to Know About Eating and Drinking*, an early Chinese cookbook (EB).

1225	Cotton manufactured in Spain (TH).
1226	Glass bottles and windows manufactured at Chiddingfold, Surrey, by Laurence Vitracius (1ST).
1230	Villard de Honnecourt sketched a water-driven sawmill (MM).
1234	Freemen of Newcastle were granted a charter to mine coal (HFB).
1236	Gilbert Sandford built the first London public water conduit from Tyburn (RF).
1238	Water-powered paper mill built in Xativa, Valencia, Spain (MM).
1250	Goose quill used for writing (TH).
1253	Linen first manufactured in England (TH).
1255	Eleanor of Castile introduced Spanish carpets to England (EB).
1259	Soap manufacture done at Sopar Lane, London (1ST).
1260	William Beukelszoon, Biervleit, Holland, first cured herring by gutting and salting (DID).
1266	Sicilian weavers fled to Lucca, Italy, following French invasion and unrest in Sicily (EB).
1268	Roger Bacon proposed spectacles (DID).
	Spinning wheel used in Paris (RF).
1270	Marco Polo mentioned Chinese matches (DID).
1272	First spinning wheel was probably used, Bologna (UZ).
1273	Francesco Borghesano, Bologna, built a hydraulic spinning machine (UZ).
1275	Marco Polo reported that the Chinese wore eye-glasses (WB).
1277	Velvet first mentioned by Joinville (DID).
1279-1368	Chinese developed hard porcelain (EB).
1280	Salvine degli Armati, Florentine physicist, devised a means for magnifying objects (WABI).
1283	First European mechanical clock made by the Austin canons and set up in Dunstable Priory, Bedfordshire (1ST).
1285	Spectacles invented by the Italian monk Nicolas Bullet (DID).
1289	Sandro di Popozo's *Traite de Conduite de la Famille* mentioned spectacles (1ST).
1299	Sugar from Morocco to Durham, Great Britain (1ST).
14th c	Papaya discovered in the West Indies (BBB).
	Closed hook iron sewing needles used (EB).
1300	Alum discovered at Rocca, Syria (DID).

Chinese used sandstone and unglazed porcelain as filters (DID).

> 1300   Arabs introduced the sweet orange to Italy (BBB).

Bernardo Buontalenti invented ice cream (EDE).

Crackowe, pointed shoe with a chain to hold up the toe, worn (WB).

1300-1400   Folding chairs, stools, trestle tables, boxes with compartments, and collapsible beds used (EB).

1306   Marco Polo reported widespread use of coal in China (EB).

1307   Inventory of Edward I, England, mentioned seven forks (WABI).

1315   Florence captured Lucca and took Sicilian weavers to Florence (EB).

Majorcan philosopher Raymond Lully identified ammonia (DID).

Lyons silk industry established by Italian immigrants (TH).

1320   Lace first made in France and Flanders (DID).

1323   Water-driven bellows used for a furnace (MM).

1328   Sawmill invented (see 300, water-driven sawmill described by Ausonius); metal technology improved blade life (TH).

1335   First recorded equal hours clock at Saint Gothard, Milan (MM).

1347   Chimneys destroyed in Venetian earthquake (HFB).

1350   German tile makers made stove tiles (EB).

1350-80   Earliest alarm clock at Warzburg, Germany, used to help monks track religious observances (RMRB).

1352   Hugh of Provence painted by Tommasco da Moderna wearing eye-glasses (EB).

1359   Herring first cured in England (DID).

1366   French parliamentary decree forced Paris butchers to slaughter and clean near a running stream (MM).

1368   Charles V planted strawberry vines at the Louvre (BBB).

1374   First French cookbook of importance, *Le Viander*, published (EB).

1380   First European blast furnace in operation (MM).

Inventory of Charles V, France, mentioned twelve forks (WABI).

1384   Ordinance of Pewters' Guild, London, founded (EDE).

1389   Inventory of the duchess of Touraine mentioned two forks (WABI).

1390   Chef of Richard I, wrote *The Form of Cury*, which also contained some apple recipes (EB).

1394	One of the early French cookbooks published, *La Menagier de Paris*; it also contained apple recipes (EB).
15th c	Fugger family, Germany, founded stores (EB).
	"Eyed" sewing needles made in The Low Countries (EB).
	Wealthy Europeans use "hot box" irons (irons had a compartment for hot coals or a hot brick) (EOOET).
	Spanish introduced rice to Italy (WB).
1400	Beer was imported to Winchester from Flanders (1ST).
	Improved felt-making process developed (WB).
>1400	Drawers developed for chests (EB).
1415	London streets first lit by lanterns (DID).
1425	Hops first mentioned in England (DID).
>1436	Wife of Louis XI had three handkerchiefs (RM).
1449	Three roller wooden sugar mill invented by Pietro Speciale, Sicily, water or oxen driven (DID).
	Thomas Brightfield, St. Martin's Parish, built a water closet flushed by water piped from a cistern (CAD).
1450	Coffee known and used as stimulant in Abyssinia (DID).
>1450	Carpetmakers guild organized in Paris (WB).
1455	Cast-iron water pipes used in Dillenburg Castle, Germany (RF).
1462	"Pocket clock" mentioned in a letter to the Marchese di Manta (1ST).
1463	Silver fork mentioned in the will of John Baret of Bury (1ST).
	Beroverio, Murano, Italy, developed Venetian glass (WABI).
1468	Spaniards introduced rice to Italy (EB).
1475	Cookery book, *Honesta Volupate*, Bartholomaeus Platina, Venice, published (OCLC).
1482	Carpets recorded in the accounts of the duke of Bedford (1ST).
1483	Bone pins used in England (DID).
1484	Arculanus discussed the use of solid gold leaf fillings (WABI).
1490	Cast-iron stoves used in Alsace (WB).
>1490	Leonardo da Vinci designed a water-filled tube as part of a cylindrical glass chimney for a lamp (EDE).
1492	Columbus's crew discovered maize in Cuba (EB).
	Portuguese found the banana on the coast of Guinea and took it to the Canary Islands (SSVFN).
>1492	Columbus discovered the peanut in Haiti (USAIR).
1493	Columbus carried orange seeds from the Canary Islands to the West Indies (BBB).

Pineapple discovered by the Spanish at Guadeloupe Island (BBB).

1498   Chinese invented the toothbrush (1ST).

16th c   Mango discovered growing in Southeast Asia (BBB).

Cashews, native to Central and South America, taken by the Portuguese to India (SSVFN).

1500   Jacob Nufer claimed to have performed the first successful cesarean on his wife (1ST).

Da Vinci built the first mechanical fan for ventilation using water power (WB).

Bluing for bleaching introduced into England from Holland (DID).

>1500   Pile carpets made in England (EDE).

Spanish brought the peach to Mexico and Florida (USAIR).

Arabs roasted whole coffee bean to make coffee (BCT).

1500-1700   Mughal Indians painted and printed fabrics (EB).

1503   Pocket handkerchief used (TH).

1504-9   Peter Henlien devised the watch (EB).

1508   Herman Schinkel, Delft, printed and sold wallpaper according to trial testimony (1ST).

1509   First known wallpaper made by Hugo Goes, Steengate, York, used at Master's Lodging, Christ College (1ST).

1516   Oranges introduced to Panama (BBB).

Dyestuff indigo imported to Europe (TH).

Leonardo da Vinci invented the spinning-flyer, the first continuous movement in textile manufacturing (EB).

Spanish priest Friar Tomas de Berlanga took the banana from the Canary Islands to Santo Domingo (SSVFN).

1517   Coffee imported to Europe (TH).

Raphael portrait of Pope Leo X depicted concave eyeglass lens for nearsightedness (EB).

1519   Bernal Diaz reported Aztecs served plums to Spaniards (BBB).

1520   Chocolate imported to Spain from Mexico (TH).

Vanilla imported to Spain from Mexico (RM).

1523   Use of the wheeled plow noted by Fitzherbert (DID).

Pope Clement VII had a bath with hot and cold water taps and hot air heat (CAD).

1524   Soap first made commercially in London (it had been made earlier at Bristol) (DID).

1527   Reference made to French Stocking Knitters Guild (1ST).

1530   Bottle corks mentioned in Palsgrave's French-English dictionary (1ST).

General use of the spinning wheel in Europe (TH).
Sulphur matches first mentioned in England (DID).
Pietro Martyre d'Anhiera described rubber balls in *De Orbo Novo*, Alcala, Spain (1ST).

1533   Johann Jurgen built a flyer spindle/rotary swift that included a treadle, using da Vinci design, leading to the popular Saxony wheel (DBE).
Four pair "knytt" hose recorded in the accounts of Sir Thomas L'Estrange (1ST).
English built rotary fans to ventilate mines (WB).
Petite Catherine de Médicis popularized high heels in Paris (CDIS).

>1533   Catherine de Médicis popularized drawers, long trouserlike underwear (GBL).

1535   First glass made in the western hemishpere (WB).

1538   Cieza de Leon in 1553 in *Cornica de Peru* recorded potato cultivation in Colombia and Peru circa 1538 (WB).

1539   Sir Anthony Ashley introduced cabbage into Great Britain (1ST).

1540   Brass pins imported to England from France (DID).

1543   Metal pins made in England (DID).

1544   Tomato introduced to Italy from South America (EB).

1545   Fine steel "Spanish" needles made by an Indian in Cheapside, London (DID).
Bernard Palissy made his first enameled ware (DID).

1548   Guinea pepper plants grown in England (TH).

1550   Blasius Villafranca's *Methodus Refrigerandi*, Rome, described saltpeter-in-water process for cooling (1ST).
Sealing wax used for first time (TH).

1553   Carding machine constructed, but English Parliament forbade its use (UZ).
Cieza de Leon, Seville, Spain, mentioned the potato in *Cornica de Peru*, as grown in Colombia (WB).

1554   First recorded coffeehouse opened in Constantinople (1ST).

1555   Olaus Magnus described the use of splints in the *History of the Goths* (FFH).

1556   Tobacco introduced into Europe from Brazil by Thevet Angolene, France (TH).

1558   Jean de Lery, minister, Fort Coligny, Rio de Janiero, mentioned sugar (DID).

1559   Venetian book described beverage drunk by Arab traders, thought to be tea (RM).

1560	Portuguese Jesuit Father Gasper da Cruz described tea (BCT).
>1560	Leloir described *les calecons*, drawer-type undergarment worn by French ladies (WC).
1564	Starch introduced into England from Flanders by Dinghen van der Plasse (CD).
1565	Sir John Hawkins introduced sweet potatoes and tobacco to England (TH).
	Konrad von Gesner, Zurich, described pencils in his treatise on fossils (1ST).
1567	Ivan Petroff and Boornash Yalysheff, two Cossacks, brought word of tea to Russia (BCT).
1568	Bottled beer invented by Alexander Nowell, dean of St. Paul's, London (TH).
1571	Linen damask first made in England (DID).
1572	Hand fan introduced into England from France (DID).
1573	Potatoes itemized in accounts of Hospital de la Sangre (1ST).
	First German cane-sugar refinery began operation at Augsburg (TH).
1575	Soft paste porcelain manufactured by Bernado Buontalenti, Florence (1ST).
1578	Conrade brothers opened a faience pottery in Nevers (1ST).
1579	Loom for weaving a number of ribbons at once developed in Danzig, but the inventor was killed and the machine destroyed (UZ).
	Spanish planted orange seedlings south of St. Augustine, Florida (EB).
1580	Coffee imported to Venice from Turkey (TH).
1581	François Rousset, personal physician to the duke of Savoy, published reports of successful cesarean surgury (EMH).
1582	Peter Morice installed a water-powered pump at London Bridge for London water supply (EB).
1584	Pencils made from graphite dug at Borrowdale, Cumberland (1ST).
1585	Bartholomew Newsam constructed the first English traveling and standing clocks (TH).
1586	Sir Thomas Harriot brought potatoes to England from Colombia (1ST).
	Sir John Harrington designed and installed the first water closet in his country home at Kelston near Bath (1ST).
1587	Marmalade recorded in accounts of Wollston Hall, England (1ST).

First beer brewed by Europeans in the New World made in the Roanoke Colony, Virginia (HS).

1588   Giovanni Battista della Porta designed an incubator after those used in ancient Egypt (DID).

1589   Reverend William Lee, Cambridge, invented the stocking frame, the first knitting machine (TH).

Juoan de Alega's *Libro de Geometrica y Traca* contained small-scale illustrations for cutting cloth (DBE).

1590   First English paper mill built at Dutford, England (TH).

1592   Windmills used in Holland to drive mechanical saws (TH).

1593   Galileo devised the thermoscope, the first thermometer (EB).

1594   Sir Hugh Platt suggested steam heat for greenhouses (HFB).

1595   Jan Hugo van Lin-Schooten, Dutch navigator, published a journal of his travels that described tea customs (BCT).

1596   Tomatoes introduced into England (TH).

1598   Lin-Schooten's *Travels* published in English; contained first printed reference to tea in English (BCT).

Artificial teeth of wax, powdered coral, and pearls made by Jacques Guillemeau, France (DID).

1599   Anthony Sherley, English adventurer, mentioned coffee in an account of his expedition to Persia (BCT).

17th c   Limes discovered in Americas and were used by sailors to treat and prevent scurvy (BBB).

1600   Caspar Lehman, jewel cutter to Emperor Rudolf, began cut-glass process (TH).

Wigs and dress trains were fashionable (TH).

East India Company obtained charter (USAIR).

ca. 1600   Ear trumpet used to enhance hearing (WB).

>1600   China began cultivation of peanut from Malay Peninsula plants (USAIR).

The Dutch developed tables with leaves (EB).

1601   King Henry IV set up 200 workmen from Flanders to make tapestries in Gobelin Paris factory (TH).

1602   Peruvian guano used as a fertilizer in Portugal (DID).

1605-17   Fynes Morrison in his Itinerary noted that Italian women wore drawers as underwear (WC).

1606   Chocolate imported to Italy (BBB).

Society of Apothecaries and Grocers and the Fruiterer's Company were founded in London (TH).

1608   Cultivated strawberries first mentioned by Sir Hugh Plat (BBB).

Forks introduced to Great Britain from Italy by Thomas Coryate (1ST).

First U.S. glass made at Jamestown, Virginia (WB).

1609   Chinese tea introduced to Europe (TH).

Caspar Lehmann of Prague granted a patent on his cut-glass process (1ST).

Tin enameled ware made at Delft (TH).

Louise Boursier, French midwife, published a book on obstetrics (WL).

Jamestown colonists grew carrots (HS).

1610   First shipments of Japanese and Chinese teas from Java to Europe (BCT).

Jeremiah Trautman, Wittenberg, performed cesarean surgury (EMH).

1611   Jamestown colonists raised cattle brought from Europe (RM).

1613   Hugh Myddleton constructed the New River, cut to bring water to London (TH).

1614   Glass industry established in England (TH).

1615   Merchant Adventurers granted a monopoly for the export of English cloth (TH).

Envelope used in Geneva (WABI).

British East India Company agents drank and acquired tea (BCT).

1616   Pineapple introduced to Italy (BBB).

1618   Mineral spring discovered at Epsom, Surrey (DID).

Czar Alexis presented with several chests of tea by the Chinese ambassador (BCT).

1619   David Ramsey and Thomas Wildgoose patented a plowing machine (DID).

Water pipes laid to bring water to individual homes in London (EB).

1620   Pierre Dupont and Simon Laudet started manufactory (for carpets) (DBE).

Mayflower cargo list recorded a wooden mortar and pestle for grinding coffee powder (BCT).

1621   Potatoes planted for the first time in Germany (TH).

1622   Potatoes imported to Virginia (RM).

1625   Italian physician Santorio used a water thermometer for medical purposes (WABI).

1626   Lace first made in England at Great Marlow, England (DID).

1627   William Brouncher, John Aprice, and William Parham
       invented a two-man plow (DID).

1629   Apple seeds and propagation wood arrived in America from
       England (RM).

       Thomas Beard, the first U.S. shoemaker, came to Salem,
       Massachusetts (WB).

1630   English colonist sampled popcorn at first Thanksgiving
       (WG).

       Sash windows installed at Raynham Hall, Norfolk, England,
       by Inigo Jones (1ST).

>1630   High heels imported from France to the United States
       (CDIS).

1631   Calico introduced into England by the East India Co. (DID).

       London Clock Makers' Company incorporated (TH).

1632   Jean Rey, French physician, independently devised a liquid
       thermometer (EB).

1633   Wind sawmill used at the Strand, London (TH).

       Thomas Johnson, London, exhibited bananas in his shop
       window (LST).

1634   William Parham, John Prewitt, Ambrose Prewitt, and
       Thomas Derney invented an engine for a plow (DID).

       Covent Garden Market opened in London (TH).

1636   Tea drunk in Paris for the first time (TH).

       John van Berg invented a threshing machine (DID).

1637   Waterproof umbrella listed in the inventory of King Louis
       XIII, France (1ST).

>1637   Conopios, disciple of Cyrill, patriarch of Constantinople,
       fled to Oxford and introduced coffee there (BCT).

1638   English settlers built a cotton mill in the United States (EB).

       Kopke, a German family, began shipping port from Oporto,
       Portugal (WJ).

1639   First apples grown in America from European stock picked
       at Beacon Hill, Boston (RM).

1640   Coke made from coal (TH).

       Massachusetts law voided marriage if woman used high-
       heeled shoes to induce marriage (CDIS).

1641   Cotton goods manufactured in Manchester (TH).

       Ferdinand II, Grand Duke of Tuscany, developed the sealed
       glass alcohol thermometer (EB).

       William Pynchon, Springfield, Massachusetts, founded the
       first U.S. meatpacking plant (WB).

1642   First cooking utensils made in the United States at Saugus,
       Massachusetts (EL).

1643	Coffee drinking popular in Paris (TH).
	Chain of Japanese pharmacies founded (EB).
1647	Captain Brocas produced the first commercially successful U.S. wine (HS).
1648	Mirrors and chandeliers manufactured at Murano near Venice (TH).
1649	Correspondent of Sir Ralph Varney asked the latter to purchase a toothbrush in Paris (1ST).
1650-1775	Jackboots, heavy, thigh-high cuffed boots, worn by gentlemen and soldiers (WB).
1650	Tea first drunk in England (TH).
	Sir Richard Westin advocated the cultivation of turnips (TH).
	First coffeehouse in England opened in Oxford, England (TH).
	Leather upholstery used for furniture (TH).
	Otto von Guericke invented the air pump (EB).
1651	Thomas Garway held the first public sale of tea in England (1ST).
	Frederick William I threatened to cut off the noses and ears of peasants who did not grow potatoes (PB).
1652	First London coffeehouse opened by Pasqua Rosee (BCT).
1654	Guericke demonstrated the air pump (1ST).
1656	First fountain pen reported from Paris by two Dutch travelers (1ST).
	Christian Huygens made the first pendulum clock (1ST).
1657	Chocolate drinking introduced in England (TH).
	First stockings manufactured in Paris (TH).
	Fountain pens manufactured in Paris (TH).
1659	Dutch settlers brought tea to U.S. (RM).
1660	Chocolate and vanilla were brought to France with Maria Theresa (RM).
	Croatian regiment marched in Paris with bright handkerchiefs around their necks; *cravat* is French for Croatian (RM).
	Water closets imported from France to England (TH).
	French merchants in Marseilles drank coffee (BCT).
	Samuel Pepys wrote in his diary: "I did send for a cup of tea (a China drink) of which I never had" (BCT).
1662	Charles II and Queen Catherine, England, popularized tea drinking (USAIR).
1663	Samuel Pepys used "silver reservoir pen" presented to him by William Coventry (1ST).

1664	Dr. Robert Hooke suggested the production of artificial silk (DID).
>1664	Frederick William forced German peasants to plant potatoes to alleviate food shortages caused by war (RM).
1665	Road-coach springs introduced into England (DID).
1666	First cheddar cheese made (TH).
	Cornelius Drebbel of Alkmaar, Holland, invented an incubator (Athenor) fitted with a thermostat (DID).
1667	Feltmakers Company, London, incorporated (TH).
1668	East India Company imported its first shipment of tea to England (USAIR).
	English Parliament forbade the importation of Dutch tea and granted a monopoly to the East India Company (USAIR).
1670	Parisian Café Procope introduced ice cream to the general populace (RM).
	Hudson Bay Company granted a charter (EB).
	Hannah Wooley wrote first cookbook by a woman, *The Queenlike Closet or Rich Cabinet* (EB).
	Coffee introduced into Germany (BCT).
	Dorothy Jones, Boston, granted a license to sell coffee (HS).
1671	First Marseilles coffeehouse opened (BCT).
1674	Lead glass made by George Ravenscroft of England (WB).
1675	Samuel Morland's water pump was in commercial production and used in his own London home (CD).
	Elector of Brandenberg served coffee (BCT).
>1675	Rice cultivation introduced to South Carolina (WB).
1677	Edward Melthorpe and Charles Milson invented a machine for hulling barley and pepper (DID).
	Ice cream was a popular dessert in Paris (TH).
	Robert Hooke recorded Sir John Hoskins's way of washing fine linen in a bag strained by a wheel (DBE).
	Johann Kunchel described aqueous solution of ammonia (DID).
1678	First German coffeehouse opened in Hamburg (TH).
	Sir John Carew, Beddington, Surrey, had an automatic flushing lavatory (1ST).
	French botanist Marchand demonstrated to the Academie des Sciences how to culture mushrooms (RD).
1679	Denis Papin, France, demonstrated the steam digestor with safety valve, the first steam pressure cooker (DID).
	London merchant opened a Hamburg coffeehouse (BCT).

1680 Blue Anchor Tavern opened in Philadelphia; it later became one of earliest U.S. restaurants (RM).

Robert Boyle used sulphur-tipped splints drawn through phosphorus-impregnated paper (DID).

Earliest surviving example of flocked wallpaper used at Worcester (EB).

>1680 Pecans cultivated in El Paso, Texas (SSVFN).

1681 London streets lit by suspended oil lanterns (DID).

1682 Weaving mill with one hundred looms established in Amsterdam (TH).

1683 Wild boars became extinct in England (1ST).

First coffeehouse in Vienna opened and Viennese coffee served (TH).

1684 Modern thimble invented by Nicolas van Benschoten, Amsterdam (1ST).

Tea grown in Java as an ornamental shrub by German physician Andreas Cleyar (BCT).

1685 French Huguenots began silk manufacture in Great Britain (TH).

1686 John Finch, John Newcomb, and James Butler invented a wire screen sifting machine for grain processing (DID).

Ice cream bought for James II was recorded in accounts of the lord steward (1ST).

1687 Coffee grinder invented (WABI).

1688 Plate glass cast for the first time by Louis Lucas, France, and used for mirrors (WB).

1689 Coffee drinking reached America (BBB).

1690 Denis Papin, France, devised a pump with a piston raised by steam (TH).

Tea licensed for sale in Massachusetts (BCT).

First identifiable English silver teapots (500).

Toothbrushes sold in London (1ST).

1693 Iron-clad roller sugar mill introduced by George Sitwell (DID).

Public drinking fountain erected at Hammersmith, London, by Sir Samuel Morland (1ST).

1694 Lamp posts erected in the City of London by Convex Light Company (1ST).

1695 Thimbles manufactured by John Lofting, Islington (1ST).

Epsom salts first prepared from Epsom Springs by Nehemiah Grew (DID).

1696 Jonathan Swift mentioned an umbrella in *A Tale of a Tub* (1ST).

	Ann, duchess of Hamilton, ordered a mangle from Edinburgh (CD).
1698	Thomas Savery, London, patented the first working steam engine for raising water (CD).
1699	Edmund Heming patented a device for sweeping the streets of London (EL).
1700	Bathroom with hot and cold running water installed at Chatsworth, Derbyshire (1ST).
	Commode became a popular piece of furniture (TH).
18th c	Spanish missionaries introduced almond trees into California (NFFOF).
	Bookcases with adjustable shelves devised in England (EB).
>1700	Wool cloth filter for sugar refining first mentioned (DID).
	Bureau à cylindre, a French desk with solid, curved wooden cover, used (USAIR).
	Secretaire à abattant, a French desk with hinged flat cover, drawers, and pigeonholes, used (USAIR).
	Gateleg tables developed in England (EB).
1701	Reference by William Burnaby in his play *The Ladies' Visiting Day* to contraceptive sheath (1ST).
	Royal charters granted to weavers in Axminster and Wilton for making carpets (TH).
	Jethro Tull devised the seed drill (1ST).
1702	Corkscrew mentioned in an analogy in a scientific article on the tadpole (RM).
	Steam engine for draining mines manufactured by Thomas Savery (1ST).
1704	Jeweled watch movement patented by Facio de Duillier and Peter Debaufre (1ST).
1706	Twinings Tea and Coffee Merchants began selling tea (USAIR).
1707	E. W. von Tschirnhaus and J. F. Bottger discovered the secret of making "hard" porcelain like Chinese (TH).
	Denis Papin invented the high pressure boiler (TH).
1707-10	Oranges planted in Arizona (BBB).
1710	Porcelain factory established in Meissen, Saxony (TH).
	Bidet recorded as used by Mme de Prie, noted Parisian beauty (1ST).
1712	First practical steam engine erected by Thomas Newcomen, Tipton, Staffordshire (used for thirty years) (DBE).
	Calico printery established by George Leason and Thomas Webber, Boston, Massachusetts (FFF).

1712-40	New England whalers discovered spermaceti wax from whales made excellent candles (CD).
1714	Dutch botanical garden gave coffee plant to Jardin des Plantes, Paris (BCT).
	Gabriel Daniel Fahrenheit devised the mercury Fahrenheit thermometer (EB).
	British patent issued for the typewriter (EB).
1715	Thomas Martin invented a machine for cleaning maize (DID).
	Collapsible-shaft folding umbrella manufactured by Marius, Paris (1ST).
1716	Bifocals first suggested (EDB).
1717	Silk throwing machine invented by John Lombe, copied from an Italian machine (DID).
	John Desaguliers improved Thomas Savery's engine (CD).
1719	Irish migrants began potato cultivation in New Hampshire (RM).
	John Akerman made cut glass in Great Britain (1ST).
	Lombe Brothers built the first British textile factory, the Derby Silk Mill (EB).
1720	Mustard first sold in London in paste form (1ST).
	First tinware made in United States at Berlin, Connecticut (EL).
	Wallpaper fashionable (TH).
	Broccoli was introduced into England by Italians (USAIR).
1722	Horse-drawn threshing machine devised by Due Quet, France (DID).
1723	French naval officer, Gabriel Mathieu, of Martinique stole coffee seedling from Paris Jardin (BCT).
1725	George Woodroffe invented a machine for cleaning wheat (DID).
	Basile Boucher improved the "draw-boy" on a loom, which was important for pattern weaving (EB).
1726	Portable bidet for officers on campaign offered in Paris (CAD).
	First bimetallic temperature responsive device, a "grid-iron" pendulum built in England for a clock (EA).
1727	Coffee first planted in Brazil (TH).
	Coal gas first obtained from the distillation of coal by Dr. Stephen Halcs (DID).
	Wife of French governor of French Guiana gave coffee seedlings to Brazilian officer (BCT).
1728	Perforated card used to control the loom "draw-boy" (EB).

	Lewis Paul invented and patented two carding machines (DID).
ca. 1728	Pierre Fauchard, France, published his *Le Chirurgien Dentiste*, describing dental practices and his dental crown (WABI).
1730	Stilton cheese first served at the Bell Hotel, Stilton, England (RM).
	Disney Stanyforth and Joseph Foljambe, U.K., patented the Dutch or Rotherham plow (FT).
1731	Thomas Ryley and John Beaumont invented a fodder (chaff) cutting machine (DID).
	Michael Menzies invented a threshing machine with rotary flails driven by water wheel (DID).
1733	John Kay patented the flying shuttle loom; it increased loom speed by eight times (TH).
	First baby carriage built at Chatsworth by William Ken for the third duke of Devonshire (1ST).
1736	Coffee trees transplanted from Arabia to West Indies (DID).
	First American cookbook, *The Compleat Housewife*, complied by Mrs. E. Smith, Williamsburg, Virginia (WG).
1737	Meiffen invented a horse-drawn threshing machine (DID).
1738	First verifiable cesarean operation performed in the British Isles by Mary Donally at Charlemont, Ireland (1ST).
	First clocks made in the Black Forest district (TH).
1739	Caspar Wistar reestablished U.S. glassmaking at Salem, New Jersey (WB).
1740	Cookery school established by E. Kidner, London (1ST).
	"Pen without end" invented by Jean F. Coulon de Thevenot, France (DID).
	Franklin stove developed for more efficient heating (RC).
	Steel smelting using "crucible" process improved by Benjamin Huntsman (TH)
1741	Artificial mineral water prepared by Dr. W. Brownrigg using carbonic acid gas (1ST).
1742	Cotton factories established in Birmingham and Northampton (TH).
	Anders Celsius devised the Celsius or centigrade temperature scale (EB).
	Faneuil Hall, Boston, was built as a market and meeting place (WK).
1743	Sheffield plate, originally used for buttons, developed by Thomas Boulsover (1ST).

East Indian yarns imported to England for finer cotton goods (TH).

1745   Colonel Cook applied steam heating to heat a home (DBE).

1746   John Roebuck invented a method of mass producing sulphuric acid (used in bleach manufacture) (EB).

Modern type of eyeglass frame invented by French optician, Thomin (WABI).

1747   Beet sugar extracted from sugar beetroot by Andreas Marggraf, Berlin (TH).

1748   Brandy first distilled from potatoes (DID).

Dr. Cullen used a vacuum hand pump to produce liquid refrigeration (DBE).

Steel pens introduced by Johann Jansen, Aachen, Germany (1ST).

Ball valve for water supply systems was introduced, essential for modern plumbing and control (1ST).

François Fresneau described a waterproof overcoat and boots coated with latex sap (WABI).

1750   Johann H. G. von Justy, Germany, suggested use of enameling for cooking utensils (EL).

Machine developed in Holland for reducing rags to fiber for papermaking (EB).

> 1750   Tambour, English desk with a cover of wooden strips glued to canvas or linen backing, popular (USAIR).

Magalhaens, Portugal, invented an eraser (WABI).

1751   Sugarcane planted in Louisiana (WG).

1752   Ben Franklin installed the first lightning conductor, Philadelphia (1ST).

1754   John Smeaton invented the triangular sugar mill with three rollers (DID).

Oil cloth first made by Nathan Taylor at Knightsbridge, London (DID).

First iron-rolling mill established at Fareham, Hampshire, England (TH).

1755   Grapefruit found in Barbados; it may have been a mutant of pummelo from Southeast Asia (SCI).

1756   First chocolate factory in Germany established (TH).

Porcelain factory established at Sevres (TH).

Cotton velvet first made at Bolton, Lancashire, England (TH).

Dilute sulphuric acid used for bleaching by Hulme (DID).

1757   Brocade first made at Lyons, France (DID).

1758 Ribbing machinery for the manufacture of hose invented by Jedediah Strutt (TH).

Lace-making machine invented by Strutt (DID).

William Bailey invented a washing machine, an improvement on "machine washing-tubs" then in use (CD).

Blast furnace installed by John Wilkinson, Bilston, Staffordshire, England (1ST).

1759 Angélique le Boursier du Coudray, descendant of Louise, published a book on midwifery (WL).

1760 Josiah Wedgewood founded his pottery works at Eturia, Staffordshire, England (TH).

First silk hats imported from Florence (TH).

George Dixon lit a room in his house at Cockfield County, Durham, using coal gas (DID).

The modern tablespoon with a rounded back adopted (EB).

1761 Robert Henchliffe, Sheffield, England, made cast steel scissors for domestic use (EB).

Steam pumping used for the London water supply (EB).

1762 De Malassagny, France, invented a threshing machine (DID).

M. Moreau devised a safety razor that used a toothed guard (CAD).

Cast iron converted into malleable iron at Carron Ironworks, Stirlingshire, Scotland (TH).

Rubber solvent, turpentine, discovered by François Fresneau, Marennes, France (1ST).

1764 Spinning jenny invented by John Hargreaves (AO).

James Watt invented the steam condenser, the first step to efficient steam engine (TH).

1765 Potato widely used foodstuff in Europe (TH).

Chocolate factory established in Massachusetts Bay Colony using cacao beans from the West Indies (BBB).

John Milne designed a sifting machine driven by wind or water for grain processing (DID).

Lazzaro Spallanzani, Italy, suggested preserving food by means of hermetic sealing (TH).

First restaurant opened in Paris by A. Boulanger (EB).

1767 Soda water prepared by Richard Bewley (1ST).

Casper Faber of Stern, Germany, developed and marketed the first Faber pencils (AMERW).

1769 Oranges planted in California (BBB).

Richard Arkwright patented the water frame, a water-powered spinning machinery (EB).

Spaniards introduced domestic pigs to California (CDIS).

Description des Arts et Metiers issued by M. Garsault, Paris, had diagrams for cutting cloth (DBE).

1770   James Edgill invented a spiral-knifed fodder cutting machine (DID).

James Sharp invented a bean-spitting mill and winnowing machine (DID).

Sixteen-spindle spinning jenny patented (AO).

John Champion of London patented an incubator (DID).

India rubber eraser sold in London (1ST).

Hand plunger designed specifically for manipulating wash sold (WB).

J. Priestly mentioned an eraser (WABI).

A. Duchateau, a French pharmacist, attempted to make dentures from mineral paste (WABI).

>1770   Cast iron ovens began to appear in England (CD).

Women an Anglo-Saxon countries wore pantaloons as underwear (GBL).

1771   Sir Richard Arkwright operated the first spinning mill in England (TH).

First synthetic dye for cotton invented by Dr. R. William (DID).

1773   John Fleming patented the Wallerer wheel for sugar mills (DID).

Richard Arkwright produced all-cotton calico (EB).

1774   Hugh Oxenham, England, patented a mangle (CD).

C. W. Scheele discovered chlorine (DID).

Joseph Priestly isolated ammonia gas (DID).

John Wilkinson's boring mill facilitated the manufacture of steam engine cylinders (TH).

1775   George Robinson designed a man- or horse-operated dressing machine for grain (DID).

Silk winding machine invented by Leture, Lille (DID).

Charles F. Weisenhall patented a double-pointed needle (EDE).

Alexander Cummings patented elements of the modern valve water closet (CAD).

James Watt perfected the steam engine (TH).

Pierre-Simon Girard invented the water turbine (TH).

>1775   Audiphone or dentiphone hearing aid used (WB).

1777	Central heating installed at Chateau de Pecq by the architect M. Bonnemain (1ST).
	S. Miller invented a circular saw (DID).
1778	Eli Whitney at age thirteen is supposed to have devised an apple parer (FHTC).
	Improved water closet constructed by Joseph Bramah, Yorkshire (TH).
	Robert Barron, England, patented a double-acting tumbler lock (EB).
	Dubois des Chement, Paris, produced durable dentures using plaster impressions (WABI).
	Joseph Bramah patented his water closet design, improving on Cummings's 1775 design (1ST).
1779	Samuel Crompton's mule with 1,000 spindles operated by one person (EB).
	Steam-driven factory equipment installed at James Pickard's button factory, Birmingham (1ST).
	Dancers in Paris wore drawers for dancing (GBL).
1780	Thomas Robinson patented the open range in England (CD).
	Thomas Turner introduced Willow pattern china (1ST).
	Casein was discovered by C. W. Scheele (later used in making plastics and nylon yarns) (DID).
	Cylindrical steel pens cut like quills made by Harrison, Birmingham (DID).
	Felice Fontana produced water gas (TH).
	Phosphorus matches used in England (DID).
	"Phosphorus candle" of phosphorus-impregnated paper in glass tube introduced to France (DID).
	Zadoc Benedict, Danbury, Connecticut, opened the first U.S. hat factory (WB).
>1780	Camembert cheese first produced near Vimoutiers, Normandy (RM).
1780-82	Cast iron boilers for heating water made in England (CD).
1781	Joseph Sterling, South Woodstock, Vermont, developed an apple parer (FHTC).
	Thomas Jefferson listed *tomatas* in his garden journal at Monticello, Virginia (USAIR).
1782	First luxury restaurant, Le Grande Taverne de Londres, Paris, opened (EB).
	Josiah Wedgewood developed a pyrometer to check temperatures in pottery furnaces (TH).

Antonine Germondy of Lyons patented a hand- or water-operated carding machine (DID).

James Watt invented the double-acting rotary steam engine (TH).

Saint-Louis, France, successfully fabricated crystal glass (WABI).

1783 Copper-cylinder calico printing developed by Henry Bell (TH).

1784 Andrew Meikle, Scottish millwright, invented the threshing machine (TH).

George Washington's expense ledger recorded the purchase of a "cream machine" that used ice (RM).

Eighty-spindle spinning jenny patented (AO).

Fournier de Granges, France, improved Germondy's carding machine (DID).

Simon Pla, France, invented a mule-powered carding machine (DID).

Aime Argand invented the Argand lamp, an iron chimney with hollow wick was used to reduce flickering, smoke, and smell (DBE).

Joseph Bramah developed the Bramah lock using a small metal tube for a key (EB).

Henry Cort introduced the first puddling process for manufacturing wrought iron (TH).

Benjamin Franklin invented bifocals (EB).

Massachusetts law penalized the seller of diseased, corrupted, or unwholesome provisions (MMGT).

Embossing machine for the blind devised (EB).

1785 Soda water manufactured by H. D. Rawlings, London (1ST).

Bench-type apple parer made by Daniel Cox and Sons, West Woodstock, Vermont (FHTC).

James Watt and Matthew Boulton installed a rotary motion steam engine in a cotton spinning factory (1ST).

E. Cartwright patented the first power-driven loom (EB).

Gosselin, Amiens, constructed a fold-up, pocket umbrella (RM).

First human artificial insemination conducted by M. Thouret on his wife (1ST).

C. L. Berthollet invented chemical bleach, Eau de Javal, which was adopted by the Lille bleachers (TH).

Hulme, U.K., used dilute sulphuric acid as a bleach (DID).

Christophe-Phillippe Oberkampf invented the first machine for printing wallpaper (EB).

1786   Hall, New York, opened the first ice cream company in the United States (RM).

Earliest attempts at internal gas lighting, Germany and England (TH).

Ezekiel Reed, American inventor, built a nail-making machine (TH).

1787   James Cooke described the modern-type chaff cutter (DID).

Ice first produced by artificial means (DID).

M. van Marcum and A. Paeto van Troostwijk liquefied ammonia (DID).

Levi Hutchins, Concord, New Hampshire, developed the modern alarm clock (RMRB).

Thomas Jefferson introduced spaghetti to the United States (HS).

1789   First steam-driven cotton factory operated in Manchester, England (TH).

Lance added glass chimney to Argand lamp (DID).

Andrew Pears made Pears's Transparent Soap (CD).

1790   Edmund Card invented a comber carding machine (DID).

Thomas Saint patented the sewing machine, but there is no evidence the machine was ever constructed (1ST).

Mrs. Johanna Hempel patented the first English filter (DID).

First steam-powered rolling mills built in England (TH).

Wristwatch itemized in the accounts of Swiss watchmakers Jacquet-Droz and Leschot, Geneva (1ST).

Thomas Jefferson grew peanuts in Virginia (USAIR).

Pineapples first planted in Hawaii (HS).

John Greenwood invented the dental drill (HS).

>1790   Count Rumford developed an improved cooking stove and double boiler (HFB).

Count Rumford further improved the Argand lamp (FFH).

1791   Antonio Mendez made sugar from Louisiana cane (WG).

Serrazin, France, invented a carding machine for wool and felting (DID).

Sand-rising water filter invented by James Peacock (DID).

Nicolas LeBlanc, France, patented a process for the manufacture of sodium carbonate (used in soap manufacturing) from salt (EB).

1791-93   Sir Samuel Brunswick, Maine, devised a circular saw (DID).

1792   Thomonde, Paris physician, invented a cotton carding machine (DID).

Illuminating gas used in England for the first time (TH).

Murdoch, Scottish engineer, used coal gas to light his home (EB).

Richard Bragg's *The New Art of Cookery* contained a recipe for tomato catsup (EOOET).

1793   Eli Whitney invented the cotton gin (TH).

Hannah Wilkinson, Pawtucket, R.I., manufactured a cotton sewing thread that was superior to previously used linen thread (FFF).

1795   Nicolas-François Appert, France, designed a preserving jar for food (TH).

Silk winding machine was invented by Biard, Rouen (DID).

Nicolas Jacques Conte, St. Cenari, France, invented a pencil of pulverized graphite and potters' clay (DID).

Joseph Bramah invented the hydraulic press (TH).

Matthew Boulton used steam to heat a friend's home (HFB).

French developed a hot air dehydrator for vegetables (EB).

1796   Susanna Carter, Philadelphia, published *The Frugal Housewife: Or the Complete Woman Cook* (FHTC).

Amelia Simmons, Hartford, Connecticut, published *American Cookery* (FHTC).

Louis Martin patented a carding machine for wool (DID).

Iron aqueduct built at Longden upon Tern, Shropshire, England (1ST).

Count Rumford developed the basic principles for fireplaces and chimneys for effective heating (HFB).

James Parker rediscovered puzzolina hydraulic cement, patented by John Smeaton as Roman (hydraulic) cement (DID).

The Taylor's Complete Guide contained instructions for cutting cloth (DBE).

1797   Meat roaster developed by William Strutt, Derby (CD).

Tellie, France, improved Martin's carding machine (DID).

Nathaniel Briggs, New Hampshire, patented a washing machine (FFF).

George Jee devised an improved mangle, reducing the number of operators and strength to operate (CD).

John Etherington, London, introduced the top hat (USAIR).

Charles Newbold, Burlington County, New Jersey, farmer, developed a cast iron plow (FFF).

1798   Nicolas-Louis Robert, France, constructed a moving screen belt for continuous papermaking (EB).

1799   Joseph Boyce invented a machine for cutting wheat and corn (DID).

Combined lamp/heating unit patented by Philippe Lebon, France (1ST).

Charles Tennant of Glasgow improved the Berthollet process by inventing bleaching powder (DID).

Dr. Hinkling invented an enameling process for saucepans (DID).

Pochan, France, invented a clothes dryer (WABI).

1800 Thomas Wigful, Norfolk, invented the post threshing machine (DID).

Robert Mean invented a machine for cutting standing corn and grass (DID).

Moritz Frederick Illig, Germany, discovered paper could be sized in vats using rosin and alum (EB).

Richard Trevithick constructed the light press steam engine (TH).

French *biggin*, drip pot for making coffee and named for the inventor, introduced to U.S. (EL).

Jean-Baptiste Reveillon and Joseph Dufour were popular designers of French wallpapers (EB).

Jean Baptise de Belloy, archbishop of Paris, devised the first percolator (dripped through to pot below) (BCT).

> 1800 Trappist monks of Notre Dame de Port du Salut manufactured Port du Salut cheese (RM).

Count Rumford improved de Belloy's coffee pot by adding a device to compress grounds (BCT).

1801 First sugar beet factory was built by F. Achard, Silesia (DID).

Edouard Adam, France, invented an apparatus for distilling brandy from wine (DID).

Joseph-Marie Jacquard demonstrated an improved loom (EB).

Biscuits were called crackers after Josiah Bent opened his bakery, Milton, Massachusetts (HS).

1802 Richard Trevithick and Vivian, Cornwall, patented a steam-driven roller sugar mill (DID).

Albert Winsor, Germany, gave a dinner party with food cooked by gas (DID).

George Bodley, Exeter iron founder, patented a closed top cooking stove (EH).

Vegetable-charcoal filter introduced (DID).

Descroiselles, Rouen pharmacist, invented the cafeollete, a form of drip coffee maker (WABI).

> 1802 Antoine Cadet created a porcelain coffee pot (WABI).

1803   Improved methods of cloth take-up from loom devised (EB).
Bryan Donkin invented the first two-piece steel pen nib
(DID).
Joseph Wise invented the metal-nibbed "perpetual pen"
(DID).
Moses Coates, Chester County, Pennsylvania, patented the
first mechanized apple parer (LR).
Robert Ransome, England, developed the tempered cast
iron plow (OFTAM).
Thomas Moore, Baltimore, Maryland, applied the term
refrigerator to his improved ice box (FFF).

1804   Nicolas Theodore de Saussaur, France, discovered saltpeter
promoted the growth of cereals (DID).
Nicolas-François Appert, France, began commercial
application of his technique for canning food (1ST).
More than 500 restaurants were operating in Paris as a
consequence of the French Revolution (see text for
fuller explanation) (EB).
Filtered water supply used in Paisley, Scotland (1ST).
Chester Chronicle reported "drawers of light pink now to ton
among our dashing belles" (DID).

1805   Steam "cane-engine" built by James Cook of Glasgow (DID).
Oliver Evans, U.S., proposed machine for creating ice (WB).
Jacquard introduced the jacquard loom (EB).
Gasworks established by National Gas Light and Heat
Company, London (1ST).
Matches tipped with potassium chlorate and sugar dipped in
sulphuric acid to ignite invented by Chanel (DID).
Semi-elliptical laminated springs were patented by Obadiah
Elliot (DID).
Thomas Plucknett patented a lawn mower (FFF).
Antonine-Augustin Parmentier, France, first produced
powdered milk (WABI).

1806   Frederic Tudor, Boston, shipped ice to Martinique, later to
Charleston, New Orleans, and India (WJ).

1807   Fruit flavored carbonated soft drinks were manufactured by
Townsend Speakman, Philadelphia (1ST).
Benjamin Silliman, Yale College, began producing bottled
soda water commercially (EB).
Bertel Sanders, a Dane, perfected a means of locking
together button parts for mass production (EDE).
Illig published his method for vat sizing of paper (EB).

Henry and Seely Fourdrinier developed an improved
version of Robert's papermaking machine (EB).

Gas street lights in London (TH).

Mme Tallien popularized long trousers with lace as
underwear (GBL).

1808   Pineapple cheese first produced in Troy, Pennsylvania (RM).

Samuel Clegg used the Argand lamp (invented 1784) as a
gas burner (DID).

Florida Water, fragrant toilet water, was introduced
(UNITD).

Camus patented hook and eye fasteners (PM).

Robert Ransome, England, patented the bolted plow;
removable parts made replacement easier (OFTAM).

1809   Joseph Hawkins patented a method for preparing imitation
mineral waters (EB).

Bobbin-net machine invented (DID).

John Dickinson, English papermaker, patented the cylinder
paper machine (EB).

First two British patents were granted referring to the
fountain pen (DID).

M. Chevreul, France, isolated fatty acids--olive, palmitic,
and stearic acids--leading to improved soap (EB).

S. Crittenden, Guilford, Maine, patented a parer and corer
(LR).

Appert, France, reduced milk to one third its original
volume and sealed it in a container (HH).

1810   Leslie combined a vessel containing sulphuric acid with an
air pump to produce 1 1/2 pounds of ice (EDE).

Peter Durand patented the idea of using vessels of metal or
tin for preserving food and licensed it to Donkin & Hall
(1ST).

Closed range manufactured in Britain (CD).

Peltier Instant Light Box matches were introduced from
France to England (DID).

Lewis Mills Norton, Troy, Pennsylvania, patented his
process for making pineapple cheese (FFF).

Sarah (Sister Tabitha) Babbitt devised a circular saw (see
also 1777, 1791) (HDI).

1811   Salmon described a reaping machine with clippers and
delivery system (DID).

J. N. Reithoffer, Vienna, manufactured rubber goods (1ST).

James Hume patented a mechanical sweeper (EDE).

1812   Donkin & Hall established a cannery at Bermondsey (1ST).

Philippe Gerard invented a machine for spinning flax (TH).

James and Patrick Clark, Paisley, Scotland, manufactured cotton sewing thread (1ST).

William Harvey, Boston, patented the Nurse lamp and warmer (FFH).

James Mallory, New York, patented a lamp (FFH).

Cylinder filter invented by Paul, Geneva (DID).

John Lorain, Phillsburg, Pennsylvania, farmer, successfully crossed flint and dent varieties of Indian corn (EB).

1813　Felt first made from shoddy at Botley, Yorkshire (DID).

1814　Patent steam kitchen and range advertised in the *Morning Post*, supplied one to fourteen gallons of boiling water (CAD).

Multiple sand filter invented by Ducommon, France (DID).

LeBlanc's soda process was first used at a plant in St. Helen's, Lancashire (CD).

John Linebach developed a cotton seed hulling machine (WB).

1815　R. V. Effinger established a cheese factory at Bern, Switzerland (1ST).

James Syme, Scotland, invented a process for making waterproof cloth (DID).

Gas meter devised by William Clegg (1ST).

Pressure filter invented by Count Real, France (DID).

1816　Fishtail or batswing gas burner introduced (DID).

François Derosne, Paris, credited with making the first phosphorus friction match (EB).

1817　Gasworks, public utility, Manchester Corporation, established in Manchester, England (1ST).

Porcelain teeth produced in the U.S. (WABI).

1818　Animal-charcoal filter first used (DID).

Jeremiah Chubb, Portsmouth, England, developed his pick-proof lock (EB).

Right and left shoe lasts, wooden forms for shaping the shoe, developed (EB).

Henry Sands Brooks opened a clothing store in New York (later Brooks Brothers) (HS).

1819　First factory-scale manufacture of eating chocolate in Vevey, Switzerland, by François-Louis Cailler (1ST).

Joseph Colin, Nantes, canned sardines (1ST).

Jordan Mott's (New York) "nut coal" burning stove, used then useless anthracite coal, reducing the fuel cost (RC)

John Schaffer invented the Penographic pen with quill-shaped brass nib (DID).

Compressed air filter invented by Hoffman, Leipzig (DID).

Amalgam fillings of mercury and silver were used for dental fillings (WABI).

Rodney and Horatio Hanks, Mansfield, Connecticut, manufactured silk thread (FFF).

John Jethro Wood, Poplar Ridge, New York, developed a plow with interchangeable parts using Newbold's 1797 cast iron plow (FFF).

1820   Thomas Hancock, London, patented a process for the manufacture of rubber (1ST).

J&J Clark introduced cotton reels (spools), Paisley, Scotland, 1/2d deposit (1ST).

Braces were first made of India rubber by T. Hancock, London (DID).

Joseph Gisllot, Birmingham, invented a new steel pen nib (DID).

Professor Robert Hare, in the *American Journal of Science and Arts*, described "burning fluid," camphene (FFH).

Cambacres discovered braided wicks for candles made self-snuffing candles (FFH).

William Underwood, Boston, established a U.S. canning factory (WB).

Frederick C. Accum, German chemist, of London published *A Treatise on Adulteration of Foods* . . . (MMGT).

>1820   Robert Johnson demonstrated the tomato was edible by eating one in public and surviving (USAIR).

Large-scale peach cultivation began in the U.S. (USAIR).

1821   First manufactured raincoat offered for sale by G. Fox, London (1ST).

Fredonia, New York, used natural gas for street lamps (1ST).

Thomas Johann Seebeck, Germany, discovered that junctions of dissimilar metals generate an electric current (EB).

Sir Humphrey Davy, England, discovered that the electrical resistance of a metal depends on temperature (EB).

1822   Differential carding machine was invented by Asa Arnold, U.S. (DID).

Thomas Hancock made rubber sheets (1ST).

1823   Macintosh patented Syme's method of making waterproof cloth (1ST).

Mansfield tinsmith introduced differential roving winding gears into carding machine (DID).

Hawkins and Mordan patented the ever-pointed pencil (DID).

Hawkins and Mordan invented pen nibs of horn and tortoise shell (DID).

William de Vere and Henry S. Crane patented a process for illumination by water gas (DID).

Faraday succeeded in liquefying chlorine (1ST).

1824   Rubber galoshes advertised by W. Goodyear, Boston (1ST).

Bag or stocking filter invented by Cleland (DID).

Portland cement patented by Joseph Aspdin, Yorkshire (EI).

L. J. Vicat, France, invented cement made from chalk and clay (DID).

Noah Cushing, Quebec City, patented a washing and fulling machine (LR).

Lemuel Wellman Wright invented a pin-making machine (DID).

First department store, Belle Jardiniere, Paris, was founded by Pierre Parissot (WABI).

Samuel Clegg described a gas cooking device made at Aetna Iron Works near Liverpool (CD).

1825   Jefferson plum developed in Albany, New York (BBB).

Canned salmon was produced by John Moir, Aberdeen, Scotland (1ST).

Thomas Kensett, New York canner, patented tin-plated cans in the U.S. (WB).

M. Chevreul and J. Guy-Lussac patented improved manufacturing methods for candles using stearic acid (EB).

Bottled gas marketed by Provincial Portable Gas Company, Manchester, England (1ST).

Sanders's son replaced the metal button shank with a flexible canvas one (DBE).

Bakewell, Page, & Bakewell Company, Pittsburgh, produced the first pressed glass (WB).

Boston & Sandwich Glass Company founded by Deming Jarves, early maker of pressed glass (WB).

Hannah Lord Montague, Troy, New York, devised the detachable collar (FFF).

1826   Tea sold in packets by John Horniman, Ryde, Isle of Wight (1ST).

> James Sharp, Northampton, designed and installed the first
> gas cooking stove in his home (1ST).
> Otto Unverdorben obtained aniline from indigo (TH).
> Early birth control manual, Richard Carlile's *Every
> Woman's Book*, London, was published (1ST).
> John Walker discovered the sulphur friction match (1ST).

\>1826   C. A. Bruce cultivated wild Indian tea in India (BCT).

1827   James Simpson constructed a sand filter for the purification
of London's water supply (TH).

> First recorded sale of sulphur friction matches by John
> Walker (TH).
> Boston & Sandwich Glass Company made the first pressed
> glass tumbler (WB).
> Jacobius Jacobenson was appointed by the Dutch to start a
> Javanese tea industry (BCT).
> Jacques-Augustin Gandais, Paris jeweler, devised the first
> practical pump percolator (tube in handle) (BCT).
> Nicolas Felix Durant invented a pump percolator with an
> internal tube (BCT).
> Samuel Williston invented a machine for producing fabric-
> covered buttons (PM).

1828   Coenraad van Houten, Amsterdam, prepared cocoa by
extracting excess cocoa butter from cacao beans (1ST).

> John Mitchell, Birmingham, introduced machine-made
> metal pens (EB).
> Samuel Jones, London, produced the Promethean match, a
> paper-wrapped glass bead that contained acid (EB).
> Chili con carne devised by Mexicans living the area that
> later became Texas (HS).
> Forerunners of the modern safety razor developed in
> Sheffield, England, and the U.S. (EB).

1829   First cooperative stores in America founded in Philadelphia
and New York (TH).

> First hotel bathroom installed at Tremont House, Boston,
> Massachusetts (1ST).
> Municipal baths opened, St. George's Baths, Liverpool
> (1ST).
> Water filtration introduced by James Simpson, Chelsea and
> Lambeth (EB).
> Coffee mill patented by James Carrington, Wallingford,
> Connecticut (FFF).
> First sewing machine was built by Barthelemy Thimmonier
> (1ST).

D. G. Yuengling & Son of Pottsville, Pennsylvania, oldest U.S. brewing company was established (HS).

Henri Leroux, French chemist, extracted salicyn from willow bark, a first step toward aspirin (WABI).

Typographer patented in the U.S. (EB).

First wrapped soap, Old Brown London soap, sold by John Atkinson (1ST).

Ebenezer Brown, Troy, New York, began manufacturing detachable collars (FFF).

1830   Nitrates were first shipped from Peru and Chile (DID).

Matthew Boulton and James Watt designed a sugar mill (DID).

Sugar beet factory built at Utling, Essex (1ST).

Mott completed the development of stove plate from pig iron, thereby reducing the cost of stove manufacture (RC).

Cooking stoves using wood and coal were common (RC).

James Perry patented his steel slit pen (TH).

D. Hyde, Reading, Pennsylvania, patented a continuous flow fountain pen (PA).

Karl von Reichenbach discovered paraffin, Blansko, Moravia (TH).

Isaiah Jennings patented "burning fluid," camphene (FFH).

J. F. Kammer, Germany, invented matches that used yellow phosphorus, sulphur, and potassium chlorate (DID).

Edwin Budding, Stroud, Gloucester, patented an improved lawn mower (DID).

First elastic manufactured by Rattier et Guibal, Paris (1ST).

Crinoline, woven horsehair fabric, invented by Oudinot, France, to stiffen soldiers' collars (WABI).

Andrew Ure, Scottish professor of chemistry, was issued a patent on a "heat-responsive element" (EA)

Uniform cakes of soap for resale by grocers were sold in Newburgh, New York; previously grocers cut soap from large blocks as needed (FFF).

>1830   The Reverend Sylvester W. Graham devised graham crackers as a health food (HS).

1831   Cooperative wholesale society, North of England Co-operative Company, Liverpool, established (1ST).

Self-snuffing, stearic acid candles sold in Paris (CD).

Earthenware ascending filter invented by Lelage, France (DID).

Jacob Perkins, England, patented a high-pressure hot-water system (HFB).

First electric bell devised by Joseph Henry, Albany, New York (1ST).

Electric transformer built by Michael Faraday, London (1ST).

Charles Sauria, France, developed a method for making matches easy to ignite (TH).

Mechanics' Magazine described John Robinson's gas burner for cooking (CD).

1832   Heckling machine invented by Phillipe de Girard, France (DID).

Corrugated iron manufactured by John Walker, Rotherhithe, England (1ST).

Dynamo demonstrated by Hypolite Pixii, Paris (1ST).

Manufacture of friction matches well established in Europe (TH).

John I. Howe patented a pin-making machine (RMRB).

Jacobenson brought seven million seeds and fifteen workmen from China to Java, but without official permission (BCT).

1833   William Baddeley invented a plunger-filled fountain pen (DID).

Jobard, Belgium, invented a process for making water gas from resins (DID).

Washboard patented in U.S. (EL).

Rotating dynamo was demonstrated at Cambridge by Joseph Saxon (1ST).

Obed Hussey patented a reaper (six months before Cyrus McCormick patented his reaper) (MITU).

1834   Cyrus McCormick patented his reaping machine (TH).

Gas refrigerator developed by Jacob Perkins, Massachusetts (WB).

First commercially produced gas stove installed in Bath hotel, Lemington, England (1ST).

Solid carbon-block filter invented (DID).

William Hunt, New York, constructed one of the first sewing machines with vibrating arm and curved needle (TH).

Philo Penfield Stewart patented the Oberlin cast iron stove, manufacturing 90,000 stoves over thirty years (SG).

Xavier Jouvain, Grenoble, France, invented a die for cutting gloves (EB).

Melamine first developed (PM).

1835   Rubber nipples introduced for infant nursing bottles (HG).

Henry Porter patented Porter's Patent Portable Buring Fluid, camphene (FFH).

Electric lamp developed by James Bowman Lindsay, Dundee, Scotland (1ST).

Charles Babbage designed his analytical engine (EB).

Elias Durand, Philadelphia, Pennsylvania, commercially bottled soda water (FFF).

1836   Scientists discovered how to pollinate vanilla artificially, enabling cultivation outside Mexico (RM).

Gas stove factory was established by James Sharp in Northampton, England (1ST).

Electroplating patented in Great Britain by George and Henry Elkington (1ST).

American patent for the manufacture of white phosphorus matches granted to A. D. Phillips (TH).

Hooks and eyes were manufactured in Waterbury, Connecticut, by Holmes and Hotchkiss (FFF).

Dubus-Bonnel, Paris, patented spinning and weaving of glass for fiberglass (WABI).

1837   John Upton patented the first steam plow (DID).

John Lea and William Perrins, pharmacists, Worcester, England, made Sir Marcus Sandys's Worcester Sauce (1ST).

William Crompton introduced a fancy goods loom (EB).

William Fourness, Leeds, invented an exhaust fan (DID).

Electric motor patented by Thomas Davenport, Rutland, Vermont (1ST).

Sorel, France, developed a process for galvanizing metal (WABI).

1838   Baron von Liebig, Germany, reduced flesh food to 15 percent of bulk weight by desiccation (DID).

First vaginal cap or pessary developed by Dr. Friedrich Wilde, Berlin (1ST).

R. W. Mitchell, Springfield, Ohio, patented a combined corer, parer, and slicer (LR).

Samuel Sloan, Poughkeepsie, New York, established a pin-making factory that made 100,000/day (RM).

First Indian tea from C. A. Bruce arrived in London (BCT).

Furniture casters were patented by John A. Philos, and Eli Whitney Blake, New Haven, Connecticut, for use on bedsteads (patent 821) (FFF).

1839   Indian tea was imported and sold in London (1ST).

Artificial fertilizer, superphosphate of lime, used by John Bennet Lawes for cultivation of turnips (1ST).

Charles Goodyear developed vulcanized rubber which made possible the commercial use of rubber (TH).

Carl August Steinheil built the first electric clock (TH).

Clarke patented a new method for enameling kitchenware (DID).

William Underwood's bookkeeper abbreviated the word "cannister" to "can" (1ST).

1840   Roller sugar mill was patented by James Robinson (DID).

Peruvian guano was first used in England (DID).

Canned lobster was produced by Tristram Halliday, St. John, New Brunswick, Canada (1ST).

Brown's Patent Universal Cooking Apparatus, a closed range, was sold in England (HFB).

William Rider, Springfield, Massachusetts, used Goodyear process of vulcanization for shirring cloth (1ST).

Friedrich Gottlob Keller first used ground wood pulp for paper in Germany (EB).

Sir Charles Wheatstone devised an electric clock (DID).

First English machine-printed wallpapers produced by Lancashire printer (EB).

Rattier and Guibal, Rouen, France, invented suspenders (see also 1820, Hancock) (WABI).

Napier vacuum machine for coffee making developed by Scottish marine engineer, Robert Napier (BCT).

>1840   Peter Cooper devised first sweetened, flavored gelatin, but it was not successful (see also 1891) (HS).

1841   Collapsible tube patented by John Rand, American artist (1ST).

Canned anchovies produced at Deammen, Norway (1ST).

C. J. Fritzsche produced aniline oil by treating indigo with potassium hydroxide (TH).

Experimental electric arc lamps installed in le quai Conti, Paris (1ST).

Soft-water process patented by Thomas Clark in Great Britain (1ST).

Street cleaning machine invented by James Whitworth (1ST).

Public laundry opened on Frederick Street, Salford, England (1ST).

Maquet, France, began industrial manufacture of envelopes in Paris (WABI).

Elias Howe invented his sewing machine (patented in 1846) (MITU).

1842 Dirchoff, Russia, invented a milk powder (DID).

H. Benjamin froze food by immersion in a mixture of ice and brine (DID).

Bevan invented the vacuum chamber for food preservation (DID).

Dr. John Gorrie, U.S., invented mechanical refrigeration (DID).

Adam Gimbel, Indiana, opened his first store (RMRB).

The *Magazine of Science and School of the Arts* recorded baths with a unit for heating water (CAD).

First machine-made matches were manufactured (DID).

1843 Peter Spence patented a method of manufacturing alum by treating burning shale and iron pyrites with acid (DID).

John Bennet Lawes manufactured artificial fertilizer at Deptford Creek, London (1ST).

First soap powder marketed by Babbitt's Best Soap, New York (1ST).

Thomas Hancock used Charles Goodyear's vulcanization process to improve Macintosh's waterproof cloth (1ST).

Charles Thurber, Norwick, Connecticut, obtained U.S. patent 3228 on his Thurber Patent Printer, a typewriter (FFF).

1844 Wood-pulp paper process invented by Friedrich Gottlob Keller (TH).

Air conditioning installed by John Gorrie at American Hospital for Tropical Fevers, Apalachicola, Florida (1ST).

I. C. Johnson made artificial hydraulic cement (DID).

Kamptulicon (floor covering) patented by Elijah Galloway, U.S. (DID).

First public baths and wash houses opened in Liverpool, England (TH).

Gustaf Pasch, Sweden, proposed safer matches by placing some ingredients on striking surface (TH).

George Kent invented a knife-cleaning machine (DID).

Orvis issued their first catalog offering fishing equipment (ACTB).

First full-sized paper pattern printed in *Bekleidungskunst für Damen--Allgemeine Muster Zeitung*, Germany (DBE).

1845   Self-raising flour marketed by Henry Jones, Bristol, England (1ST).

Joshua Heilman, French inventor, patented a machine for combing wool and cotton (TH).

Howard invented the linen filter (DID).

E. B. Bigelow constructed the power loom for manufacturing carpets (TH).

William M'Naught developed the compound steam engine (TH).

Anton Schroetter, Vienna, patented amorphorus phosphorus matches (DID).

Tiffany issued its first catalog (ACTB).

1846   John Deere constructed a plow with a steel moldboard (TH).

Catherine Beecher published *Miss Beecher's Domestic Receipt Book* describing basic cookery techniques (WL).

Incubator repatented (DID).

Abraham Gesner, U.S., distilled kerosene from coal and sold it as lamp oil (CD).

Sponge rubber patented by Charles Hancock, London (1ST).

Nancy Johnson invented an ice cream maker (LR).

Elias Howe patented his sewing machine (TH).

Crinoline (woven horsehair) petticoats first worn (WABI).

1847   Ring doughnuts introduced by Captain Hanson Gregory, Camden, Maine (1ST).

Fry and Sons, England, produced semisweet, solid eating chocolate (BBB).

Meat extract prepared by Justus von Liebig, Royal Pharmacy, Munich (1ST).

Evaporated milk made for the first time (TH).

Latch hook knitting machine developed (EB).

Hungarian physician I. Semmelwis discovered connection between childbed puerperal fever and septicemia (TH).

First child born alive under anesthesia was Wilhelmina Carstairs (1ST).

James Young produced paraffin (kerosene) from shale oil (DID).

Stephen White, Manchester, and H. M. Paine, Worcester, Massachusetts, patented a process for making water gas (DID).

Insulated electric cables manufactured by Gutta Percha Company, London (1ST).

Safety razor with comb teeth guard invented by William Samuel Henson, Chard, Somerset (DID).

Alexander Bain, Edinburgh, invented an electric clock (DID).

Lavigne, France, patented a machine to print ribbon for making tape measures (WABI).

Gutta-percha used as a dental filling (WABI).

Disc plow developed (EB).

1848   Canned tomatoes produced by Harrison W. Crosby, steward at Lafayette College, Easton, Pennsylvania (1ST).

First canned sweet corn was sold by Nathan Winslow, Portland, Maine (1ST).

Morton Salt Company, Chicago, was founded (RM).

First large-scale department store, Marble Dry Goods Palace, New York, occupying one city block, opened (1ST).

Charles Burton, New York, devised the first modern baby carriage (RMRB).

Platinum mantle burner for gas lamps introduced (DID).

George Lloyd patented a centrifugal mine fan (DID).

Guibal invented a rotary fan and built a seventeen-foot diameter one (DID).

First oil refinery in Great Britain established by James Young, Riddings, Derby, England (1ST).

Bottger produced the first safety matches (TH).

William Young patented Nancy Johnson's ice cream maker (LR).

Joseph Wiss founded his scissors-making firm (RM).

John Curtis, Bangor, Maine, developed State of Maine Pure Spruce Gum for chewing (1ST).

1849   Electric arc lamp erected in the north tower of Hungerford Bridge (1ST).

Elisha Foote, U.S., patented a stove with a thermostatically controlled damper (HFB).

First commercial laundry in U.S. started by a failed miner, Davis, for other miners in California (WB).

Dry cleaning process discovered by M. Jolly-Bellin, Paris (1ST).

Charles P. Carter, Ware, Massachusetts, patented a screw apple parer (LR).

Safety pin patented by Walter Hunt, New York (1ST).

Safety pin patented in Great Britain by Charles Rowley, six
  months after Hunt patented his (1ST).

Machine for manufacturing envelopes patented by Jesse K.
  Park and Cornelius S. Watson, New York City (FFF).

General Merritt Hemingway produced silk thread on
  spools, U.S. (FFF).

> 1849   Charles Mattee designed a crude wooden wash wheel for
  Davis's laundry (WB).

< 1850   Springs introduced into furniture construction during the
  first half of the century (EB).

1850   Dried meat biscuit invented (DID).

Australian James Harrison and American Alexander
  Twining each developed commercial refrigeration
  (1ST).

Alexander Twinning patented ice-making equipment (EB).

Le Chevalier P. Clausser, France, invented a method of
  making short-staple flax and mixed cotton/wool goods
  (DID).

Levi Strauss made jeans (1ST).

Eau de Cologne invented by Jean Marie Farina, Colonge
  (DID).

First regular manufacture of perambulators began in
  London (1ST).

Plumber's advertisement, Boston, described servant-
  pumped hot and cold water system for the bath (CAD).

The *Journal of Gas Lighting* described Defries Magic
  Heater for heating bath water (CAD).

Joseph Glass invented chimney sweeping tools (DID).

Cylindrical hand-operated washing machine with agitated
  water invented by Joel Houghton, U.S. (DID).

Thomas Masters devised the first oven thermometer
  (FHTC).

Isaac Singer devised the continuous stitch sewing machine
  for domestic use (TH).

The banana was first brought to the U.S. (SSVFN).

California food and drug act enacted (MMGT).

Rolltop desk devised by Abner Cutler, Buffalo, New York
  (USAIR).

Cast metal fillings used for teeth (WABI).

> 1850   Large-scale cultivation of asparagus undertaken in the U.S.
  (RM).

Dr. Arthur H. Hanell studied food adulteration (MMGT).

Baking powder first became available (HS).

High-heeled shoes reappeared in New Orleans brothel worn by a French prostitute (CDIS).

World of Fashion included printed patterns (DBE).

Pressed glass lamps were popular in the U.S. (EB).

1851   Marsh incorporated a conveyer belt in a reaping machine (DID).

M. Mason, France, invented a method for drying vegetables (DID).

James Russell, Baltimore, introduced mass-produced ice cream (1ST).

First large U.S. cheese factory founded by Jesse William, Rome, New York, to produce cheddar cheese (RM).

Linus Yale patented his Yale lock (1ST).

Robert Newell, New York, displayed his parautopic lock at the Great Exhibition, London (EB).

Tubular steel furniture, rocking chair, exhibited by R. W. Winfield, London (1ST).

Wire-nail-making machine was made by Adolph Felix Browne, New York (DID).

John Gorrie obtained first U.S. patent for a mechanical refrigerator (EB).

Pierre Carpentier, France, patented a process for ribbing galvanized metal sheets (WABI).

Amelia Bloomer tried to popularize Turkish pantaloons for women (EB).

Lord Kelvin completed development of his principles for heat pumps (EB).

John Nicols, Lynn, Massachusetts, adaped Elias Howe's sewing machine for sewing shoes (WB).

Brigantine *La Ville de Paris* arrived in San Francisco loaded with luxury goods (NB).

Simon Lazarus opened a men's clothing store in Columbus, Ohio (MP).

Delmar Kinnear Patent Lamp patented (FFH).

Isaac Singer patented his sewing machine, an improvement of the earlier (1846) design by Elias Howe (MITU).

Richard H. Pease, Albany, New York, produced Christmas cards (FFF).

1852   Potash found at Strassfurt (DID).

Lord Kelvin and Professor Rankine proposed open-cycle refrigeration (DID).

Bower's Registered Gas Cooking Stove with insulated oven and cooking top sold (1ST).

Marshall Field Department Store established in Chicago (RC).

Nicolas Taliaferro, Augusta, and William Cummings of Murphyville, Kentucky, patented a self-heating iron (LR).

Patent issued to Cyrus T. Moore, Concord, New Hampshire, on an improved corn straw broom (LR).

Public lavatories with water closets opened in London (1ST).

Soda pulp for paper made from wood in England (EB).

John E. Lundstrom and brother, Jonkoping, Sweden, made matches proposed by Pasch in 1844 (DID).

German butchers' guild devised hot dog in Frankfurt (BW).

> 1852 South African farmer sold ostrich feathers in local market, leading to eventual export of feathers (WJ).

1853 First potato chips prepared at Moon Lake House Hotel, Saratoga Springs, New York, by George Crumb (1ST).

James F. and Edwin P. Monroe patented the egg beater (KC).

D. M. Smith, Springfield, Vermont, patented the spring clothespin (LR).

First domestic gas-fire produced (DID).

Dr. Stonehouse invented a charcoal air filter (DID).

First hotel to offer private baths was the Mount Vernon Hotel, Cape May, New Jersey (1ST).

Charles Gerhardt, University of Montpellier, France, first derived acetylsalicylic acid (aspirin) (WABI).

1854 Sewing machine to stitch buttonholes was patented by Charles Miller, St. Louis, Missouri (FFF).

Water-softening plant was installed at Plumstead Waterworks, England (1ST).

John Snow traced cholera outbreak to contamination of well water by privy vault forcing use of sewer (EB).

Gesner patented his process for making kerosene (WB).

John H. and George W. Austen, New York, invented a kerosene lamp (1ST).

Heinrich Goebel invented the first form of the electric light bulb (TH).

Spongial gold replaced gold-leaf filling for teeth (WABI).

Daniel Halladay, Connecticut, developed a self-governing windmill (CDIS).

City of Paris Department Store opened in San Francisco by Felix and Emile Verdeu of Paris (NB).

Farmers Institute, Massachusetts, provided first example of later cooperative extension service work (WB).

Walter Hunt, New York City, patented the (detachable) paper collar (FFF).

1855   Self-propelled rotary cultivator built by Robert Romaine, Montreal (1ST).

R. W. Bunsen produced a gas burner that was later adapted for heating, cooking, and lighting (TH).

Georges Audemars, Switzerland, obtained the first patent for producing artificial silk (rayon) (DID).

Samuel Yost, Connorsville, Indiana, developed a rotary washboard washing machine (LR).

Filtration of all river-water supplies for London made compulsory (EB).

Water desalinization plant described by S. Sidey, instructor at War Office Revictualling Station (1ST).

London sewers modernized (TH).

London municipal lavatories opened (1ST).

Austrian engineer Franz Koller developed tungsten steel (TH).

Aluminum manufactured by Henry Deville, Salindres Foundry, Paris (1ST).

John E. Lundstrom, Sweden, patented the safety match (1ST).

First practical programmed computer built by George Scheutz, Stockholm, and exhibited in Paris (1ST).

Snap fastener invented by Paul-Albert Regnault, France, for fastening gloves (WABI).

Grunwald, Germany, carried out industrial tests to produce powdered milk (WABI).

Adhesive gold used by Robert Arthur, U.S., in dental fillings (WABI).

George Vanderbilt installed the first bathroom in a private home in the U.S. in his New York home (HS).

J. B. Jolly, Paris, established the first dry cleaners using turpentine as a cleaning fluid (WABI).

Glass oven doors used (CD).

1856   Gail Borden patented condensed, canned milk (RM).

Aerated bread invented by Dr. Dauglish (DID).

Butter factory was established by R. S. Woodhull, Campbell Hall, Orange County, New York (1ST).

William H. Perkin prepared the first aniline dye (TH).

Pettit & Smith adapted the Bunsen burner (1855) for
domestic heating and commercially marketed it (1ST).

James Young, Bathgate, Glosgow, Scotland, introduced the
first British kerosene lamp (1ST).

Dr. Gesner, London, produced kerosene by distillation
(DID).

Davis patented a lard-burning lamp (FFH).

Sir Henry Bessemer introduced the converter into his
process for making steel (TH).

Sir William Siemens made ductile steel for boiler plating
(TH).

Crinoline petticoats were replaced by the steel framed
underskirt invented by Tavernier, France (WABI).

W. H. Dwinelle popularized the dental crown in the U.S.
(WABI).

William Filene opened his first store in Salem,
Massachusetts (MP).

1857   James Harrison, Geelong, Australia, used Perkins
refrigeration process to transport meat by ship (DID).

R. A. Stratter, Philadelphia, patented a mangle (LR).

Rumanians produced 2,000 barrels of oil from hand dug
wells (WB).

James Miller Williams, Canada, dug an oil well and
established a refinery at Oil Springs, Ontario (WB).

First toilet paper made by Joseph Gayetty, New York, with
unbleached pearl-colored manila hemp paper (RM).

Lyons, France, introduced permanent electric street lights
(1ST).

Lubricating oil manufactured by Price's Patent Candle
Company, Battersea, London (1ST).

1858   Condensed milk advertised by Gail Borden, Burrville,
Connecticut (1ST).

John Mason invented the threaded-top Mason jar used for
home canning (WB).

First U.S. patent for a can opener was issued to Ezra J.
Warner, Waterbury, Connecticut (EL).

First rotary type, mechanical washing machine was patented
by Hamilton E. Smith (DID).

Hiram H. Herrick, East Boston, Massachusetts, patented
carpet sweeper that swept in only one direction (LR).

Lucius Bigelow patented an improved carpet sweeper (EL).

Pencil with attached eraser patented by Hyman Lipman,
Phildelphia (1ST).

Crystal Palace Bazar, first London department store,
    opened (WABI).

Lyman Reed Blake, Abington, Massachusetts, invented a
    machine for sewing the soles to shoes (WB).

Rowland Hussey Macy opened the first Macy's (HS).

1859   Hydraulic pressure was used in a sugar mill by Jeremiah
    Howard (DID).

First commercial ice produced by Alexander Twinning,
    Cleveland (EB).

Ferdinand Carre, France, used ammonia in refrigerator
    (EB).

Wire egg beater patented by J. F. and E. P. Monroe and
    was one of the more successful models (LR).

Sir Joseph Whitworth invented a carpet sweeper based on
    his 1842 road sweeper (DBE).

First commercial oil well at Titusville, Pennsylvania, dug
    (WB).

Moses G. Farmer, Salem, Massachusetts, installed electric
    lamps in his parlor that were powered by a galvanic
    battery (1ST).

R. L. G. Plante developed the first practical electric storage
    battery (TH).

Robert Cheesebrough discovered oil-rig workers used oil
    residue to treat cuts and burns (Vaseline) (CDIS).

William Goodale, Clinton, Massachusetts, obtained U.S.
    patent 24734 on a paper-bag-manufacturing machine
    (FFF).

George B. Simpson, Washington, D.C., patented an electric
    stove (patent 25532) (FFF).

1860   Goodenough introduced a machine for making horseshoes
    (DID).

C. Degiton, Philadelphia, patented a potato paring knife
    (LR).

First U.S. patent for the corkscrew issued to M. L. Byrn,
    New York (RM).

Gas-ring fire introduced (DID).

Frederick Walton invented cork linoleum (TH).

Wax match-making machine invented by Louis Perrier,
    Marseilles (DID).

Charles Hill Morgan invented the first commercially
    feasible automatic machine to make paper bags (HS).

The Reverend Henry Moule, U.K., developed the earth
    closet toilet (CAD).

> 1860   Linus Yale, Jr., devised the Yale cylinder lock (EB).

    1861   Basic slag first used as a source of phosphorus for fertilizer (DID).

Meat refrigeration plant established by Thomas Mort, Darling Harbour, Sydney, N.S.W. (1ST).

Isaac Salmon discovered the addition of calcium chloride to canning water bath reduced canning time (RM).

Gilbert Van Camp, Indianapolis, began canning food products for sale (RM).

Mrs. Beeton's *Book of Household Management* published (TH).

    1862   William Morris's first wallpaper designs available (EB).

Morrill Land Grant Act granted every state 30,000 acres for each senator and representative (WB).

U.S. Department of Agriculture established (WB).

    1863   First canned peaches produced by Cutting Company, California (1ST).

Viennese devised the croissant to celebrate their defeat of the Ottoman Turks (BBB).

Timothy Earle, Smithfield, Rhode Island, patented a metal strip egg beater (LR).

Tilden's universal sifter was patented (flour sifter) (KC).

Ebenezer Butterick developed the first packaged paper dressmaking pattern (TH).

Hamilton E. Smith invented the first self-reversible washing machine, an improvement on his 1858 machine (DID).

Johann Faber, Stein, Germany, patented the screw-top ever-pointed pencil case (DID).

Open-hearth steel furnace was developed by Matin brothers in France based on Siemens process (TH).

West Indian avocado introduced into Florida (USAIR).

    1864   Louis Pasteur developed a pasteurization process that better preserved wine (TH).

Canned asparagus produced by William Hudson, Hunter's Point, Long Island (1ST).

Meat extract manufactured at Frey Bentos, Uruguay (1ST).

Self-filling water boiler available (HFB).

Turtle nursing bottle introduced for infant feeding (ECTL).

George Lemoine, France, developed phosphorus sequisulfide matches, a nontoxic form of phosphorus (EB).

William Filene opened a wholesale business in New York City (MP).

1865 Dr. Joseph Lister applied carbolic acid as an antiseptic and used it in an operation (DID).

Thaddeus Lowe invented an ice machine (TH).

Union Stockyards opened in Chicago (TH).

Methylated spirit gas pressure stove invented by B. Whangton (DID).

Earnshaw's Patent flour scoop and sifter patented (KC).

James H. Nason, Franklin, Massachusetts, patented a coffee percolator (PA78).

Pasteur cured the silk worm disease and saved the French silk industry (TH).

Schutzenberger was aware of process reactions to make rayon acetate (Celanese) (DID).

Soap in liquid form patented by William Sheppard, New York (FFF).

Carpet sweeper in general use (TH).

First oil pipeline laid in Pennsylvania (TH).

S. F. Pierre licensed his paper-bag-making process to printers (WB).

1866 Parkesine, nitrocellulose-based plastic (like celluloid), produced by Alexander Parkes, Birmingham (1ST).

Gregor Mendel published his work on plant genetics in Austrian journal (EB).

J. Osterhoudt, New York, invented the key opening can (WABI).

Dr. James M. Sims, Women's Hospital, New York City, performed human impregnation by artificial insemination (FFF).

1867 Armour meat packing factory opened in Chicago (TH).

Reinforced concrete patented by Joseph Monier, France (EI).

U.S. patent issued to Tilghman for sulphite pulping (EB).

Barbed wire patented by Lucien B. Smith, Kent, Ohio (1ST).

Monosodium glutamate prepared in German laboratory (MB).

English physician Thomsd Allbutt devised the modern glass medical thermometer (WABI).

Morris Rich opened his first Atlanta store (MP).

Arm and Hammer Baking Soda introduced (HS).

Burnham and Morrill Company, Portland, Maine, canned sweet corn (HS).

Scientific American described a British typewriter (EB).

Ebenezer Butterick founded the E. Butterick Company, New York City, to manufacture his paper patterns (FFF).

Averill Paint Company, New York, offered premixed paints based on patent 66773 granted to D. R. Averill of Newburg, Ohio (FFF).

1868   McIlhenny Co., Avery Island, Louisiana, introduced Tabasco Sauce (EP).

James Highen, Melbourne, Australia, printer, patented a sheep shearing machine (EI).

J. A. Hagg introduced an atmospheric gas burner (DID).

M. Kelly patented Kelly's Diamond for barbed wire (1ST).

Filter pump invented by Robert Wilhelm Bunsen (DID).

William Filene financially ruined in Jay Gould's Black Friday (MP).

Many paper products available including paper plates, cups, collars, vests, napkins, and towels (DH).

Luther Sholes patented the first practical U.S. typewriter (EB).

>1868   William Filene opened a new store in Lynn, Massachusetts (MP).

1869   Celluloid patented by James W. Hyatt, Albany, New York (1ST).

Margarine patented by Hyppolyte Mege-Mouries, Paris (1ST).

George H. Hartford formed Great Atlantic and Pacific Tea Co., A&P, and opened a chain of stores (RM).

Vitreous enamel panels for stoves patented (CD).

Commercial gas stove rented for six shillings/annum in England (DID).

Whirlwind, the first vacuum cleaner sold, used a hand crank to create suction (WJ).

Joseph Campbell and Abraham Anderson of Camden, New Jersey, successfully canned tomatoes (USAIR).

Union Co., Pennsylvania, devised successful paper bag making machine combining the best features of others (WB).

Cornelius Swartout patented the first U.S. waffle iron (HS).

William S. Hadaway, Jr., New York City, patented an electric stove with a ring-shaped, spiral-coiled burner for uniform heat conduction (FFF).

Shaker elder recorded use of a steam-operatred laundry, including a Parker wash mill (washing machine) and centrifugal dryer (HDI).

English patent for desalinization granted (EB).

English built desalinization plant at Aden (EB).

1870 William W. Lyman, U.S., patented the first can opener to use a cutting wheel, not a spike or knife blade (EL).

F. A. Walker offered tin and copper measures for home use including cooking (FHTC).

F. A. Walker sold a potato masher (FHTC).

James McCall began manufacture of packaged paper dress patterns (ET).

Hellyer's Optimus, improved valve water closet introduced (CAD).

Horse-drawn mowing machine introduced (DID).

Alexander Livingston, Reynoldsburg, Ohio, developed a smooth, red tomato (USAIR).

Thomas Adams, Staten Island, developed Black Jack, the first flavored, chicle-based chewing gum (1ST).

Charles Hires, Philadelphia pharmacist, tasted herb tea and acquired the recipe for Hires Herb Tea (UNITD).

Dutch farmers in Kalamazoo, Michigan, began growing celery for commercial consumption (HS).

B. F. Goodrich, Akron, Ohio, manufactured rubber hoses (EOOET).

>1870 Wealthy Americans exchanged Christmas cards (FORB).

Rubber condoms sold (EOOET).

1871 First margarine factory began operation at Oss, Holland (1ST).

Mary F. Potts, Ottumwa, Iowa, patented a double-pointed sad-iron (a flatiron with a removable handle) with spring-loaded clamp-on-handle (LR).

Toilet roll devised by Seth Wheeler, New York (1ST).

Thomas Lipton opened Lipton's Market, Glasgow, Scotland and used coupons called Lipton's Pounds (TIME).

F. W. Rueckheim, Chicago, developed Cracker Jack snack food (CDIS).

Sir William Siemens, Germany, proposed an electrical resistance thermometer (EB).

First hot dog, called a dachshund sausage, was sold at Coney Island, but without a bun (HS).

B. F. Goodrich, Akron, Ohio, manufactured rubber clothes wringers (EOOET).

B. F. Goodrich, Akron, Ohio, manufactured rubber
    preserving rings (EOOET).

B. F. Goodrich, Akron, Ohio, manufactured rubber bottle
    stoppers (EOOET).

1872   Samuel R. Percy, New York, patented dried milk in the U.S.
    (FFF).

Wolseley, Australia, developed the first practical sheep
    shearing machine (EI).

Marcel, French hairdresser, invented the marcel wave (WB).

U.S. patent issued to Silas Noble and James P. Cooley,
    Granville, Massachusetts, for toothpick-making
    machine (WJ).

Montgomery Ward, Chicago, began catalog sales (WABI).

Alexander Graham Bell began experiments to help deaf
    that led to the development of the telephone (WB).

William and Andrew Smith packaged Smith Brothers's
    Cough Drops in the first factory-sealed packages (HS).

Luther C. Crowell, U.S., devised the square bottomed
    brown paper bag with longitudinal folds (U.S. patents
    123811 and 123812) (HS).

Thomas Edison developed the first electric typewriter (EB).

1873   Washington navel orange, originally from Baia, Brazil,
    planted in California (BBB).

Campbell's soups offered for sale (RM).

James Sargent, Rochester, New York, devised a lock based
    on a Scottish lock patent for a time-lock (EB).

Joseph F. Glidden, DeKalb, Illinois, designed a
    commercially successful barbed wire (WB).

Luther Burbank developed the Burbank potato (EB).

Chester Greenwood, Farmington, Maine, devised the first
    earmuffs (WJ).

Charles Phillip, Glenbrook, Connecticut, devised and
    marketed Phillips' milk of magnesia (EOOET).

1874   DDT first synthesized by Othmar Zeidler, Strasbourg (EB).

H. Solomon introduced a pressure-cooking method for
    canning food; it reduced sterilization time (TH).

Robert Green, Philadelphia, invented the ice cream soda
    (RM).

Monroe improved his egg beater and assigned the patent to
    Dover Stamping Co. (LR).

Jacob Vollrath, Sheboygan, Wisconsin, made enameled cast
    iron ware (SS).

First gas iron patents were issued in the U.S. (EL).

Machine for manufacturing barbed wire fencing introduced (EB).

Louis Prang, Roxbury, Massachusetts, sold Christmas cards in the United Kingdom (FFF).

1875 B&M canned baked beans produced by Burnham and Morrill, Portland, Maine (1ST).

Daniel Peter, Switzerland, developed milk chocolate (1ST).

Mellin's Food and Murdock's Liquid Food sold as infant or invalid supplement (HG).

London's main sewerage system completed (TH).

British Parliament enacted food legislation (MMGT).

Union Company changed its name to Union Bag and Paper Co. and began manufacture of paper bags (WB).

First Agriculture Experiment Station was established at Wesleyan University, Middletown, Connecticut (WB).

Louis Prang, Roxbury, Massachusetts, sold Christmas cards in the United States (FFF).

1876 Carl von Linde introduced the first successful refrigerator using ammonia in the compression system (EB).

Hires Herb Tea (later Root Beer) was sold nationally in dry concentrate form (UNITD).

H. J. Heinz sold catsup nationally (RM).

Melville R. Bissell, Grand Rapids, Michigan, patented the carpet sweeper (1ST).

Halcyon Skinner, U.S., perfected the power-driven Axminster carpet loom (WB).

Charles Darwin published the results of his studies on cross- and self-fertilization of plants (EB).

Tinfoil wrapped bananas sold for ten cents at the American Centennial Exposition in Philadelphia (SSVFN).

The rolltop desk shown at American Centennial Exposition, Philadelphia (USAIR).

First I. Magnin store was opened in San Francisco (NB).

Gustave de Laval, Sweden, patented the cream separator (WABI).

1877 Railway refrigerator cars built for Chicago meatpacker Gustavus Smith (1ST).

First frozen meat shipped from Argentina to Europe (1ST).

John Jossi, Wisconsin, invented American Brick cheese (RM).

W. E. Brock, New York, patented a popular hand apple parer and corer (LR).

Chester Greenwood patented ear muffs (WJ).

1878   Appleby invented a knotting device for a reaping machine (DID).

Sugar cubes were produced in Great Britain by Henry Tate, Silvertown Refinery, London (1ST).

Fur farming began in Canada (TH).

Ivory Soap (first called White Soap) marketed by Procter & Gamble, Cincinnati, Ohio (FFF).

Electric street lighting was installed in London (TH).

Swan and Edison independently developed the incandescent electric light bulb (1ST).

John Peet introduced Assam tea seed from India to Java (BCT).

1879   Saccharin discovery reported by Constantine Fahlberg and Ira Remsen, Johns Hopkins University, Baltimore (1ST).

Australian frozen meat sold in London (TH).

Milk bottles introduced by Echo Farms Dairy Company, New York, from bottle devised by Louis Whiteman (1ST).

Bread slicing machine manufactured by Summerscales, Keighley, Yorkshire (MH).

Rotary masher for vegetables manufactured by George Kent Co. (MH).

First variety chain store, Woolworth's, Utica, New York, opened (1ST).

Julien Bradford, Portland, Maine, devised an electric heat and vapor governor for spinning shops (HONEY).

James Ritty, Dayton, Ohio, devised his "incorruputible cashier," the first cash register (MITU).

Richard S. Rhodes, Riverpark, Illinois, patented the Audiophone in the U.S. (patent 219828) (FFF).

<1880   William Filene opened two Boston stores (MP).

St. Louis physician developed peanut butter for patient intolerant of other forms of protein (RM).

Jewish immigrants ate Kosher dried beef shaved and mixed with onions to form hamburg patties (HS).

1880   Listerine antiseptic introduced (HS).

Bones used as fertilizer (DID).

Baked beans in tomato sauce first sold in U.S. (1ST).

First canned fruits and meats available in stores (TH).

Scotch and Irish still used hot stones in certain locales for heating and cooking (CD).

Can opener patented by William E. Brock, New York (KC).

Holly, Lockport, New York, installed a well-insulated steam-heating system (HFB).

First commercially produced incandescent electric light bulbs manufactured at Menlo Park, New Jersey (1ST).

New York streets first lit by electric lights (TH).

Carnegie developed the first large steel furnace (TH).

Hires Herb Tea sold as a liquid concentrate (later Hires Root Beer) (UNITD).

First hearing aid patent was registered to Francis D. Clarke and M. G. Foster (DBE).

Sherwin-Williams Company, Cleveland, Ohio, began manufacturing standarized premixed paints (EOOET).

James Ritty improved his cash register, making it more accurate (see 1879 and 1884) (FFF).

Kampfe Brothers, New York, developed the Star Safety Razor using a fixed (not removable) blade (FFF).

1880-90   Bordeaux mixture, copper sulphate and lime, used on vines to combat downy mildew (DBE).

1881   Incandescent electric street lamps were installed in Newcastle upon Tyne (1ST).

Dr. Aletta Jacobs, Amsterdam, offered birth control advice; she recommended the use of the diaphragm (WL).

First electric power station providing current for public and domestic use operated at Godalming, England (1ST).

1882   Electric iron patented by Henry W. Seely, New York (1ST).

Electric fan for commercial production developed by Dr. Schuyler Wheeler, Crocker & Curtis, New York (1ST).

First Christmas tree lit with electric light bulbs installed by Edward C. Johnson, New York City (1ST).

Edison designed the first hydroelectric plant (TH).

Miles Nervine, a sedative, marketed (WABI).

Jan Ernst Matzeliger invented a machine for shaping and fastening the leather over the shoe sole (WB).

Vogue first issued paper patterns (EDB).

Edward S. Moore, Salem, Massachusetts, installed a solar hot box at the Peabody Museum (BP).

Paul Beirsdorf, Hamburg, Germany, devised an adhesive bandage, Hansaplast (WABI).

H. Kroger opened his first grocery store (BW).

1883   Mary Lincoln published the first *Boston Cooking-School Cookbook* (WL).

First man-made fiber, cellulose-based artificial silk (rayon), developed by Joseph Swan, but was not used (1ST).

Clothes pounder with multiple cones for manipulating wash patented by James McCalla, Galton, Ohio (LR).

Galvanized portable garbage container introduced by Eugene Poubelle, Paris prefect of police (1ST).

First U.S. Tea Act passed (MMGT).

Albert Butz, Minneapolis, began working on a thermo-electric regulator for the coal furnace (HONEY).

"Visible writing" typewriter developed (EB).

Charles Stilwell, U.S., developed a machine for making a self-opening, pleated, flat-bottomed grocery bag (patent 279505) (EOOET).

1884   Bananas regularly imported from Canary Islands to Great Britain by Elder Dempster & Company (1ST).

Evaporated milk patented by John Meyenberg, St. Louis, Missouri (1ST).

First tea shop opened in England; earlier tea was sold in coffee houses (RM).

Modern pedestal water-closet with oval-shaped portrait-frame seat exhibited by Jennings at Health Exhibition, U.K. (1ST).

Electric transformer used in Metropolitan Railway, England, electric lighting system (1ST).

Practical steam turbine designed by Charles Parsons, Gateshead, to drive the dynamo in local lamp works (1ST).

Lewis E. Waterman developed a practical fountain pen (PA).

Carl F. Dahl invented kraft pulp in Danzig (EB).

First artificial insemination with donor in place of husband performed by Professor Pancoast, Philadelphia, Pennsylvania (1ST).

Carl Zeiss, Jena, Germany, developed boron-silicon glass later developed by Corning into Pryex glass (WABI).

Cemented bifocals devised (EB).

Bust improvers advertised (WC).

John Patterson, founder of National Cash Register Company, began to sell cash registers after having bought the rights from James Ritty (MITU).

1885   Evaporated milk manufactured by John Meyenberg's Helvetia Milk Condensing Company, Highland, Illinois (1ST).

First cream crackers manufactured by William Jacob, Dublin (1ST).

Exchange Buffet, first self-service cafeteria, opened in New York (1ST).

First appearance of the can opener in Army and Navy Stores catalog (DBE).

Comte Hilaire de Chardonnet patented a process for making artificial silk (rayon) (DID).

Karl Auer von Welsbach invented the incandescent gas mantle (TH).

Moxie soft drink was introduced initially as a nerve tonic in the northeastern U.S. (WJ).

Lady's suit invented by Redfern of England (WABI).

Albert Butz formed the Butz Therm-Electric Regulator Company, later renamed Honeywell (HONEY).

1886   Coca-Cola invented by Dr. James Pemberton, Atlanta, Georgia (1ST).

Dr. Pepper (soft drink) invented (BBB).

Charles M. Hall, U.S., and Paul Heroult, France, independently discovered the electrolytic process for aluminum (FHTC).

Krafft, Germany, discovered detergent properties (further discoveries were made in 1896) (DBE).

Henri Moissan isolated fluorine (WB).

Susan K. Knox patented a fluting machine (LR).

Benjamin Waddy Maugham produced the first gas "geyser" for rapid heating of water (CAD).

Hydroelectric installations begun at Niagara Falls (TH).

Kieselguhr filter was patented by Heddle and Stewart (U.K.) and Weischmann (U.S.) (DID).

Griswold Lorillard, Tuxedo Park, New York, devised the tuxedo, evening wear for men (WJ).

U.S. Oleomargarine Act passed (MMGT).

Butz patented the thermostatically controlled damper/flapper (HONEY).

Marvin Chester Stone devised the paper drinking straw (HS).

George K. Anderson, Memphis, Tennessee, patented the typewriter ribbon (patent 349026) (FFF).

Excelsior Needle Company, Wolcottville, Connecticut, manufactured uniform sewing needles by machine at a lower cost (FFF).

1887   Malted milk introduced by Horlick's, Racine, Wisconsin (1ST).

Paper napkins used at John Dickinson & Company annual
dinner, Castle Hotel, Hastings (1ST).

Electrically heated curling tongs available in Berlin
according to *Scientific American* (EL).

Electric heating system patented by Dr. W. Leigh Burton
(1ST).

Coin-in-the-slot gas meter patented by R. W. Brown,
Birmingham (1ST).

HUB vented nursing bottle for infant feeding patented
(ECTL).

First contact lens developed by A. E. Fick (DID).

U.S. Tea Act replaced 1883 Tea Act (MMGT).

John Meyenberg granted a second patent on his sterilization
process for evaporated milk (HH).

Hatch Act provided financial support for the agricultural
experiment stations (WB).

P. J. Towle, Minnesota grocer, devised Log Cabin pancake
syrup (HS).

Dr. John Roe, U.S., performed cosmetic rhinoplasty; he
published a paper in 1887 and second in 1891 on the
procedure (WABI).

1888   Wolseley sheep shearing machine used at Dunlop Station,
Australia (EI).

Electric switch socket designed by Sir David Salomons,
Tunbridge Wells, England (1ST).

Working-class dwellings were built with bathrooms, Port
Sunlight, Cheshire, England (1ST).

John L. Loud, U.S., patented and manufactured a ballpoint
pen, but the pen was not successful (DID).

Nikola Tesla created the electric motor (TH).

First Sears catalog issued, offering only watches (TIME).

Marvin Chester Stone, Washington, D.C., patented the
paper drinking straw (WABI).

Jacob L. Wortmen, Philadelphia, Pennsylvania, patented the
manifold copy typewriter ribbon (patent 376764) (FFF).

Dr. William Champion Deming, New York City, devised
the "hatching cradle," an incubator for infants (FFF).

1889   First gasoline powered agriculture tractor built by Charter
Engine Company, Chicago, Illinois (1ST).

First electric oven installed at Hotel Bernina, Samaden,
Switzerland (1ST).

James H. Northrop's new automatic loom in operation
automatically supplying filling and thread (ET).

First electric sewing machine was produced by Singer in Elizabethport, New Jersey (1ST).

Rubber heels introduced by the Aberdeen Rubber Sole & Heel Company, Scotland (1ST).

Mrs. W. A. Cockran perfected the first dishwashing machine (1ST).

Electric heating system was introduced commercially by Burton Electric Company, Richmond, Virginia (1ST).

D. T. Bostel, Brighton, introduced the wash down water closet (CAD).

Punched card system created by Herman Hollerith (TH).

Raffaele Esposito, Naples, Italy, devised Pizza alla Magherita in honor of Queen Magherita (RD).

Lysol disinfectant first introduced for cleaning (CDIS).

Borsch developed an improved means of welding bifocal lenses together (WABI).

1890 Disc plow invented (DID).

Packaged gelatin invented by Charles Knox (RM).

Egg lifter for poached eggs patented by Ann E. Smith (KC).

First aluminum saucepan made by Henry W. Avery, Cleveland, Ohio (1ST).

Matthai-Ingram offered a graduated measuring cup (FHTC).

Schutzenberger, Germany, first made rayon (DID).

Gas-mantle combined with Bunsen's gas burner to create a lamp (DID).

Thomas Lipton began selling tea (TIME).

U.S. Meat Inspection Act passed (MMGT).

French firm Bourjois launched makeup packaged in an elegant carton (WABI).

Second Morril Act provided further funding for land grant colleges (WB).

U.S. patent 421802 issued for a concealed floor socket for two-pronged electric plug (TAC).

Front stroke typewriter developed by John N. Williams (EB).

1891 Van Camp's, Indianapolis, canned their first baked beans with tomato sauce (1ST).

First commercially produced electric oven sold by Carpenter Electic Heating Manfacturing Company, St. Paul, Minnesota (1ST).

E. H. Whitney and J. L. Kirby, Cambridge, Massachusetts, patented an egg beater incorporating principles still used (LR).

Electric kettle marketed by Carpenter Electric
Manfacturing Company, St. Paul, Minnesota (1ST).

W. L. Judson, U.S., invented the clothing zipper (TH).

Stevens dishwasher, Cleveland, advertised (FHTC).

First baby incubator devised by Dr. Alexander Lion, Nice
(1ST).

U.S. Cattle Inspection Act passed (MMGT).

Clarence M. Kemp, Baltimore, Maryland, patented the first
Climax solar water heater (BP).

George A. Hormel, Austin, Minnesota, founded the
Hormel Meat Packing Company (USAIR).

>1891   May Wait invented Jell-O; by 1900 it was sold nationally
(RM).

1892   First canned pineapple sold (TH).

William Painter patented the pry-off bottle cap (RMRB).

Emil Frey produced Liederkranz cheese (RM).

Coin-operated gas meters installed by South Metropolitan
Gas Co. led to general adoption of gas (1ST).

Thermos flask invented by James Dewar, Cambridge,
England (1ST).

Pittsburgh Reduction, later Alcoa, made first stamped and
cast aluminum cookware (EL).

First artificial silk (rayon) factory established at Besançon,
France (1ST).

Viscose process for rayon patented by Edward Bevan and
Charles Cross (DID).

Dr. Washington Sheffield, New London, Connecticut,
introduced toothpaste in a tube (1ST).

Electric iron demonstrated by J. J. Dowsing at Crystal
Palace (1ST).

First book matches patented by Joshua Pusey, Lima,
Pennsylvania (1ST).

Asa G. Candler bought the rights to Coca-Cola syrup (WJ).

Beecham's Tooth Paste, U.K., was marketed in a tube
(1ST).

R. E. Bell Crompton and J. H. Dowsing, England, patented
an electric radiator using a cast iron plate (WABI).

James Mitchell devised a machine for extruding cookie
dough, later used for Fig Newtons (EOOET).

First U.S. portable typewriter patented by George
Blickensderfer, Stamford, Connecticut (patent 472,692)
(FFF).

1893   First ready-to-eat breakfast cereal, Shredded Wheat,
produced by Henry D. Perky, Denver, Colorado (1ST).

First electric kitchen shown at the Colombian Exposition,
Chicago (WB).

Electric toaster marketed by Crompton Company,
Chelmsford, England (1ST).

Electric central heating introduced by American Electric
Heating Corp. (1ST).

Steam-driven mowing machine introduced (DID).

Thomas Lipton entered his tea in Chicago's World Fair and
received first prize (TIME).

Coca-Cola trademark registered with the U.S. Patent Office
(WJ).

Wire reinforced glass fabricated by Leon Appert, France,
for use in casings and windows (WABI).

U.S. Supreme Court ruled that the tomato was a vegetable
for import and taxation purposes (USAIR).

Charles Hire renamed Hires Herb Tea; Hires Root Beer
marketed (UNITD).

English electric wall socket manufactured (TAC).

Welch's Grape Juice, formerly Dr. Welch's Unfermented
Wine, introduced at the Chicago's World Fair (HS).

Waldorf Salad devised by Oscar Tschirky (HS).

Disc plow reintroduced to the U.S. (see 1847) (EB).

Felix Hoffman, Bayer chemist, developed aspirin (WABI).

Visible writing typewriter developed by Herman L. Wagner,
Brooklyn, New York (patent 497560); later sold to John
T. Underwood (FFF).

1894   Charles Post developed coffee substitute, Postum (PA78).

Coca-Cola bottled by Joseph Biedenham, Vicksburg,
Mississippi (1ST).

Milton Hershey, Lancaster, Pennsylvania, introduced
Hershey Bars (BBB).

Susan Stavors created Minute Tapioca by using a coffee
grinder to process tapioca (WM).

First electric kettle shown in Chicago (TT).

Word frankfurter devised for hot dog, assumed to have
come from Frankfurt, Germany (CDIS).

First full line Sears catalog offered (TIME).

House Furnishing Review (January 1894) described machine
similar to Silex coffee maker (EL).

1895 Charles Post developed Grape-Nuts, an early
breakfast/health food and offered the first cents-off
coupon (LB).

First flaked breakfast cereal, Granose Flakes, prepared
from wheat by Dr. John Kellogg, Battle Creek,
Michigan (1ST).

Northrop added the first bobbin changing to his loom (ET).

King C. Gillette invented the successful safety razor (TH).

Michael J. Owens, Toledo, Ohio, patented a glass-blowing
machine for automatic bottle production (WABI).

J. W. Foster and Sons, England, oldest athletic shoe
company, established (see 1979, manufacture of
Reeboks for export to the U.S.) (BUSWK).

Montgomery Ward catalog depicted "breast support," early
bra (MONTW).

Frederick A. Chapman, Philadelphia, patented a cut-out
block and box for electric power outlets in walls and
ceilings (TAC).

First U.S. pizzeria opened on Spring Street in New York
(HS).

E. F. Brooks and W. H. Congers, Pasadena, California,
bought marketing rights to Clarence Kemp's Climax
heater (BP).

Montgomery Ward catalog offered a steam or gas heated
ironer (MONTW).

Fig Newton cookies sold by the Kennedy Biscuit Works,
Cambridge, Massachusetts (EOOET).

Underwood Typewriter Company founded in New York
(FFF).

1896 Ice cream cones first produced by Italo Marcioni, New
Jersey (1ST).

Chinese ambassador Li-Hung-Chang's chef devised chop
suey to appeal to Chinese and American tastes (1ST).

Leonard Heischfield introduced Tootsie Rolls candy (BBB).

Fannie Merritt Farmer's *The Boston Cooking-School
Cookbook* published; it used standardized measures
(EB).

Chlorinated water supply introduced in Pola, Italy (1ST).

Niagara Falls hydroelectric plant opened (TH).

Book matches manufactured by Diamond Match Company,
Barberton, Ohio (1ST).

U.S. Filled Cheese Act passed (MMGT).

Wahlin Company, New York, manufactured the New American Cream Separator derived from Laval's 1876 machine (WABI).

Edison applied for a patent on the fluorescent light (FD).

1897   Sundaes sold because soda water prohibited from sale on Sunday: ice cream with syrup only sold, Red Cross Pharmacy, Ithaca, New York (RM).

Sears offered aluminum cooking utensils in its catalog (SS).

Joseph McCreery, Toledo, Ohio, patented an air conditioning spray process (WB).

Hypochlorite of lime used to disinfect water supplies in England (EB).

Motorized garbage trucks used by Cheswick Vestry (1ST).

Inverted gas mantle invented by Kent (DID).

First gasoline-powered mower built by Benz Co., Stuttgart, and by Coldwell Lawn Mower Co., Newbury, New York (1ST).

Ludwig Anthelin, Germany, discovered carbon tetrachloride for cleaning (WABI).

Seventy-one almond growers formed a co-op, which later became the California Almond Growers Exchange (NFFOF).

Dr. John Dorrance, Campbell's, invented a process for condensing soup (RC).

Ice cream sundae eaten at the Red Cross Pharmacy, Ithaca, New York (FFF).

1898   George Cobb, Fairport, New York, developed the double-seamed, solderless can (RM).

Kellogg Corn Flakes developed by William Kellogg, manufactured by Sanitas Food Company, Battle Creek, Michigan (1ST).

Pepsi Cola invented (BBB).

Coffee vending machines installed, Leicester Square, London, by Pluto Hot Water Syndicate (1ST).

Weaveable, dyeable viscose rayon yarn developed by C. H. Stearn and C. S. Cross, Kew, Surrey (1ST).

J. R. Clark Co., Minneapolis, introduced the first ironing board with legs (EL).

Practical use natural gas used for lighting of Heathfield Station, Sussex (1ST).

E. D. Cahen and H. Sevene patented and used Lemoine's discovery of a nontoxic phosphorus compound (EB).

Campbell's Soup adopted the red and white label (UNITD).

Pharmacist created Beeman's Gum for treating heartburn (WJ).

Forerunner of Whirlpool-Seegar offered a reciprocating tub washing machine (AHLMA).

Frank Walker, Los Angeles, patented a recessed model of a roof-top solar water heater (BP).

1899 First U.S. household refrigerator machine patent was granted to Albert T. Marshall, Brockton, Massachusetts (FFF).

Oxo (flavoring) cubes produced at Fray Bentos, Uruguay (1ST).

Motorized milk van introduced by Eccles Co-op, Lancashire, England (1ST).

Angus Campbell (Scotland) invented the spindle-type cotton picking machine (DID).

Spinnerets for Bevans and Cross 1892 process patented (DID).

John S. Thurman, U.S., patented the pneumatic carpet renovator, a vacuum cleaner (PA).

Califont, multipoint pressure geyser was developed by Ewart (CAD).

Aspirin in powder form introduced commercially by Bayer AG, Leverkusen, Germany (1ST).

Captain Lorenzo Dow Baker founded the United Fruit Company, which exported bananas to the U.S. (SSVFN).

J. L. Borsch invented a more economical procedure for welding the two lenses of bifocal glasses (WABI).

First Lake Placid Conference on Home Economics held at Lake Placid, New York (led to American Home Economics Association) (ER).

Safety heel, of rubber, patented by Humphrey O'Sullivan, Lowell, Massachusetts (patent 618128) (FFF).

1900 Eaterie in New Haven, Connecticut, pioneered the hamburger (TT).

Boker Coffee offered quarter, half, and one teaspoon measuring spoons (FHTC).

Ready-made clothing was sold (RC).

Horse and Pony mower introduced by Ransomes, Sims, and Jefferies, Suffolk, manfacturers of agricultural machinery (TT).

Scientists in Netherlands, Germany, and Austria discovered the importance of Mendel's work (EB).

Canadian hybrid Marquis wheat released to farmers (EB).

Horticulturist George Cullen, Florida, developed an avocado strain that produced fruit of a uniform size (USAIR).

Campbell's condensed soup won a prize at the Paris Exposition (UNITD).

Milton Hershey devised the wrapped candy bar (UNITD).

Ducoudum Ironer, French industrial ironer, used (SG).

Women used vaseline and cucumber as a face cream (HS).

Unsuccessful electric razors were devised in the U.S. (EB).

>1900   Harvard beets devised, named for Harvard color, crimson (CDIS).

Charles Francis Jenkins invented the conical paper cup (HS).

L. L. Bean issued its first catalog (ACTB).

1901   Satori Kato, Chicago, sold powdered, instant coffee at Pan American Exposition, Buffalo, New York (ADBE).

American Can Company founded (RM).

Peanut vending machine installed at Pan-American Exposition, Buffalo, New York, by Mills Novelty Company (1ST).

First synthetic vat dye, Indanthrene Blue, successfully made, a fast dye that did not fade or wash out (TT).

Sew-on press-studs (snaps) invented in Paris (TT).

Gillete patented the safety razor (1ST).

Vacuum cleaner devised by Hubert Cecil Booth (1ST).

First coin-in-slot electric meter approved for use in Britain (TT).

Peter Cooper-Hewitt, U.S., demonstrated the forerunner of the fluorescent lamp tube using liquid mercury (TT).

First Christmas tree lights sold commercially by the Edison General Electric Co., Harrison, New Jersey (1ST).

Electronic hearing aid first devised by Miller R. Hutchinson, but was bulky and difficult to use (WABI).

Campbell (U.S.) and Wimmer (Denmark) dried concentrated milk on trays to make powdered milk (HH).

Quaker Oats Company established to market rolled oats cereal (BUSWK).

Hearst sport cartoonist T. A. Dorgan coined the term *hot dog* for dachshund sausages (HS).

Will B. Otwell, Macoupin County, Illinois, founded the Boys Corn Club (EAA).

1902   Alexander Kerr perfected the Economy lid with a rubber sealing gasket fixed to the lid for home canning (WB).

Early automat offered food for a "nickel in a slot," Horn & Hardart Baking Co., Philadelphia (TT).

Frank Clarke, Birmingham gunsmith, patented first automatic tea-making machine (TT).

Electric kettle introduced into Britain (TT).

W. H. Walker, U.S., patented man-made textile yarn, acetate rayon, later used for women's underwear (TT).

Vacuum Cleaner Co. Ltd. produced a prototype based on Hubert Booth's design (1ST).

Alfred R. Wolff designed air conditioning for Carnegie Hall and other New York City buildings (WB).

Willis H. Carrier designed a scientifically based air conditioning system (WB).

Emile Fourcault, Belgium, developed a machine to draw a continuous sheet of glass (WB).

First electric light using an osmium metal filament, more efficient, but too expensive, developed (TT).

Von Steinbuchel, Graz, introduced Scopolamine morphine for childbirth (TT).

U.S. Renovated or Process Butter Act passed (MMGT).

Just patented a drum milk drier that was later improved by Hatmaker (HH).

Barnum's Animal Crackers first marketed (HS).

British newspaper advertisement offered "patent bust improvers" (DBE).

J. C. Penney opened his first store in Kemmerer, Ohio (HS).

O. J. Kern, Winnebago County, Illinois, founded the Boys Experimental Club (EAA).

1903   Flavor and texture of margarine improved by hardening vegetable oils (TT).

Italo Marcioni, New Jersey, patented a mold for ice cream cones (TT).

M. J. Owens, U.S., began bottle production with fully automatic machine that lowered cost and wages (TT).

Anonymous Frenchman invented enrober machine for coating candies with chocolate (RM).

Gillette began manufacture and sale of his razor at American Safety Razor (1ST).

Earl Richardson, California, developed an iron with a "hot point," which later became the Hotpoint iron (EL).

General Electric introduced the detachable cord for irons (EL).

Laminated glass invented by Edouard Benedictus, France (WB).

Ambrose W. Straub, St. Louis, patented a peanut butter making machine (USAIR).

Dr. John Harvey Kellogg made peanut butter for his patients, Battle Creek, Michigan (USAIR).

O. J. Kern, Winnebago County, Illinois, founded the Girls Home Culture Club (EAA).

1904   Canada Dry Ginger Ale was introduced (BBB).

Processed cheese manufactured by J. L. Kraft, Chicago, Illinois (1ST).

Ice cream cones were reintroduced in St. Louis by E. A. Hamwi, who rolled waffles in shape of cone (1ST).

Thomas Sullivan, New York, sent customers samples of tea in silk bags, first tea bags (WB).

Iced tea served by Richard Blechynden of England at Louisiana Purchase Exposition, St. Louis, Missouri (WB).

Hamburger sandwich popularized at the Louisana Purchase Exposition, St. Louis, Missouri (RM).

Vacuum flask, originally designed for lab work, first used in the home (TT).

Leon Guillet, French scientist, made the first stainless steel, but failed to capitalize on it (TT).

Dr. Scholl, former shoe salesman and Illinois medical school graduate, patented the arch support (TT).

Gas used for first time to power central heating and large-scale hot water supply, Clampton, England (TT).

Laclede Gas Light Co. began experiments in central house heating in St. Louis, Missouri (TT).

Initial efforts to harness steam from volcanic region made at Lardarello, Tuscany, Italy (TT).

George Parker invented the lever-fill fountain pen (TT).

Folding metal push chair developed, a boon to parents and servants (TT).

C. W. Post introduced Post Toasties (LB).

Harvey Hubbell invented a separable plug for electrical appliances with a screw part for light socket (TAC).

Hot dogs first sold with buns at the Louisiana Purchase Exposition, St. Louis (HS).

O. H. Benson, later with the U.S. Department of Agriculture, used the three leaf clover which later became the 4-H four leaf clover sign (EAA).

1905   Carl Linde patented ammonia compression refrigerator (DID).

Wafer ice cream sandwich was devised by Lewis using a rectangular brass slide to form the ice cream (1ST).

General Electric range marketed (HFB).

Rayon viscose yarn manufactured commercially by Courtauld, Coventry (1ST).

Elastic rubber replaced traditional whalebone and lacing in foundation garments (TT).

Portable vacuum cleaner for domestic use was marketed by Chapman and Skinner, San Francisco (1ST).

Electric light built with tantalum metal filaments marketed by Siemens and Halske, Berlin (TT).

Vick's Magic Croup Salve, later renamed VapoRub, introduced by Vick Chemical, North Carolina (TT).

Marvin Chester Stone invented a machine for making paper drinking straws (WABI).

First U.S. patent issued for using ionizing radiation to kill food bacteria (ACTB).

MacLachlan developed a spray processing for faster drying of milk, skim milk, eggs, or blood (HH).

First clock controlled thermostat to turn down heat introduced (HONEY).

Max Kiss, New York, introduced Ex-Lax, a chocolate-based laxative (EOOET).

Hyppolyte Morestin, France, performed mammary hypertrophy (WABI).

1906   G. Washington, U.S., developed refined soluble coffee (DBE).

U.S. Pure Food and Drug Act triggered by Upton Sinclair's *The Jungle,* based on Chicago Stockyard (1ST).

Freeze-drying invented by A. d'Arsonval and F. Bordas, Paris, for preserving biological materials (TT).

Cornflakes moved from health food market to general use (TT).

Milk cartons introduced by G. W. Maxwell, San Francisco (1ST).

A. L. Marsh patented nickel-chromium resistance wire, used in the modern cooking element of the electric stove (TAC).

First demonstration of the permanent wave machine
developed by Karl L. Nessler, London (1ST).

Tungsten filament used in commercially produced electric
light bulb (TT).

Jack Sapirstein established American Greeting Co. (FORB).

E. Baumann, Paris, devised the folding baby carriage
(WABI).

U.S. Meat Inspection Act supserseded 1890 Meat
Inspection Act (MMGT).

Gold, porcelain, and embossed steel used for dentures in
the U.S. (EB).

1907   Bakelite invented by Dr. Leo Baekland, Belgium (1ST).

Silk-screen print process used by Samuel Simon,
Manchester (1ST).

First household powdered soap, Persil, marketed by Henkel
& Cie, Dusseldorf (1ST).

Frank Griffin Carter invented an inlaid soap with insert of
abrasive marble flour (DID).

Thor, an electric washing machine, designed by Alva J.
Fisher, manufacted by Hurley Machine Corp., U.S.
(1ST).

Murray Spangler sold the rights to his domestic upright
vacuum cleaner to Hoover (TT).

David E. Kenney, New York, granted a patent on a suction
cleaner originally applied for in 1901 (EL).

Strip Lighting, Moore Tubing, used to illuminate Savoy
Hotel court (1ST).

Scott Paper Company introduced Sani-Towels, which later
became ScotTowels (UNITD).

Bullock's Department Store, Los Angeles, opened by John
Bullock (NB).

I. C. Merrel, I. S. Merrel, and Gere patented improvments
to milk spraying process for making dried milk (HH).

Al Neiman and Herbert Marcus opened their first Dallas
store (MP).

Annette Kellerman wore a one piece, skirtless bathing suit
at Revere Beach, Boston (HS).

First Maytag wringer washing machine manufactured
(FORB).

1908   First gasoline engine tractor with crawler tracks produced
by Holt, California; better traction, larger tool (TT).

Cellophane patented by Dr. Jacques Brandenberger, Zurich
(1ST).

Paper cups introduced by the Public Cup Vendor Company, New York (1ST).

Self-threading sewing machine needle introduced, invented by Frank Griffin Carter, Brighton, Sussex (TT).

Fritz Haber synthesized ammonia (TH).

W. Hoover introduced the vacuum cleaner for home use (TT).

First oscillating electric fan produced in U.S. by Eck Dynamo & Electric Co. (1ST).

Coolidge, U.S. scientist, developed a prototype of the durable tungsten filament with his new process (TT).

Willsie and Boyle built a solar-powered water pump, but it was uneconomic (TT).

Middle Eastern oil first struck in commercial quantities (1ST).

Titanium dioxide extracted by two U.S. chemists made possible purer white products, especially paints (TT).

George H. Beal discovered that self-fertilization weakened, but cross-breeding restored, plant vigor (EB).

Kikunae Iheda, Japanese chemist, discovered the flavor-enhancing properties of monosodium glutamate (first discovered in 1867) (MB).

Hydrox cookie first sold in U.S. by Sunshine Biscuit Company (CDIS).

Melitta Bentz, Germany, devised the Melitta coffee filter (WABI).

Fred Walker & Company began the manufacture and sale of Vegemite (WJ).

Landers, Frary & Clark introduced an electric percolator, Universal Percolator, with cold water pump (EL).

Fused bifocals introduced (EB).

Blackstone offered an improved agitator washing machine (AHLMA).

Forerunner of Whirlpool-Seegar offered the Dolly washing machine (AHLMA).

1908-17   Sheet window glass drawing machine invented by I. W. Colburn and developed by Libby Owens Sheet Glass Co. (WB).

1909   Nitrogen fertilizer discovered by Fritz Haber, Karlsruhe, Germany (1ST).

First commercial manufacture of Bakelite undertaken (TH).

General Electric made the first electric toaster (TT).

Synthetic rubber developed by F. Hofman, Bayer Co., Elberfeld, Germany (1ST).

Brassiere for dressy occasions mentioned in *Vogue* (1ST).

P & O Line fitted with two steam-heated electric washing machines and two electrical centrifugal dryers (DID).

General Electric introduced the first travel iron (EL).

Sears offered low-cost homes in its catalog; the program lasted until 1937 and included plans, paint, building materials, and nails (WJ).

Edouard Benedictus, France, patented laminated glass (WABI).

Eugene Schueller successfully tested hair dye (WABI).

Shackwell, St. Louis, Missouri, independently discovered freeze drying (WABI).

Mrs. Ann Bridges and Mrs. Sutton acquired rights to a coffee maker and used new Pyrex glass (EL).

New City of Paris store built to replace the store destroyed in the 1906 earthquake in San Francisco (NB).

American Home Economics Association established (EOA).

William J. Bailey introduced the Day and Night insulated solar water heater (BP).

First successful portable typewriter introduced (EB).

Benjamin D. Chamberlin, Washington, D.C., first filed a patent on a machine to make glass light bulbs (patent 491812) (FFF).

1910   Holt, California firm, developed the first self-propelled gasoline-powered combine harvester (TT).

Glass-lined railway milk tanker used by Whiting Milk Company on the Boston & Maine Railroad (1ST).

Carron electric range sold (TT).

Synthetic rubber manufactured by Bayer Co., Leverkusen, Germany (1ST).

First stockings made from synthetic fiber of rayon, Bamburg, Germany (1ST).

First stamped, nickel-plated steel shell introduced for General Electric irons (EL).

New convection electric heater developed that was easier and safer to install and use (TT).

U.S. heavily taxed white and yellow phosphorus matches (EB).

Diamond Match Co. bought U.S. patent for Lemoine's compound, but was unable to use it (EB).

One-piece bifocals developed (DID).

Joyce Hall established Hallmark, greeting card
    manufacturer of Kansas City (FORB).

Eugene Schueller founded the French Harmless Hair Dye
    Co. (later L'Oreal) (WABI).

Saccharin used commercially as a sweetener (ACTB).

Bentron and Emerson of the Carl Zeiss Company
    developed one-piece bifocals (WABI).

William Atkinson, Boston architect, experimented with a
    "sun box" (BP).

1911   Westinghouse introduced the first electric frying pan (EL).

Electric heater based on H. J. Dowsing's patent using
    incandescent lamps marketed by General Electric Co.,
    Britain (TT).

William Armstrong Fairburn, U.S., adapted French red
    phosphorus formula for production in the U.S. (EB).

General Baking Co. was formed to offer white bread; the
    company later became General Host (FORT).

Nabisco introduced Oreo cookies (CDIS).

Beiersdorf, Germany, invented Nivea cold cream (WABI).

Procter & Gamble introduced Crisco shortening (WJ).

Barlow and Seeling, later Speed Queen, offered a wringer
    washing machine (AHLMA).

1912   Alpha-Beta established self-service grocery stores in
    California (1ST).

Belling introduced the electric geyser, a rapid immersion
    heater, but with heavy power consumption (TT).

Zippers used in clothing by Firma Vorweck & Sohn,
    Wuppertal-Barmen, Germany (1ST).

Girdle introduced (DBE).

Otto Titzling invented his brassiere but failed to patent it
    (PA).

Heating pad, forerunner of electric blanket, was developed
    by Dr. Sidney Russell, U.S. (TT).

Electric dishwasher marketed in U.S. (WABI).

Prizes first added to Cracker Jack boxes (CDIS).

French firm Bourjois created pastel cheek makeup, the first
    mass-produced rouge (WABI).

Paul Sabatier, French chemist, discovered nickel made a
    good hydrogenation catalyst (WB).

Army and Navy Stores catalog advertised "golf-shape
    knickers" (WC).

William Atkinson, Boston architect, published *The
    Orientation of Buildings or Planning for Sunlight* (BP).

R. Belling, Enfield, England, developed a fire-proof clay for radiator cores and built the Standard (WABI).

Richard Hellman, New York, sold one-pound boxes of Hellman's mayonnaise (EOOET).

1912-15   British developed activated sludge process for sewage treatment (EB).

1913   Nitrogen fertilizer manufactured by Badische Anilin and Soda Fabrik, Germany (1ST).

First domestic electric refrigerator, Domelre, manufactured in Chicago (1ST).

Cellophane manufactured by La Cellophane, Paris (1ST).

Stainless steel first cast by Harry Brearley, Sheffield, England (1ST).

Gidden Sundback, Hoboken, New Jersey, patented "separable fasteners," an improved and successful zipper (1ST).

Brillo pads sold for cleaning pots and pans (TT).

A. Richler, Belgium, discovered synthetic chemical detergents (TT).

Sheet glass of high quality produced as Fourcault machine started commercial operation (TT).

Nitrogen-filled electric bulb developed by Irving Langmuir, General Electric, improved the life of tungsten filament (TT).

Ostrich feathers became South Africa's fourth largest export; the market collapsed shortly after (WJ).

C. E. Grey and Jensen patented further improvements to the milk spray-drying process (HH).

Maytag made a hand-operated washing machine (AHLMA).

U.S. Imported Meat Act (MMGT).

Richard Hellman, New York, sold glass jars of mayonnaise (EOOET).

1914   Ammonia produced at Oppau, Germany, on large scale using Haber process for fertilizers and explosives (TT).

Aluminum foil bottle caps (for milk) produced by Josef Jonsson, Linkoping, Sweden (1ST).

Two-piece Economy lid with a separate screw band invented, a boon to home canners (WB).

Protein plastic casein made from skim milk used to make plastic knitting needles and crochet hooks (TT).

Brassiere patented by Mary Phelps Jacob (TT).

Liquid chlorine used in water supply systems (EB).

First major sewage works used bacteria in decomposition of waste opened at Manchester, England (TT).

Eugene Sutter developed a hair dryer as part of a permanent wave machine (WABI).

Short panties, petticoats with legs, introduced (GBL).

Smith-Lever Act provided for federal, state, and county funding of the cooperative extension service (WB).

Mrs. Cochran's company introduced a domestic dishwashing machine (EOOET).

Corning Glass Works, Corning, New York, began using Chamberlin's machine further modified by others to make glass light bulbs (FFF).

1915   Ford developed a farm tractor (TH).

U.S. grocery store installed a turnstile and checkout (TT).

Thermostatic oven regulators for gas ovens developed (1ST).

Pyrex glass produced by Corning Glass Laboratories, U.S., resistant to chemicals and heat, used for cooking (TT).

Long-life stainless steel designed specifically for cutlery patented (TT).

Manchester Yellow, common dye ingredient found to moth proof wool, by E. Meckbach, German Dye Trust (TT).

Lipstick in a metal tube was manufactured by Maurice Levy, U.S. (1ST).

John Fisher, U.S., developed a prototype of the agitator washing machine (TT).

Enameled baths began to replace cast iron baths (TT).

Mechanical pencil invented (TT).

Volcanite used for molding teeth fillings (TT).

Aspirin tablets first retailed by Bayer (1ST).

Alexander Samuelson, Root Glass Company, Terre Haute, Indiana, designed the wasp-waisted Coca-Cola bottle (CDIS).

First clock controlled thermostat to both lower and restore furnace temperature introduced (HONEY).

National Electric Light Association met in New York City to establish some uniformity in appliance plugs and outlets (TAC).

Upton Machine Company, later Whirlpool-Seegar, was marketing the Cataract Rocking Tub washing machine (AHLMA).

Willis Carrier founded Carrier Corporation to market his air conditioner design (ACTB).

1916   Clarence Saunders launched a small chain of self-service grocery stores, Piggly Wiggly (WG).

Margaret Sanger opened a birth-control clinic, Brooklyn, New York (1ST).

Mechanical home refrigerator sold for $900 (ACTB).

Double-shelled enameled bathtubs were mass produced in the U.S., replacing cast-iron tubs (ACTB).

Gas-filled tungsten filament incadescent bulb introduced with thirteen times more light than carbon bulbs (ACTB).

Stainless steel introduced to the U.S. (ACTB).

Market Square, the first planned shopping district in the U.S., opened in Lake Forest, Illinois (WK).

1917   Ford began production of cheap, durable frameless tractors, increasing farm mechanization (TT).

Shortage of soap in Germany led to marketing of Nekal, first commercial detergent for textile industry (TT).

M. Hammer and C. Lewis, U.S., produced a fully automatic, electrically ignited oil burner for home heat (TT).

British and U.S. physicians used Scopolamine morphine to ease childbirth (TT).

Six U.S. manufacturers agreed on a standard appliance plug and receptacle for electric power (TAC).

Converse Shoe Company introduced the All-Star athletic shoe (ER).

U.S. Rubber Company introduced Keds, the first mass-marketed athletic shoes (EOOET).

S.O.S., soap-impregnated steel wool pads, were devised by Edwin W. Cox, San Francisco (EOOET).

1918   First electric food mixer available in U.S. (TT).

Lux, Sweden, produced a vacuum suitable for cleaning curtains, cars, and upholstery (TT).

Alkyd resins developed and manufactured on a large scale from petroleum to produce better paints (TT).

Titanium oxide introduced into white paint for improved color, coverage, and nontoxicity (TT).

Wood cellulose, Celucotton, used as a cotton substitute for bandages and dressings (TT).

Donald E. Jones, Connecticut, used double cross breeding to produce stronger plants with desired features (EB).

Kelvinator refrigerator designed by Nathaniel Wales, U.S., marketed (WABI).

Because of toxic fumes, carbon tetrachloride was replaced
by trichlorethylene for dry cleaning (WABI).

Landers, Frary, & Clark, New Britain, Connecticut,
introduced an electric waffle iron (EL).

Efficient power take-off mechanisms developed for tractors
(EB).

Murphy bed developed and patented by Lawrence Murphy
(CDIS).

1919   Canned grapefruit introduced by Yankee Products Ltd.,
Puerto Rico (1ST).

Woodward's, Vancouver, Canada, introduced self-service
food department (1ST).

Oven thermostat used in an electric range (TAC).

Belling introduced the first popular U.K. domestic electric
stove, Modernette, based on submarine unit (TT).

Stretch-spinning of rayon invented by Edmund Thiele and
Dr. Elsaessor, Germany, by cuprammonium process
(DID).

Pips, glass imperfections left by manufacturing process,
were eliminated from electric light bulbs, thereby
strengthening the bulb (TT).

Moon Pies, southern favorite, invented in Chattanoga
(CDIS).

Frigidaire refrigerator marketed in U.S. (WABI).

Sailor Jack and his dog Bingo first appeared on Cracker
Jack boxes (CDIS).

Coney Island vendor devised frozen custard dessert (WJ).

U.S. Horse-Meat Act (MMGT).

1920   Agricultural fumigants first used (DID).

London firm sold energen rolls, bread substitute for
diabetics (TT).

Tea bags produced by Joseph Krieger, San Francisco (1ST).

First practical immersion water heater with thermostatic
control introduced (TT).

Coco Chanel introduced No. 5 perfume (TT).

Magicoal, imitation coal fire, introduced by H. H. Berry
(TT).

U.S.S.R. legalized abortion on demand (1ST).

Karl Spire and Arthur Stoll extracted the alkaloid drug
ergotamine from ergot for migraine and abortion use
(TT).

C. Munters and B. von Platen, Sweden, patented a silent
refrigerator (WABI).

Earle Dickson, cotton buyer for Johnson and Johnson, devised a temporary dressing, later Band-Aid (CDIS).

Barlow-Seeling, later Speed Queen, offered a vacuum washing machine (AHLMA).

Armstrong Electric, Huntington, West Virginia, introduced a waffle iron with a heat-indicating light (EL).

Electric office typewriter marketed (EB).

>1920 Cartier introduced personalized silver-plated swizzle sticks (CDIS).

Race electric hair dryer introduced by Racine Universal Motor Co. (EL).

Asphalt tiles developed in the U.S. (EB).

Disposable cartridge pens available (WB).

1921 Aerial crop dusting initiated by Ohio Agricultural Experiment Station; first plane flown by J. Macready (1ST).

Bouillon cubes imported from Europe to U.S. (TT).

First chocolate-covered ice cream bars, called Eskimo Pies, were marketed by Christian K. Nelson, Onawa, Iowa (1ST).

Peter Paul Halijian, New Haven, Connecticut, introduced Peter Paul Mounds candy bar (BBB).

Otto Schnering introduced Baby Ruth candy bar, named for Grover Cleveland's daughter (BBB).

Swan electric kettle, with double the heating power of others, introduced by Bulpitt and Sons, England (TT).

International Silver Co., Meriden, Connecticut, produced knives with stainless steel blade (EL).

Artificial, or "cultured," pearls produced (TT).

Van Heusen's stiff, starchless collars introduced (TT).

First iron with embedded Calrod heating element introduced (EL).

Friedrich Bergius hydrogenated coal to oil (TH).

Dr. Marie Stopes established Mothers' Clinic that dispensed birth-control information in London (1ST).

Johnson and Johnson introduced the Band-Aid, first stick-on bandage (TT).

White Castle restaurant chain established in Wichita, Kansas, and began selling its hamburgers (FORB).

Connecticut Agricultural Experiment Station sold the first commercially produced corn hybrids (EB).

Kimberly-Clark marketed Kotex sanitary napkins (WABI).

X-rays used to kill trichinosis eggs in pork (ACTB).

1922   Canned baby food manufactured by Harold H. Clapp,
        Rochester, New York (1ST).
        First wrapped ice cream brickettes manufactured by
        Thomas Wall (1ST).
        Butterfingers candy bar introduced by Schnering (BBB).
        Best Stove, Detroit, introduced oven type toaster; in which
        bread was held vertically (EL).
        Joseph Block, U.S., suggested the use of a whistle on tea
        kettle to German manufacturer, Gebruder Hansel,
        Westphalia (EL).
        A. Clevel discovered a new class of dyes for coloring rayon
        fibers (TT).
        Nocord Electric Company introduced a cordless electric
        iron (EL).
        Country Club Plaza, Kansas City, included a group of
        buildings developed and managed as unit; it included
        parking (WK).
        Leo Gerstenzang began to develop the idea for Q-Tips
        (CDIS).
1923   Regulo, the first commercially successful thermostat for
        domestic ovens, was used by Davis Gas Co. (TT).
        Jacob Schick patented the electric razor (TH).
        Cellular concrete used to build low-cost housing at
        Scheveninger, Holland (TT).
        Chicago Crawford Avenue Power Station used excess steam
        to drive turbines to regenerate heat (TT).
        Self-winding watch patented in Switzerland by John
        Harwood (1ST).
        Air-O-Mix, Delaware, marketed Whip-All, a portable mixer
        (WABI).
        European wild boar introduced into the Carmel Valley,
        California (CDIS).
        U.S. Filled Milk Act (MMGT).
        Marconi Co. launched an electronic valve operated hearing
        aid, the Otophone, that weighed 16 pounds (DBE).
        Clarence Birdseye patented his process for quick-freezing
        foods (HS).
        Converse Shoe Company introduced the high-top Chuck
        Taylor basketball shoe (ER).
1924   Soybeans harvested for the first time by combine harvester
        (TT).
        Insecticide first used (TH).

CO_2 process first used in manufacture of ice cream, New York (DID).

Aga Stove invented by Gustav Galen, Swedish physicist, using solid fuel; later models use gas or oil (TT).

D. A. Rogers, South Minneapolis, announced the first automatic toaster with a dial to regulate timing (EL).

Kleenex, U.S., introduced paper tissues (1ST).

Spin dryer used for washing clothes made by Savage Arms Corp., Utica, New York (TT).

Westinghouse introduced nonadjustable, automatic iron featuring Klixon disc thermostat (EL).

Dead-bolt locks used for homes (TT).

Spiral binding, Spirex, invented by Staale (TT).

Cosmetic mammoplasty available (TT).

Henry A. Wallace (later Secretary of Agriculture and U.S. Vice-President) sold cross-bred corn (EB).

Mexican avocado introduced into California (USAIR).

Wheaties breakfast cereal marketed (UNITD).

1925   California agriculturists conducted earliest successful hydroponics experiments (TT).

First commercial crop dusting service offered by Huff Daland Dusters Incorporated, Georgia (1ST).

Battery cages for higher egg production by laying hens introduced, U.S. (TT).

More efficient glass bottle manufacturing machine introduced, better than the 1903 Owens machine (TT).

Potato chips manufactured commercially in Albany, New York (WG).

Howard Johnson opened his first ice cream shop, Wollaston, Massachusetts (RM).

Star Can introduced, the first can opener with a toothed can wheel to rotate can against cutting wheel (EL).

Hiram M. Berry invented the barbed-spindle cotton picking machine (DID).

Internally frosted glass bulbs introduced by British manufacturers for improved diffuse light source (TT).

Pierre and Jacques Guerlain created Shalimar (WABI).

Dr. Suzanne Noel began performing cosmetic face lifts (WABI).

American Home Economics Association requested the U.S. National Bureau of Standards to investigate standard measures for home (JHC).

Zalmon Simmons, U.S., marketed his innerspring
Beautyrest mattress (EOOET).

Benjamin D. Chamberlin, Washington, D.C., received U.S.
patent 1551935 on a glass bulb manufacturing machine
(see also 1909, 1914) (FFF).

>1925   Igepon detergents invented by Platz and Daimler (DBE).

1926   First factory ships for freezing and processing fish used by
Britain and France (TT).

Gas refrigerator introduced to the American market by
Electrolux, Evansville, Indiana (FFF).

B. von Platen and George Munters, Stockholm, Sweden,
patented the absorption refrigerator in U.S. (FFF).

Moisture-proof cellophane invented by William Hale
Church and Karl Edwin Pringle, DuPont (DID).

Hotpoint developed Calrod, a superior heating element for
electric stoves (TAC).

Pop-up electric toaster was marketed by McGraw Electric
Company, Minneapolis, Minnesota (1ST).

Waters Genter Co., Minneapolis, introduced Toastmaster,
the first successful automatic ejector toaster (EL).

Zippers replaced buttons on jeans (TT).

Eldec Co., New York City, introduced "steeming iron"
imported from Switzerland (EL).

Globar electric heater introduced by Carborundum Co. that
used a silicon-carbide rod (TT).

Machine to draw glass directly from the furnace developed
by Pittsburgh Plate Glass (WB).

Mass production of safety glass was developed by Libby
Owens Glass Company (WB).

Mies van der Rohe tubular steel chair manufactured by
Thonet, Germany (1ST).

Stainless steel pens introduced (EB).

First seed company devoted to the commercial production
of hybrid corn established (EB).

General Electric manufactured a hermetically sealed
household refrigerator (WABI).

Erik Rotheim, Norway, invented the aerosol (WABI).

First White Tower restaurant opened in Milwaukee,
Wisconsin (WT).

Kraft bought an interest in Fred Walker & Company,
manufacturer of Vegemite (WJ).

Ironer on wheels offered in mail-order catalog (SG).

Leo Gerstenzang began marketing Baby Gays, later renamed Q-Tips, for cleaning baby's ears, nose, etc. (CDIS).

Speed Queen manufactured the Vacuum Dasher washing machine (AHLMA).

1927 A. J. Hosier, British farmer, invented the milking bail or portable milking shed (TT).

First silent electric refrigerator manufactured by Electrolux, Britain (TT).

American Can Co. discovered canning with liquid reduced spoilage and vitamin loss (RM).

Continental Can Company founded (RM).

Wall-mounted can opener marketed by Central States Manufacturing Company, St. Louis, Missouri (1ST).

Polar Ware Company, Sheboygan, Wisconsin, made the first stainless steel cookware (EL).

John and Mack Rust invented a mechanical cotton-picker capable of harvesting a bale of cotton per day (TT).

Liberty Gauge and Instrument Company, Cleveland, introduced an adjustable-temperature automatic iron (EL).

First acrylic polymer produced commercially by Rohm and Haas, Germany; it was used as lacquer and for glass (TT).

Electric blanket marketed by Thermega, London (1ST).

Albert W. Hall added improvements to fluorescent lights (TH).

Hautier, France, patented the first controlled low pressure cooker (WABI).

Imedia hair dye developed offering more shades and using organic materials for better coverage (WABI).

Englishman Frederick Hurten Rhead was named art director for Homer Laughlin China Company (he later designed Fiestaware dinnerware) (USAIR).

T. G. N. Haldane, England, built a heat pump and used it in his office in London and in a home in Scotland (WABI).

Fremont Canning Co., later Gerber, began marketing canned baby food (ACTB).

1928 Machine for boning and cleaning kippers first used, Fleetwood, England (TH).

Franklin Mars developed Milky Way candy bar (BBB).

Rhythm birth control method devised by Dr. Harmann Knaus, Berlin (1ST).

Self-winding watch manufactured by Harwood Self-Winding
Watch Company, London (1ST).

Progresso Quality Foods Company founded (BW).

H. Biesenberger, Vienna, contributed to techniques for
mammary reduction plasty (WABI).

Broccoli introduced to the eastern U.S. from California by
the D'Arrigo brothers, vegetable growers (HS).

Dr. George Nicola Papanicolaou, U.S., devised the Pap
smear test for cervical cancer, but it was not accepted
for broad use until 1941 (HS).

1929　　D. J. Kennedy, U.K., patented the padded bra (PA).

Phillipe de Brassiere, France, used Titzling brassiere design
in dressy fabrics and became famous (PA).

First do-it-yourself hair coloring introduced (TT).

German chemist discovered fluorescent brightening agents,
later used in washing powders (TT).

General Electric introduced an iron with Button Nook to
facilitate ironing around buttons (EL).

Foam rubber produced at Dunlop Latex Development
Laboratories, Fort Dunlop, Birmingham (1ST).

American manufacturers made aluminum furniture (TH).

Machine-blown electric bulb developed by Corning Glass
Works (WB).

First ocean thermal power plant operated at Mantanzas
Bay, Cuba (TT).

Howard Johnson opened his second restaurant (CJ).

C. Munters and B. von Platen devised a condensor for their
refrigerator, the Electrolux (WABI).

Sink waste disposal unit developed (WABI).

Panty girdle introduced (COLLR).

<1930　　Rudolph Boysen developed the boysenberry, a cross of the
blackberry, red raspberry, and loganberry (HS).

1930　　Corn hybrids produced in U.S. by cross-breeding (TT).

Poultry farmers, U.S., built special sheds for rearing young
fowl at a lower cost (TT).

Clarence Birdseye, Springfield, Massachusetts, sold
individually packaged products of frozen foods (1ST).

Wonder Bread, the first packaged, sliced bread, introduced
(TT).

Mars introduced Snickers candy bar (BBB).

Michael Cullen operated King Kullen food stores, Long
Island, New York, as first supermarkets (1ST).

Electric range with automatic timing control introduced (TAC).

Gas range with automatic oven and top burner lighting introduced (TAC).

First electric kettle with safety shut off available from General Electric, England (TT).

Proctor Electric introduced the first automatic toaster with heat sensor to monitor toast cooking (EL).

Beardsley and Wolcott, Waterbury, Connecticut, offered "drop down" toaster, bread inserted at top, drops down (EL).

J. Walter Reppe, Germany, made artificial fabrics from acetylene (TH).

The Draper high speed loom introduced, improving loom speed by 20 percent (ET).

Hoover introduced the first small, hand-held vacuum (TT).

First domestic water heater to work efficiently was the Progas instantaneous water heater (TT).

Thermopane, insulated window glass, invented by C. D. Haven, U.S. (WB).

State supported birth-control clinics were established in Bangalore and Mysore, India, by Mysore Government (1ST).

Polystyrene commercially developed by I. G. Farbenindustrie, Germany, used for toilet articles, cases (TT).

Polyvinyl chloride developed by W. L. Semons of B.F. Goodrich, U.S. (TT).

Vannevar Bush built the first analog computer (EB).

Sunbeam, Chicago, marketed Mixmaster, the first stationary mixer (WABI).

McNary-Mapes amendment to the 1906 U.S. Food and Drug Act (MMGT).

E. Schwarzmann, Vienna, contributed to techniques for mammary reduction plasty (WABI).

Twinkies snack cake introduced by James Dewar (HS).

Waters Genter Co. introduced their Waffle Master, an automatic waffle iron (EL).

> 1930   Thin latex rubber condoms introduced (EOOET).

Large still for commerical and industrial use built at Aruba (EB).

Bar coding system devised for identifying items, later adopted for supermarket use (WJ).

Women's briefs, underwear, introduced (GBL).

Decorative linoleums marketed (EB).

Vacuum-tube hearing aid that weighed four pounds developed (WB).

1931   Diesel caterpillar tractor manufactured by Caterpillar Tractor Co., Peoria, Illinois (1ST).

Kinetic Chemical Company produced Freon, non-toxic, low boiling liquid for use in refrigeration and aerosol sprays (TT).

Gas range with time control introduced (TAC).

Preston C. West, Chicago, patented a portable electric can opener, but it was apparently never made (EL).

Bunker introduced, first can opener to use pivoted handles to hold can with key-type handle for cutting wheel (EL).

Refillable toothbrush with removable bristles introduced (TT).

First electric razor manufactured by Schick Incorporated, Stamford, Connecticut (1ST).

General Electric introduced a cast-aluminum soleplate iron (EL).

Glass building bricks introduced (DID).

Miles Laboratories, Indiana, introduced Alka-Seltzer (TT).

Electrolux began refrigerator production in Sweden (WABI).

Servel (Swedish Electrolux) began refrigerator production in U.S. (WABI).

Patent request for window air conditioner filed (WABI).

Knapp-Monarch Company introduced Therm-A-Magic percolator that shut off after making coffee (EL).

Highland Park Shopping Village, Dallas, shopping center with stores facing each other instead of the road (WK).

Mercury switches introduced into thermostats for heating (HONEY).

Most U.S. electric appliances had converted to standardized plug first adopted in 1917 (TAC).

1932   Mars introduced Three Musketeers candy bar (BBB).

Stephen J. Poplawski developed a blender from his mixer (EL).

Wrinkle-free fabric announced after fourteen years of development by R. S. Willows, Manchester, England (1ST).

Synthetic sponge manufactured by Novacel, France (1ST).

First practical dishwasher marketed in U.S. (TT).

Room air conditioners available (WB).

Foam rubber used for seats in Shakespeare Memorial Theater (1ST).

Dual water immersion heaters for bath or sink were introduced by Hotpoint, Bocker, and Bray (TT).

Calgon detergent developed (WABI).

1931-32   Royal Institute of British Architects devised the helidon for aiding proper solar orientation (BP).

1933   Polyethylene produced at Northwich Cheshire Laboratories of Imperial Chemical (ICI) by R. O. Gibson (1ST).

Lithiated Lemon, English soft drink, promoted in the U.S. as 7-Up, "mixer" drink (TT).

First all-electric tea maker introduced by Goblin; it included an electric light and clock (TT).

Samson United, Rochester, New York, designed a toaster to accommodate rolls and bread (EL).

Dreft, synthetic detergent for wool and silk, marketed for home use by Procter & Gamble (FFF).

First soapless shampoo, Drene, introduced by Procter & Gamble in U.S. (1ST).

Tennessee Valley Authority established, major electrification project (TT).

First electric clock incorporated in heating thermostat (HONEY).

Vinyl abestos floor tiles introduced at the Chicago World's Fair (EB).

1934   Refrigeration process for meat cargoes developed (TH).

Phthalocyanine dyes prepared (TH).

First thermostatically controlled iron sold in U.S. by Sunbeam; it used a double-pole automatic thermostat (EL).

First launderette was the Washateria, opened at Fort Worth, Texas, by J. F. Cantrell (1ST).

Persplex commercially produced by Imperial Chemicals Industries for use in airplane windows and light fittings (TT).

First coiled tungsten filament electric bulb produced, U.S. (TT).

Swiss lab developed freeze-dried coffee (WABI).

Amendment on seafood inspection added to the 1906 U.S. Food and Drug Act (MMGT).

Nabisco introduced Ritz Crackers (CDIS).

Cornuts, toasted corn snack food, first made (CORN).

1935 Kreuger Beer, Newton, New Jersey, sold beer in easy-carry cans (TT).

Cheap, imitation wool called Merinova and resistant to moths, heat, and mildew, manufactured of casein in Italy (TT).

Zippers introduced on men's trousers (1ST).

Sink waste disposal unit marketed by General Electric Corporation, Bridgeport, Connecticut (1ST).

Fluorescent lighting first shown publicly in Cincinnati by General Electric Company (1ST).

Iceland introduced legalized abortion on medico-social grounds (1ST).

M. Perraud, Jif-Waterman, invented the ink cartridge (WABI).

A. Edwin Stevens produced the Amplivox, the first practical hearing aid at 2 1/2 pounds (DBE).

First thermostat styled by industrial designer introduced by Honeywell (HONEY).

Synthetic resins used for false teeth in Germany (EB).

Libby-Owens Corning manufactured thermopane windows (BP).

Cooper & Sons, Wisconsin, introduced Jockey Briefs based on a French men's swimsuit (EOOET).

1936 First artificial insemination association established in Denmark; it led to improved cattle (TT).

Inexpensive "baby" combine harvester marketed by Allis-Chalmers (TT).

Upright Hoover--"beats as it sweeps as it cleans"--introduced (TT).

First commercial convection heater equipped with electric fan introduced in the U.S. by Belling (TT).

First practical application of fluorescent lighting used in Washington, D.C. (1ST).

British scientists extracted the basic ingredients of milk of magnesia from dolomite and sea water (TT).

Commercial tampons available (EBM).

Hoover Dam completed (TH).

Borden adopted Elsie the Cow as symbol (CDIS).

Faberware Coffee Robot, a vacuum-type brewer, with thermostatic control that kept coffee warm indefinitely, was introduced (EL).

Fiestaware dinnerware introduced by Homer Laughlin China Company (BUSWK).

	D. D. Peebles patented a process to reprocess dried milk for better reconstitutability (HH).
1937-39	Otto Bayer discovered polyurethanes (TT).
1937	After thirty-eight years as a health food, peanut butter was improved and marketed nationwide (TT).
	Rice Krispies breakfast cereal launched (TT).
	Kitchen Aid introduced the first electric coffee grinder (EL).
	James Kennedy developed the process used to make the first copper-clad bottom on stainless steel cookware (EL).
	Toast-o-Lator, New York, introduced a toaster using a conveyer device (EL).
	Nylon developed at Dupont Laboratories and patented by William Caruthers (TH).
	Philips, Holland, introduced an electric razor with a circular cutting head (TT).
	Durabilt, first folding travel iron, introduced by Winsted Hardware, Winsted, Connecticut (EL).
	Steem-Electric Company, St. Louis, introduced Steem-Electric, first U.S.-made steam iron (EL).
	Scotch tape developed by Richard Drew, 3M Company (TT).
	Margaret Rudkin founded Pepperidge Farm Bakery (WJ).
	Harry H. Holly, U.S., developed a hamburger patty machine (CDIS).
	Humpty Dumpty Stores, Oklahoma City, provided shopping carts--a basket on a folding chair on wheels (WABI).
	Earl Hass patented the tampon and founded Tampax Co. (patent filed and tampons sold in 1936) (WABI).
	Nabisco baked more than twenty-nine milllion Ritz Crackers per day (CDIS).
	Bendix manufactured the automatic washer (AHLMA).
	Kraft marketed Kraft macaroni and cheese (CDIS).
	Hormel marketed Spam, canned spiced ham and pork shoulder (CDIS).
1938-41	Peter Schlumbohm invented the Chemex coffee maker (EL).
1938	Nestlé's introduced Nescafé, first commercially successful instant coffee (1ST).
	Alfred Vischer, Chicago, designed a pressure cooker as interlocking pan and lid that eliminated bolts (TT).
	Roy J. Plunkett, Dupont, discovered Teflon (WJ).
	First nylon toothbrush marketed in U.S. (1ST).
	Fiberglass developed by Owens-Illinois Glass Co. (WB).

General Electric and Westinghouse began commercial production of fluorescent lamps (1ST).

Ladisalas Biro invented a ballpoint pen, the first successful such pen (TH).

Contact lens of methylmethacrylate developed (DID).

Fred Waring patented the blender developed by Fred Osius (EL).

U.S. Food, Drug, and Cosmetics Act replaced the 1906 Food and Drug Act and amendments (MMGT).

Massachusetts Institute of Technology staff began two decades of research into solar heating (BP).

Jockey shorts for men manufactured by Lyle and Scott Ltd., U.K. (WC).

Henry W. Wright published the results of his study of the sun's heating effects (BP).

1939   DDT used as an insecticide by Dr. Wiesmann to combat Colorado Beetle outbreak in Switzerland (1ST).

Bird's Eye introduced precooked frozen foods (1ST).

Polyethylene manufacture begun by Imperial Chemical Industries (1ST).

Hand-held electric knife marketed in the U.S. (TT).

First hand-cranked can opener with "positive single-action," no separate piercing lever, made by Vaughan Manufacturing (EL).

Nylon yarn produced commercially by Dupont at Seaford factory (1ST).

Met-L-Top, all metal ironing table with adjustable legs, introduced (EL).

Disprin, soluble aspirin, introduced by Rickett and Colman (TT).

Nuclear fission discovery published by Otto Horn, Chemical Institute, Berlin (1ST).

Rotolactor, a milking machine, shown at the New York World's Fair (CDIS).

General Electric introduced a dual temperature refrigerator with frozen foods in separate compartment (WABI).

Julian S. Kahn, New York, invented a disposable spray can (WABI).

Artificial impregnation of a rabbit performed by Dr. Gregory Pincus, Clark University, Worcester, Massachusetts (FFF).

Warner Co., Bridgeport, Connecticut, sold bras with cup sizes (HS).

Hamilton, U.S., marketed an automatic clothes dryer (AHLMA).

1940   Freeze drying adapted for food preservation (TT).

Mars introduced M&M's, no-mess chocolate candy (BBB).

Swing-A-Way, first wall-mounted can opener with mounting permitting it to swivel right or left (EL).

First nylon stockings marketed using Dupont nylon yarn (1ST).

Massachusetts Institute of Technology staff built a solar house (BP).

Howard Johnson opened his first turnpike restaurant in Pennsylvania (CJ).

Lipton's Chicken Noodle Soup (dried soup mix) introduced (TIME).

Automatic dishwasher developed in the U.S. (WABI).

Richard and Maurice McDonald opened their first restaurant in Pasadena, California (HS).

1941   Aerosol insecticides developed by L. D. Goddhue and W. N. Sullivan, U.S. (TT).

Packham potato harvester sped up crop production, but potatoes still had to be sorted by hand (TT).

Heinz developed a self-heating can for lifeboat and wartime use, but high production costs made it impractical for commercial use (TT).

Polyester fiber (Dacron) discovered by E. Whinfield, Calico Printers Association, Accrington, England (TT).

Fluorescent brightening agents patented by I. G. Farbenindustrie (TT).

Polyurethane commercially produced by I. G. Farbenindustrie (TT).

1942   Peter Halijian introduced the Almond Joy candy bar (BBB).

Foam glass developed by Pittsburgh Corning (WB).

First electronic brain or automatic computer developed in the U.S. (TH).

Dannon yogurt introduced to the U.S. (WJ).

Red Fiestaware dinnerware made with uranium oxide was discontinued because of World War II (BUSWK).

First round thermostat for home heat introduced; reintroduced after the end of the war in 1952 (HONEY).

1943   DDT used in the United States (TT).

Bernard E. Proctor, Massachusetts Institute of Technology, demonstrated the use of irradiation to preserve foods (IEEE).

Ladisalas Biro patented his ballpoint pen (1ST).

Raymond Davis Cady devised a waterproof match for use in the Pacific (EB).

1944 British extracted oil from sunflower seeds for use in margarine and other foods (TT).

Mark I computer completed by Howard Aiken, Harvard (EB).

Bruno Court, French Company, registered solid stick perfume (WABI).

Dr. George Lof, Boulder, Colorado, used solar-heated air for home heating (BP).

Reynolds Metal Company demonstrated aluminum foil for home use in New York City (full production began in 1947) (MITU).

1945 Herbicide 2,4-D patented in the U.S. (TT).

Bertha West Nealey with Ruth Bigelow established the Constant Comment Tea Company (WM).

Microwave oven patented in the U.S. by Perry LeBaron Spencer (1ST).

Tupperware Corporation founded (TT).

Brightening agents added to washing powders (TT).

Joseph C. Mulder, Indiana University, investigated stannous fluoride for preventing tooth decay (NY).

Fluoridated water supply introduced in Grand Rapids, Michigan (1ST).

First successful commercially produced ballpoint pens produced by Eterpen Company, Buenos Aires (1ST).

L'Oreal Laboratories, France, patented the cold permanent wave (WABI).

U.S. airline began serving precooked frozen dinners (WABI).

>1945 Lipton bought rights to use improved tea bag producing machine to manufacture Flo-Thru tea bags (TIME).

1946 Fisherman used ultrasonic equipment to detect shoals of fish (TT).

Instant Milk Company, U.S., patented the agglomeration process that improved dried skim milk (TT).

Lurex, aluminum-based yarn, produced more cheaply than other metallic yarns such as lamé (TT).

First bikini swimsuit designed by French couturier Louis Reard (1ST).

Procter and Gamble introduced Tide, first "built" detergent (TWA).

Rotary clothes hoist invented in Australia (TT).

Simons Company, Petersburg, Virginia, marketed an electric blanket in the U.S. with thermostatic control (FFF).

Timex watch introduced (TT).

Electronic brain (computer) built at University of Pennsylvania (TH).

Gaggia, Italy, invented the espresso machine for coffee (WABI).

Folding ironer marketed by Earle Ludgin & Company, Chicago (SG).

Belding Hemingway Corticelli, Putnam, Connecticut, sold commercially nylon threads, Monocord and Nymo (FFF).

1946-47　E. W. Flosdorff, U.S., demonstrated the use of freeze-drying techniques for foods, including coffee, orange juice, and meat (WABI).

1947　Microwave oven manufactured by Raytheon Co., Waltham, Massachusetts (1ST).

Kenneth Wood, England, designed Kenwood Chef, the forerunner of the food processor (WABI).

Christian Dior introduced the "New Look" with long skirts (WABI).

Robert Ricci marketed L'Air du Temps perfume (WABI).

Christian Dior marketed Miss Dior perfume (WABI).

1948　Instant tea marketed in the U.S. (WB).

Teflon first commercially produced by DuPont (TT).

Swingmaster, the first can opener with magnetic lid-lifter, introduced by Steel Products Manufacturing (EL).

DuPont introduced Orlon (polyester) (DID).

Solar house built by Dr. Maria Telkes using glass-covered, black-painted metal plates to heat air (BP).

First large, land-based desalting plant built in Kuwait with capacity of 1.2 million gallons/day (EB).

Transistor developed by John Bardeen and Walter Brattain, Bell Laboratories, Murray Hill, New Jersey (TH).

Georges de Mestral developed the idea for the Velcro fastener (WABI).

Adidas, athletic shoe manufacturer, established in West Germany (WJ).

Dishpans made of polyethylene introduced (1ST).

Burton Baskins and Irving Robbins merged their ice cream chains to form Baskin-Robbins (HS).

McDonalds, operated by the McDonald brothers, became a self-service restaurant (HS).

Heinz introduced a barbecue sauce (FF).

> 1948 Vinyl floor tiles became generally available (EB).

1949 Prepared cake mixes introduced by General Mills and Pillsbury, U.S. (TT).

Sunbeam introduced a toaster with a bimetallic sensor control; bread lowered itself automatically (EL).

Rival Manufacturing offered the first can opener with removable cutting wheel for easier cleaning (EL).

Procter & Gamble began sponsorship of Dr. Mulder's research on a fluoride toothpaste (NY).

Tufted carpet first made in the U.S. (TT).

Cyclamate first marketed as a nonnutritive sweetener (ACTB).

Charles Lubin introduced Sara Lee baked goods (HS).

1950 Sulzer weaving machine introduced, modern automatic shuttleless loom (TT).

Necchi zigzag sewing machine marketed in the U.S. (TIME).

Small contact lens covering only cornea developed (DID).

Home use portable typewriters available (RB).

> 1950 Fusing used in clothing manufacture (EB).

Bikini panties patterned on the bikini swimsuit introduced (GBL).

1951 Terlyene (Dacron, polyester) products manufactured in Great Britain (1ST).

DuPont produced the first orlon (polyester) garments (TT).

Dacron (polyester) men's suits marketed by Hart, Schaffner, and Marx (FFF).

Sears offered "bonnet" hair dryer in its catalogs (EL).

White fluorescent lighting tubes produced using Peter Ranby's discovery of halophosphate (TT).

Electric power produced by atomic energy at Arcon, Iowa (TH).

First commercial application of transistors (1ST).

First commercially manufactured computer, UNIVAC I, installed at the U.S. Census Bureau, Philadelphia (1ST).

Lacoste shirts first sold in the U.S. (HS).

1952 Acrilan, acrylic fiber, announced by Chemistrand Corp., U.S. (TT).

Clinical tests began with Crest toothpaste (NY).

Contraceptive pill of phosphorated hespidin produced (TH).

Institute of Margarine, Moscow, invented powdered butter (WABI).

Harold Ridley, English physicist, invented a polymethylmethacrylate lens for implantation in the eye (WABI).

Round thermostat reintroduced by Honeywell for home heating systems (HONEY).

Sonotone Corp. manufactured and sold a 3 1/2 ounce transistorized hearing aid (EDB).

Herman Kirsch of Kirsch Beverages Incorporated, College Point, New York, introduced NoCal, a sugar-free soft drink (FFF).

Nylon stretch yarn introduced from Switzerland by Heberlein Patent Corporation, New York, and used in men's stretch socks (FFF).

1953   U.S. Army began a test program to irradiate precooked foods (IEEE).

Low-cost plastic aerosol valve mechanism developed (TT).

Karl Ziegler invented a cheaper process using atmospheric pressure for producing polyeythlene (TT).

First collapsible polyethylene tube made by Bradley Container Corporation, Delaware, for Sea and Ski (1ST).

Sunbeam marketed an automatic electric fry pan (KC).

General Electric introduced the first steam travel iron (EL).

One-step hair dye, Imedia Creme, developed (WABI).

Sidney Rosenthal applied for a patent on Magic Marker, a felt-tipped marker pen (FORB).

Transistorized hearing aid introduced (WB).

Bendix introduced the Duomatic, a combination washer and dryer (AHLMA).

Irish coffee first served in the U.S. at the Buena Vista Café, San Francisco (HS).

1954   Hybrid wheats made possible by work of Ernest Robert Sears, wheat geneticist, U.S. (TT).

First frozen TV dinners introduced (TT).

Peanut M&M's, chocolate candy, introduced (EP).

S. W. Farber, England, introduced immersible electric frying pan developed by H. K. Foster (EL).

General Electric introduced the first steam/dry iron (EL).

Polypropylene invented by Dr. Natter from petroleum by-products, less costly than nylon, for carpets (TT).

First clinical tests made of oral contraceptive pill (1ST).

First atomic power station at Oblinsk, U.S.S.R., began producing electrical current for industrial use (1ST).

First business-use computer, Lyons Electronic Office, LEO, used at J. Lyons & Company, Cadby Hall, London (1ST).

Arthur Godfrey popularized California Dip made with Lipton's Onion Soup and sour cream (TIME).

Miller Pesticides Chemical Act amended 1938 U.S. Food, Drug, and Cosmetics Act (MMGT).

Procter & Gamble began marketing Fluffo yellow shortening (FORT).

Tractor transmissions improved (EB).

1955   First domestic deep freezer capable of freezing fresh food sold in the U.S. (TT).

McDonald's restaurant chain opened in Des Plaines, Illinois (RM).

Tefal Co., Paris, manufactured nonstick saucepans (1ST).

General Electric introduced a toaster with lower drawer for heating buns (EL).

Automatic kettle with thermostat, which automatically switched kettle off after boiling, introduced (TT).

Triacetate fibers, Arnel and Celanese, with luxurious feel and appearance at lower cost, introduced (TT).

Jesse E. Harmond, U.S., developed the continuous process for flax fiber, thereby reducing time to produce fiber (WB).

Velcro fastener patented (TT).

Wilkinson Sword introduced stainless steel razor blades (TT).

McNeil Consumer Products first marketed Tylenol pain reliever (NYT).

Freeze drying used to preserve Texas shrimp and Maryland crab (WABI).

25,000 tons of almonds grown in California (NFFOF).

Lightweight, unbreakable organic lenses for glasses made from a thermo-hardening material (WABI).

D. D. Peebles registered another patent for reprocessing milk for reconstitutability (HH).

New Hatch Act incorporated and extended earlier legislation for continued funding of agriculture experiment station (WB).

1956   First electric can openers produced by Klassen Enterprises and Udico Corp, California (EL).

Nonstick saucepans introduced to Great Britain by Habenware (1ST).

Proctor introduced a four-slice toaster (EL).

Crest toothpaste with stannous fluoride introduced (NY).

Southdale Center, near Minneapolis, was opened as the first enclosed shopping mall (WK).

Smith-Corona announced a portable electric typewriter (EB).

1957   Quick-freeze plant for trawlers developed by Scottish research scientists (TT).

Ultrasonic generator, U.S., foamed bottled and canned beer to remove air bubbles that cause cloudy look (TT).

Fast food manufacturers interested in dehydrated potato flakes for mashed potatoes (TT).

Karl Zyssert marketed a vegetable shredder (TT).

Ovenproof pyro-ceram cookingware introduced, accidentially discovered when oven curing temperature was too high (TT).

Moth-proofing of wool by dipping on the hoof confirmed (TT).

Permanent-press treatment for wool via chemical treatment developed in Australia (TT).

Hoover made the spin dryer part its of twin-tub washing machine (TT).

First steam/spray iron introduced by General Electric (EL).

Polypropylene commercially produced (TT).

New dry cell battery developed by P. R. Mallory Company, lasted several times longer than previous dry cell (TT).

Velcro fastener patented worldwide (WABI).

StaPuf liquid fabric softener introduced (WHIRL).

NuSoft fabric softener introduced (WHIRL).

U.S. Poultry Inspection Act (MMGT).

Hamilton Watch Company, Lancaster, Pennsylvania, introduced the electric watch (FFF).

1958   Massey Ferguson demonstrated new labor-saving sugarcane harvester to Australian farmers (TT).

B. A. Stout and S. K. Ries, Michigan State University, devised a crude tomato harvester (KM).

Lycra, man-made elastic, sold by DuPont; it was stronger, longer lasting than rubber (TT).

Hush Puppies entered the U.S. shoe market (EP).

Panty hose sold (DBE).

Domestic tumble dryers marketed by Parnell after
successful use in coin-operated laundrettes (TT).

Capacity of Kuwaiti desalting plant increased to 5 million
gallons/day (EB).

First bifocal contact lens developed (TT).

Frank and Dan Carney, Wichita, Kansas, opened the first
Pizza Hut restaurant (RD).

Electric lawn mower with rotating blade introduced by Wolf
(WABI).

Food Additives Amendment (Delaney Amendment) to 1938
U.S. Food, Drug, and Cosmetics Act enacted (MMGT).

Soviets used irradiation for commercial food products
(ACTB).

D. D. Peebles registered another patent relating to dried
milk reconstitutability (HH).

Stouffer's introduced the first premium-priced frozen foods
(BUSWK).

1959 Stout and Ries improved their tomato-harvesting machine
(KM).

University of California staff tested a tomato harvester
(KM).

Kew Gardens Hotel installed a microwave oven (1ST).

General Electric introduced a toaster with improved safety
using a douple-pole main switch (EL).

Kleenex dispenser box introduced (RM).

Pilkington Brothers, Britain, developed float glass, a
cheaper process for high-quality glass (TT).

Hollow clay bricks used for thermal insulation (TT).

Johnson and Johnson acquired McNeil and rights to Tylenol
pain reliever (NYT).

Rene Lelievre and Roger Lemoine invented a heating-
combing iron, Babyliss (WABI).

French Society of Opticians invented the Varilux lens for
replacing bifocals and trifocals (WABI).

1960 Stout and Ries tested a further improved model of their
tomato harvester in Florida and Michigan (KM).

Blackwell Manufacturing Company built and sold the
University of California tomato harvester (KM).

Aluminum cans used for soft drinks (BBB).

Teflon pans introduced into the U.S. by T-Fal (EL).

Robot Coupe, developer of the commercial food processor,
established in France (later Cuisinart) (NYT).

American Dental Association endorsed Crest as an effective means of protecting teeth (NY).

First fiber-tipped pen, Pentel, marketed by Japan Stationery Company, Tokyo (1ST).

IBM Selectric typewriter introduced (EB).

Bulova Watch Company introduced the Accutron, an electronic watch using a precision tuning fork (FFF).

U.S. Food and Drug Administration approved Tylenol pain reliever for sale without a prescription (NYT).

First commercially produced oral contraceptive pill, Enovid, marketed by G. D. Searle, Skokie, Illinois (1ST).

Little Debbie snack foods introduced by McKee Baking Company, Collegedale, Tennessee (CDIS).

Europe imported the automatic dishwasher from the U.S. (WABI).

L'Oreal developed Elnett hair spray (WABI).

Color Amendment to the 1938 U.S. Food, Drug, and Cosmetics Act was enacted (MMGT).

Seam-free women's stockings introduced (JM).

First Nieman-Marcus Christmas catalog (CDIS).

> 1960   Aren Byhjing Pederson, Denmark, invented Carmen, electrically heated curlers (EL).

Catalog showrooms began operation (BUSWK).

Epoxy floors developed (EB).

< 1961   Disposable diapers marketed; Pampers was the first successful mass-marketed brand (PG).

1961   Domestic-use front-opening freezers sold (TT).

Coke introduced Sprite, lemon-lime drink (MR).

Green Giant sold boil-in-bag frozen vegetables (PM).

General Electric introduced the first toaster oven using horizontal toasting, also used for frozen foods (EL).

First electric toothbrush manufactured by Squibb Company, New York (1ST).

Self-wringing mop, flat sponge with metal plate, was patented in the U.S. (TT).

Mexican and Colombian wheat crossed with Japanese varieties produced a Mexican cross-bred wheat (EB).

Campbell Soup Company bought Pepperidge Farm Bakery (WJ).

Downey liquid fabric softener marketed (WHIRL).

1962   Rachel Carson's *Silent Spring* alerted the public to dangers of DDT and other chemical pesticides (EB).

Instant orange juice available in powdered form (TT).

Colgate-Palmolive introduced Baggies plastic sandwich and
storage bags (CP).

Diet Rite Cola introduced (BBB).

Kelvinator, Seattle, demonstrated an ultrasonic dishwasher,
but it was too costly for normal use (TT).

Sunbeam introduced a spray-mist steam iron (EL).

Sacs of silicone rubber gel replaced sponge pads for breast
implants (TT).

Omnifocal lens that eliminated bifocal line developed (TT).

Gaines, first semidwarf wheat, produced (EB).

International Rice Research Institute was established by the
Rockefeller Foundation at Los Banos, Philippines (EB).

Digital Computer Corporation produced LINC, Laboratory
INstrument Computer, the first desktop computer
(PCW).

Standard electric power receptacles had a third hole added
for grounding (TAC).

1963   Alcoa introduced the tab-top can for beverages (BBB).

Tab diet soft drink introduced by Coca Cola Company
(BBB).

Pepsi followed Coca-Cola's lead in introducing Tab and
introduced Diet Pepsi (BBB).

Hoover steam and dry electric iron introduced in Britain
(TT).

Hospitals adopted Beneflex disposable bottles for feeding
infants (TIME).

Roche Laboratories introduced Valium (TT).

First prenatal blood transfusion performed by Professor
George Green, Auckland, New Zealand, on Mrs. E.
McLeod (1ST).

Robot-coupe, a restaurant food processor, patented by
Pierre Verdun, France (WABI).

Carefour, France, hypermarche, opened (WABI).

1964   New strain of miracle rice grown at Philippines
International Rice Research Institute (TT).

First freeze-dried coffee marketed by General Foods (FFF).

U.S. Food and Drug Administration approved irradiation
for potatoes, wheat, and wheat flour preservation
(IEEE).

Ronson Corporation, Woodbridge, New Jersey, introduced
the Can Do, the first portable electric can opener with
mixer (EL).

New cheap, long-lasting moth-proofing agent, Siven 55,
   developed by Fibre and Forest Institute, Israel (TT).

Permanent press fabrics marketed (ET).

Topless swimsuit designed by Rudi Gernreich, California
   (1ST).

Photochromatic glass invented by Dr. Stookey, Corning
   Glass, New York; it darkened on exposure to sunlight
   (TT).

Borden's bought Cracker Jack Company (CDIS).

Ghirardelli Square, San Francisco, a renovated chocolate
   factory, converted to shops and restaurants (WK).

Metal halide lamp introduced (MHEST).

1965   Long-life milk packaged for nonrefrigerated storage (TT).

Frost-free refrigerators sold in the U.S. (CREPT).

Microwave oven introduced for domestic use (TT).

Salton-Hotray, New York, marketed an electric bun warmer
   (EL).

Shavex electric styling comb imported to the U.S. from
   Switzerland (EL).

General Electric introduced the Teflon-coated soleplate
   iron (EL).

First electric clock with no moving parts and using a
   cathode ray display introduced (TT).

New permeable plastic for soft-contact lens developed (TT).

Mexican cross-bred wheat used internationally (EB).

Charles Hall showed his water bed at San Francisco
   Cannery Gallery (CDIS).

Mary Quant, England, created the miniskirt (WABI).

G. D. Searle scientist discovered aspartame (ACTB).

Ocean Spray introduced Cranapple juice (HS).

Gatorade, a rehydrating beverage, was developed by a team
   of University of Florida urologists (MKTGN).

>1965   Control panty hose introduced (JM).

1965-72   Fecal mitten used on Gemini and Apollo (EB).

1966   Ceramic cooking surface for domestic use available (TT).

Biodegradable liquid detergents introduced (TT).

Water-Pik developed by Dr. Gerald M. Moyer (EL).

Colgate-Palmolive introduced Handiwipes, developed from
   Johnson and Johnson's sanitary napkin fabric (CP).

Nestlé introduced Taster's Choice freeze dried coffee (HS).

1966-72   Plastic bags containing dried food for one-way rehydration
   using a water gun were used in space (EB).

1967   Railroads adopted bar codes for tracking railroad cars (WJ).

Clairol imported Carmen electric curlers to U.S. (EL).
Procter & Gamble introduced detergents with enzymes that
    digested dirt (TT).
Royal assent given to the British Abortion Act (1ST).
H. J. Heinz introduced Great American soups, but they
    were later withdrawn from the market (BW).
Imitation milk marketed in Arizona (HS).
U.S. Wholesome Meat Act (MMGT).

1968   Freeze-dried coffee marketed nationally in the U.S. (FFF).
Washable wallpaper manufactured by Dupont, U.S. (TT).
British Abortion Act became effective (1ST).
Hooleys' opened first Cub store, Fridley, Minnesota (WJ).
Dupont introduced Teflon II, improved Teflon (WJ).
U.S. Wholesale Poultry Products Act (MMGT).

1968-72   Cookies coated with methyl cellulose to prevent crumbs
    developed by NASA for use in space (EB).

1969   DDT banned in the United States (1ST).
Cross-bred tick-resistant cattle bred for the tropics
    introduced (TT).
Naxon Beanery, forerunner of the Crock-pot, introduced
    (KC).
Samsonite Corporation, Denver, introduced a can opener
    with removable can-piercing lever for easier cleaning
    (EL).
Home yogurt maker marketed (TT).
Shot-of-Steam iron with trigger-controlled ejector
    introduced by Sunbeam (EL).
Campbell's introduced Chunky soups (BW).
Alan Kay and Ed Cheadle developed FLEX, an early,
    expensive graphics desktop computer (PCW).
Dr. John Olney urged a ban on monosodium glutamate in
    baby food (MB).

1970   Self-cleaning ovens available (CREPT).
Corelle Livingware dinnerware introduced by Corning (EL).
Michael V. Zamaro, California, marketed a water-filled
    vinyl mattress, popularizing the waterbed (TIME).
Kassai, Japan, began manufacture of baby strollers
    (BUSWK).
Westinghouse demonstrated the induction cooking surface
    (IEEE).
Gilbert Levin, U.S., developed a process, PhoStrip, for
    removing phosphorus from waste water (FORT).
Pampers disposable diapers marketed nationally (WJ).

Azzaro Femme perfume marketed by Loris Azzaro (WABI).

135 million pounds of almonds grown in California (NFFOF).

Cyclamate withdrawn from the U.S. market (ACTB).

U.S. Egg Products Inspection Law (MMGT).

1971   Manny Wesber invented Canfield's Diet Chocolate Soda (CJ).

Rival Crock-pot marketed (KC).

Light, fireproof glass-reinforced cement invented by the Building Research Establishment, England (TT).

Water used for cutting, pioneered by McCartney Manufacturing Company (TT).

John Blankenbaker assembled Kenbak I, an early microcomputer (WJ).

Pierre Verdun marketed Magimix, a French food processor, for home use (WABI).

Richard Sapper, Italy, manufactured halogen lamps designed by Artemide (VIS).

1972   General Electric introduced self-cleaning steam, spray, and dry iron (EL).

Matsushita Electric and Osaka Industrial Research Institute developed an ultrasonic kerosene heater (TT).

Electronic lock and card key system used in high security buildings (TT).

Homer Laughlin China Company discontinued Fiestaware dinnerware (BUSWK).

City of Paris Department Store closed permanently (NB).

Natural breakfast cereals introduced by major food companies: Post, Kellogg, General Mills (HS).

Automatic filter drip coffee makers marketed (CT).

1973   Universal Product Code introduced for identifying grocery products (WJ).

Push through tabs on cans reduced litter (TT).

McDonald's began serving breakfast with Egg McMuffin (WJ).

U.S. Supreme Court reaffirmed a woman's right to abortion on request (TIME).

Russian power station used volcanic steam (TT).

Hershey added nutritional information to Hershey candy bar wrapper (CDIS).

Micral, French microcomputer, marketed (WJ).

Violet No. 1 food dye banned in the U.S. (MMGT).

French surgeon Valdimir Mitz demonstrated Superficial musculo-aponevrotic system for face lifting (WABI).

IBM engineers developed the first portable computer, the SCAMP, but it was not produced commercially (PCW).

Electronic memory typewriter introduced (EB).

1974 Fison's Gro-Bags, compost-filled sacks for growing fruit and vegetables, marketed in England (TT).

Bic introduced the disposable razor in Europe (NEWSW).

Half of Iceland population used hot springs heat piped to homes and offices (TT).

Frozen pizza first marketed (RD).

U.S. National Heart and Lung Authorization Bill enacted (MMGT).

Computer controlled scanners were installed in Marsh Supermarkets in Troy, Ohio (BUSWK).

1975 Cuisinart, food processor developed by Robot Coupe, introduced to the U.S. (NYT).

New faster-weaving loom using air jets to move threads introduced (TT).

Imperial Chemical Industries promoted use of "deep shaft" sewage treatment (TT).

Johnson and Johnson sold Tylenol pain reliever to the public (NYT).

Altair, first mass-produced personal computer, offered for sale. (CDIS).

General Mills introduced granola bars (FORT).

Wally Amos, Hollywood, California, opened the first Famous Amos Cookie Store (WJ).

Nabisco introduced Double Stuf Oreo cookies (CDIS).

Getty Synthetic Fuels extracted landfill gas as a fuel source (CDIS).

< 1976 Europeans developed pump dispensers for toothpaste (WJ).

1976 Plastic aluminum was developed (TT).

Silverstone, nonscratch Teflon, introduced by Dupont (WJ).

Bic and Gillette introduced disposable razors in the U.S. (NEWSW).

Apple I microcomputer developed by Apple Computer Co. (CDIS).

Red No. 2 food dye banned in the U.S. (MMGT).

Red No. 4 and carbon black food dye banned in the U.S. (MMGT).

Weinstein, U.S., developed the disposable oral thermometer (WABI).

Chill Car Industries perfected a process for instantly cooling a canned beverage (WABI).

Renovated Faneuil Hall opened containing stores, restaurants, and boutiques (WK).

Bounce and Cling-Free in-dryer fabric softeners marketed (CREPT).

1977   Commodore offered PET; Tandy offered Radio Shack TRS-80 microcomputers (CDIS).

McDonald's launched Egg McMuffin nationally (CDIS).

Convection oven first available (MI).

Warner Lambert offered home pregnancy test (FORT).

GMB Co., France, developed an individual portable solar water heater (WABI).

U.S. Food Drug Administration proposed banning saccharin use, but U.S. Congress imposed a moratorium (ACTB).

Pozel, German firm, developed an instant heating process for vending machine containers (WABI).

1978   Programmed washing machine introduced (TT).

Louise Brown, first test-tube baby, was born (TT).

Sophia Collier founded the American Natural Beverage Corporation (BUSWK).

Toxic shock syndrome associated with tampon use identified (EBM).

Dave Little, Australia, invented the Suntrac solar water heater (WABI).

1979   Throw-away toothbrushes molded from one piece of plastic made in Milan, Italy (TT).

X-10, Northvale, New Jersey, marketed a home security system (EW).

Dr. Sozio, U.S., developed a solid ceramic for dentures (WABI).

Paul Fireman discovered Reeboks in England and planned introduction to the U.S. (BUSWK).

Jell-O Pudding Pops introduced (GF).

George Kovacs, U.S., introduced his first halogen lamps (VIS).

1979-80   Leg warmers for women were fashion wear (JM).

1980   U.S. Department of Agriculture assumed responsibility for former U.S. Army food irradiation program (IEEE).

Kassai introduced its Aprica baby stroller to the U.S. market (BW).

Procter & Gamble withdrew Rely tampon from the market (EBM).

Hershey began using foil wrapper on Hershey candy bar (CDIS).

Armour introduced gourmet frozen dinners, Dinner Classics (CDIS).

West Edmonton Mall, Alberta, Canada, shopping entertainment center with amusement park addition opened (WJ).

Alistair Pilkington, England, invented Kappafloat, glass with the property of trapping solar heat (WABI).

Pillsbury began work on Toaster Strudel (FORT).

Minnetonka bought Calvin Klein Cosmetics and Obession perfume (BUSWK).

1981   Searle introduced Equal sweetener based on aspartame (Nutrasweet) (CDIS).

57 percent of U.S. homes had air conditioners (FUTUR).

California "wind farms" established in Altamont Pass (TR).

Osborne I and IBM PC microcomputer marketed (CDIS).

Wendy's experimented with a breakfast menu (CDIS).

Colgate-Palmolive introduced gel toothpaste (WJ).

Gilbert Levin, U.S., patented L-sugar, same taste as sugar, but not metabolized by the body (FORT).

Foods using Yellow No. 5 food dye began carrying a warning label (MMGT).

Aspartame approved by U.S. Food and Drug Administration for use in food and dry beverage mixes (ACTB).

Low calorie Aunt Jemima Pancake Syrup introduced (BUSWK).

R. R. Donnelley began investigations of electronic shopping (MKTGN).

Giorgio perfume launched by Fred and Gale Hayman (CDIS).

California wine cooler introduced (MKTGN).

Brother low-cost dot matrix typewriter introduced.

1982   General Foods used Nutrasweet in its products (CDIS).

Pepsi Free, caffeine-free Pepsi, introduced (MR).

Diet Coke debuted (MR).

Tylenol poisonings case led to tamper proof packaging for nonprescription drugs (NYT).

Quaker Oats Company introduced Chewy Granola Bars (FORT).

Campbell Soup Co. launched Le Menu frozen foods (WJ).

Campbell Soup Co. launched Prego Spaghetti Sauce (WJ).

Admiral introduced the Entertainer, premium-priced refrigerator with built-in wine rack (FORB).

Milk Marketing Board, U.K., developed a cheese substitute (WABI).

Uni-Charm, Japan, introduced superabsorbent disposable diapers in Japan (BUSWK).

Disposable fashion watch, Swatch, introduced (WABI).

1983   Nutrasweet used in cold milk product mixes (CDIS).

U.S. Food and Drug Administration approved general use of irradiation for preserving spices (IEEE).

Procter & Gamble introduced Duncan Hines cookies, soft and chewy inside, crunchy outside (WJ).

Caffeine-free Coke debuted (MR).

Coke adopted Nutrasweet as the sweetener in Diet Coke (MR).

Last Maytag wringer washing machine manufactured (FORB).

Space shuttle produced its water supply from fuel cells (NASA).

Hand-powered, rachet-driven trash compactor developed by Nelson and Johnson Engineering; adapted for home and camp (EB).

NASA shuttle used toilet facility with freeze drying of solid waste matter (NASA).

Kimberly-Clark introduced disposable diapers with reusable tabs (WJ).

IBM XT and Apple LISA microcomputers marketed (CDIS).

Wendy's introduced a new breakfast menu nationally (CDIS).

Vipont marketed an antiplaque toothpaste (WJ).

Dannon introduced a French-style yogurt (Custard) (WJ).

Aspartame approved by the U.S. Food and Drug Administration for use in soft drinks (ACTB).

Brightly colored Reeboks introduced to the U.S. (BUSWK).

R. R. Donnelley built a prototype Electronicstore kiosk for merchandise ordering (MKTGN).

Crisco introduced a yellow, butter-flavored shortening that led to the demise of Fluffo shortening (FORT).

>1983   Ray Ward, Mesa, Arizona, adapted shuttle water filter for home use and marketed it commercially (EB).

1984   Pepsi and Coke switched to 100 percent Nutrasweet formulations in Diet Pepsi and Diet Coke (CDIS).

Environmental Pollution Agency banned the use of
fumigant ethylene dibromide (WJ).
General Electric introduced electronically controlled
refrigerator with a sensor to detect open door (BUSWK).
Coke available on the space shuttle (MR).
First U.S. hypermarche, Biggs, opened near Cincinnati
(CDIS).
Liquid Tide introduced (WJ).
Acoustic wand for cleaning teeth devised by Joseph S.
Heyman, NASA (IEEE).
General Electric and Sunbeam marketed irons that shut
themselves off automatically (WD).
General Electric Home Minder home security system based
on X-10 components introduced (EW).
GIFT, gamete intrafallopian transfer, developed for
artificial insemination (SCI85).
Procter & Gamble patented a disposable bib with
gravitationally operated pocket to catch baby's dribbles
(WJ).
Apple Macintosh, Fat Mac, IBM AT, AT&T PC, and
Epson PX-8 personal computers debuted (CDIS).
Pepsi introduced Slice while Coke test marketed Minute
Maid orange soda in Canada (BW).
Campbell Soup Company launched Chunky New England
Clam Chowder (WJ).
Campbell Soup Company launched Great Start Breakfasts
(WJ).
Colgate-Palmolive introduced pump dispenser for
toothpaste to compete with Minnetonka's Check-up
(WJ).
Recipe Writer, software package for home use, with master
index, menus, and shopping list, introduced (AMERW).
Nabisco introduced mint-flavored Oreo cookies (CDIS).
Induction hotplate invented by Anton Seelig, Germany; it
heated food without getting hot itself (WABI).
Dole introduced Fruit'n'Juice bars (FORB).
Scott Paper Company introduced ScotTowel Jr., smaller-
sized paper towel (WJ).
CompuServe tested the Electronic Mall, an on-line
purchasing service (INFO).
1985   Experiments reported on methods to increase successful
spawning of shellfish making farming practical (WJ).

Environmental Protection Agency, U.S., banned the use of Silvex and 2,4,5-T herbicides (WJ).

La Vie de France, Virginia, introduced a low cholesterol croissant made with margarine (WJ).

Bob Greene, columnist, popularized Canfield's Diet Chocolate Soda (CJ).

Burger King began national introduction of its new breakfast menu with Croissan'wich (WJ).

Tide with enzymes and bleach introduced; Procter & Gamble noted bleach worked for wider range of temperatures (WJ).

Gillette introduced a twin-blade cartridge razor with water-soluble strip of plastic resin (WJ).

Procter & Gamble tested tartar-control formula Crest toothpaste (WJ).

Mitsubishi introduced the "wired home," a housekeeping system, to monitor appliances and heating and to entertain (EW).

Ivarsons, Minneapolis, offered prefabricated, energy-saving Swedish homes for the U.S. market (CDIS).

NASA abandoned freeze drying of toilet wastes for simpler storage system (CDIS).

Valium patent expired opening the way to generic manufacturers (CDIS).

California State office building, San Francisco, got 5 percent of its power and most of its hot water from fuel cells (USA).

NASA investigated use of photosynthetic bacteria for converting solar energy to hydrogen for shuttle (USA).

Arco developed thin-film silicon for photovoltaic cells to make fuel cells competitive with current technology (FORT).

IBM announced end of the IBM PC Jr. while Apple did the same for LISA personal computers (WJ).

Source Perrier announced plans to add citrus flavor to Perrier (ECON).

U.S. Food and Drug Administration proposed a ban on the use of sulfite to preserve fresh fruits and vegetables in salad bars (CJ).

Coke introduced a special can for use by astronauts on space shuttle (CDIS).

Coke introduced a new formula Coke nationally (WJ).

Coke reintroduced original formula Coke as Classic Coke (WJ).

Coke introduced Cherry Coke nationally (WJ).

Campbell Soup Company launched Prego Plus Spaghetti Sauce (WJ).

NASA scientists announced that the spider plant was effective in removing formaldehyde from air (RD).

Ralph Portier, Louisiana State University, developed fourteen strains of bacteria and yeasts that ate pollutants (RD).

Dr. Edward Kass identified magnesium absorbing fibers as increasing risk of toxic shock (WJ).

Campbell's introduced dried soup mixes for the second time, but the first time under Campbell label (WJ).

Improved monoclonal home pregnancy test introduced (FORT).

Monoclonal Antibodies, Mountain View, California, offered OvuStick for determining peak fertility (FORT).

Wacoal Corporation, Japan, introduced a machine-washable bra using "shape-memory" alloy wire (BUSWK).

U.S. Department of Agriculture developed electronic device to measure fruit and vegetable ripeness (BUSWK).

NASA developed Dynacoil, material used for high-resiliency running shoe by KangaRoos USA, St. Louis (BUSWK).

Prince Company, Lowell, Massachusetts, introduced Tug-n-Tie package for Dutch Maid Noodles, easy open, tie reseal (MKTGN).

Harry H. Holly developed new hamburger patty machine, Universal System, that interwove meat strands to retain moisture during cooking (CDIS).

McDonald's introduced McDLT, devised by Will May at Texas franchise; packaging change for hamburger that separated the lettuce and tomato (FORT).

American Cyanimid introduced Combat, a more effective roach poison (FORT).

American Cyanimid announced a new corn hybrid resistant to chemical weed killers (WJ).

Pillsbury introduced Toaster Strudel (FORT).

Bacteria used to clean Navy sewage tanks (CDIS).

Textile designers used computers for design work (CDIS).

463

Lem-me, a dish washer scrubber, introduced by Swiss-tex (MKTGN).

Procter & Gamble developed new superabsorbant Pampers; they absorb 700-800 times their own weight in fluid (WJ).

Brassiere du Pecheur developed a syrup for making beer by adding water (BUSWK).

T. J. Cinnamons Bakery, Kansas City, opened in January selling large, fresh-baked cinnamon rolls (TIME).

Coca-Cola test marketed in clear plastic cans (WJ).

Digital hearing aid patented by A. Maynard Engebretson and Gerald R. Popelka, St. Louis (IEEE).

R. R. Donnelley made Electronicstore kiosks available for shopping malls and department store use (MKTGN).

Laser used to develop custom orthopedic shoes (IEEE).

Christian Dior's Poison perfume launched (CDIS).

Bathwomb introduced by Water Jet Corporation with nine massage jets, facial mist, four stereo speakers, and phone (CDIS).

Peachtree Schnapps introduced (FF).

1986   Plant Genetic Systems, Belgian biotechnology company, developed a tobacco strain that kills insects (BUSWK).

Warner-Lambert reintroduced Beeman's, Black Jack, and Clove chewing gums (WJ).

Alcan Research Labs announced the MicroMatch packaging system (MKTGN).

Reynolds Metals Company introduced the Container-Mate foil closure for packaging use (MKTGN).

Campbell's Soup Company test marketed Cookbook Classic Soups, single-serve, table-ready, microwave (MKTGN).

Coca Cola introduced Minute Maid Fruit Juicee (formerly Guido's Ice Juicee) (FORB).

Fiestaware dinnerware reintroduced by Homer Laughlin China Company (BUSWK).

Japanese scientists developed a new electric toothbrush that neutralized plaque film (SMR).

Soft bifocal contact lens, Alges lens, developed by University Optical Products, Florida (CDIS).

Soft bifocal contact lens, Hydrocurve II, developed by Barnes-Hind of Revlon Vision Care (CDIS).

John Sullivan, U.S. Department of Agriculture, adapted puff drying techniques used for blue berries to mushrooms (WJ).

Procter & Gamble introduced Dribbles, a disposable bib for infants (WJ).

Dr. Beatrice Courzinet and Dr. Gilber Schaison, France, tested a "morning after pill" for contraception (TIME).

Second Tylenol tampering led Johnson and Johnson to withdraw capsules and market caplets, flattened, oval shaped tablets (WJ).

1987   Nabisco baked more than sixty milllion Ritz Crackers per day (CDIS).

Kimberly-Clark considered marketing germicidal tissues to corporate customers (BR).

Apple introduced the Open Mac with open board slots for greater flexibility and IBM compatability (WJ).

Compaq announced the 80386-based computer (WJ).

Kroger Supermarkets experimented with Service Plus, customer-operated scanner units for grocery check-out (BUSWK).

K. C. Masterpiece Barbecue Sauce, created by Rich Davis, Kansas City, was marketed nationally (FF).

Index to the Chronologies

BC dates are indicated by -, AD dates have no designation, C = century. Modern geographic names are used throughout.

A&P. *SEE* GREAT ATLANTIC AND PACIFIC TEA
ABERDEEN RUBBER SOLE CO. 1889
ABORTION 1920, 1935, 1967-68, 1973
ABRASIVE SOAP 1907
ABSORPTION REFRIGERATION 1926
ABU LU'LU'A 644
ABYDOS, EGYPT -5510
ACCUM, FREDERICK C. 1820
ACCUTRON 1960
ACETATE. *SEE* RAYON
ACETYLENE 1930
ACETYLSALICYLIC ACID 1853. *SEE ALSO* ASPIRIN
ACHARD, F. 1801
ACOUSTIC WAND 1984
ACRILAN 1952
ACRYLIC POLYMER 1927
ACTIVATED SLUDGE 1912-15
ADAM, EDOUARD 1801
ADAMS, THOMAS 1870
ADDITIVES. *SEE* FOOD ADDITIVES
ADIDAS SHOES 1948
ADMIRAL (REFRIGERATOR) 1982
ADULTERATED FOODS -300, 1820, 1850
AERATED BREAD 1856
AEROSOLS 1926, 1939, 1941, 1953
AETIUS 500
AETNA IRONWORKS 1824
AFGHANISTAN -500
AFRICA, CENTRAL -2000
AGA STOVE 1924

AGITATOR (WASHING MACHINE) 1850, 1908, 1915
AGRICULTURE -10,000
AGRIPPA -27
AIKEN, HOWARD 1944
AIR CONDITIONERS 1842, 1844, 1897, 1902, 1915, 1931-32, 1981
AIR PUMP 1650, 1654
AIR PURIFICATION 1985. *SEE ALSO* AIR CONDITIONING
AIR-O-MIX 1923
AKERMAN, JOHN 1719
AKHMIN, EGYPT 300
AKRON, OHIO 1870-71, 1930
ALARM CLOCK 1380, 1787, 1902, 1914
AL-BALARI, EGYPT -4400
ALCOA 1892, 1963, 1986
ALCOHOL THERMOMETER 1641
ALEGA, JUOAN DE 1589
ALEXANDER THE GREAT -328--326
ALEXIS, CZAR 1618
ALFRED, KING 890
ALGES LENS 1986
ALKA-SELTZER 1931
ALKYD RESINS 1918
ALL-STAR ATHLETIC SHOES 1917
ALLBUTT, DR. THOMAS 1867
ALLIS CHALMERS 1936
ALMOND 18TH C, 1897, 1955, 1970
ALMOND JOY 1942
ALPHA BETA 1912
ALSACE 1490, 1832
ALTAI MOUNTAINS -201
ALTAIR 1975
ALUM 1300, 1800, 1843

CARDING 1553, 1728, 1782, 1784, 1790-92, 1796-97, 1822-23
CARD KEY 1972
CAREFOUR (STORE) 1963
CAREW, JOHN 1678
CARL ZEISS CO. 1910
CARLILE, RICHARD 1826
CARMEN CURLERS 1960, 1967
CARNEGIE, ANDREW 1880
CARNEGIE HALL 1902
CARNEY, FRANK AND DAN 1958
CARPENTER ELECTRIC HEATING MFG. CO. 1891
CARPENTER'S SQUARE -600--500
CARPENTIER, PIERRE 1851
CARPETS -201, -100, 1100, 1255, 1450, 1482, 1500, 1620, 1701, 1845, 1876, 1949, 1954, 1957
CARPET SWEEPER 1811, 1858, 1859, 1865, 1876. *SEE ALSO* STREET SWEEPER
CARRE, FERDINAND 1859
CARRIER, WILLIAM H. 1902
CARRINGTON, JAMES 1829
CARRON ELECTRIC RANGE 1910
CARRON IRON WORKS 1762
CARROT -500
CARRUCA 1
CARSON, RACHEL 1962
CARSTAIRS, WILHEMINA 1847
CARTER, CHARLES P. 1849
CARTER, FRANK GRIFFIN 1907-8
CARTER, SUSANNA 1796
CARTIER (JEWELER) >1920
CARTRIDGE PENS 1920, 1935
CARTWRIGHT, E. 1785
CARUTHERS, WILLIAM 1937
CASEIN 1780, 1914, 1935
CASH REGISTER 1879-80, 1884, 1974, 1987
CASPIAN SEA REGION -4000
CASTERS (FURNITURE) 1838
CAST IRON 1455, 1490, 1762, 1770, 1782, 1796-97, 1803, 1830, 1834, 1915
CATAL HAYUK -6500
CATALOG SHOPPING 1844-45, 1872, 1885, 1888, 1894-95, 1897, 1900, 1909, 1912, 1926, 1951, 1960

CATARACT ROCKING TUB (WASHING MACHINE) 1915
CATERPILLAR TRACTOR 1931
CATHERINE, QUEEN (ENGLAND) 1662
CATHERINE DE MÉDICIS, QUEEN (FRANCE) 1533
CATO, MARCUS PORTIUS (THE ELDER) -2D C
CATSUP 1792, 1876
CATTLE. *SEE* LIVESTOCK
CELANESE 1865, 1955
CELERY 1870
CELLON, GEORGE 1900
CELLOPHANE 1908, 1913, 1926
CELLULOID 1866, 1869
CELLULOSE 1883, 1918, 1972
CELSIUS, ANDERS 1742
CELT (AX) -15,000
CELUCOTTON 1918
CEMENT -150, 1796, 1824, 1844, 1867, 1923, 1971
CENTIGRADE TEMPERATURE SCALE 1742
CENTRAL HEATING -350, -80, 1777, 1831, 1893, 1904
CENTRAL STATES MFG. 1927
CERAMIC COOKING SURFACE 1966
CEREALS 1893, 1895, 1898, 1901, 1904, 1906, 1924, 1937, 1972
CEREALS (GRAINS) -3500, -300. *SEE ALSO* BARLEY, CORN, OATS, RICE, WHEAT
CESAREAN DELIVERY 1500, 1581, 1610, 1738
CHAFF CUTTER 1731, 1787
CHAIRS 1400, 1851, 1926, 1929, 1967
CHALDEAN -2000
CHAMBERLIN, BENJAMIN D. 1909, 1914, 1925
CHAMPION, JOHN 1770
CHANDELIERS 1646
CHANEL (MATCHES) 1805
CHANEL NO. 5 1920
CHAPMAN, FREDERICK A. 1895
CHAPMAN AND SKINNER 1905
CHARDONNET, HILAIRE DE 1885
CHARLEMAGNE 768, 9TH C

GRANOLA BARS 1975, 1982
GRANOSE FLAKES 1895
GRAPEFRUIT 1755, 1919
GRAPE JUICE (WELCH'S) 1893
GRAPE-NUTS 1895
GRAPES -4000, -450
GRAPHITE 1584, 1795
GREAT AMERICAN SOUPS 1969
GREAT ATLANTIC AND PACIFIC
    TEA CO. 1869
GREECE -900, -500, -4TH C, -390,
    -371, -350, -300, -200, -2D C, >100,
    500, 8TH C
GREEN, GEORGE 1963
GREEN, ROBERT 1874
GREENE, BOB 1985
GREENE, CATHERINE. *SEE*
    WHITNEY, ELI (COTTON
    GIN)
GREEN GIANT 1961
GREENHOUSES 1594
GREENWOOD, CHESTER 1873,
    1877
GREENWOOD, JOHN 1790
GREETING CARDS 1851, >1870,
    1874-75, 1906, 1910
GREGORY, HANSON 1847
GREW, NEHEMIAH 1695
GREY, C. E. 1913
GRO-BAGS 1974
GROCERS AND GROCERIES 1606,
    1869, 1882, 1887, 1912, 1915-16,
    1930, 1937, 1968, 1974, 1987. *SEE*
    ALSO SUPERMARKETS
GROTTE-DU-PONTAL, HERAULT
    -4001
GRUNWALD (GERMANY) 1855
GUANO 1602, 1840
GUERICKE, OTTO VON 1650, 1654
GUERLAIN, PIERRE AND
    JACQUES 1925
GUIBAL, 1848
GUILLEMEAU, JACQUES 1598
GUILLET, LEON 1904
GUINEA 1492, 1548
GUM. *SEE* CHEWING GUM
GUTTA PERCHA CO. 1847
GUY-LUSSAC, J. 1825
GYOKI (JAPAN) 749

HAAS, EARLE C. 1937
HABENWARE 1956
HABER, FRITZ 1908-9, 1914
HADAWAY, WILLIAM S., JR. 1869
HAGG, J. A. 1868
HAIR DRESSING 1872, 1887, 1906,
    1914, 1945, 1959-60, 1965, 1967
HAIR DRYER 1914, 1920, 1951
HAIR DYES 1909-10, 1927, 1929, 1953
HAIR SPRAY 1960
HAITI 1492
HALCS, STEPHEN 1727
HALDANE, T. G. N. 1927
HALIDE LAMPS 1964, 1971, 1979
HALIJIAN, PETER PAUL 1921, 1942
HALL 1786
HALL, ALBERT W. 1927
HALL, CHARLES 1965
HALL, CHARLES M. 1886
HALL, JOYCE 1910
HALLADAY, DANIEL 1854
HALLIDAY, TRISTRAM 1840
HALLMARK CARDS 1910
HALLSTATT, AUSTRIA -6500, -601
HALOPHOSPHATE 1951
HAMBURG, GERMANY 1678-79,
    1882
HAMBURGER 1880, 1900, 1904, 1921,
    1937, 1955, 1985
HAMILTON, DUCHESS OF (ANN)
    1696
HAMILTON (U.S.) 1939
HAMILTON WATCH CO. 1957
HAMMER, M. 1917
HAMWI, E. A. 1904
HANCOCK, CHARLES 1846
HANCOCK, THOMAS 1820, 1822,
    1843
HANDIWIPES 1966
HANDKERCHIEF 4TH C, 1436, 1503,
    1660
HANDMILLS -75,000, -15,000,
    -6000
HANELL, ARTHUR H. >1850
HANKS, RODNEY AND HORATIO
    1819
HANSAPLAST 1882
HANSEL, GEBRUDER 1922
HARE, ROBERT 1820

MAIZE. *SEE* CORN
MAJORCA 1315
MAKEUP. *SEE* COSMETICS
MALASSAGNY, DE 1762
MALAY PENINSULA 1600
MALLORY, JAMES 1812
MALLORY, P. R., CO. 1957
MALLS. *SEE* SHOPPING CENTERS
MALTED MILK 1887
MAMMOPLASTY 1905, 1924, 1928, 1930, 1962
MANCHESTER, ENGLAND 1641, 1789, 1817, 1825, 1847, 1907, 1914
MANCHESTER YELLOW (DYE) 1915
MANDARIN ORANGES -220
M&M'S 1940, 1954
M&M'S, PEANUT 1954
MANGLE 1696, 1774, 1797, 1857, 1871. *SEE ALSO* IRONS, WASHERS
MANGO 16TH C
MANTA, MARCHESE DI 1462
MANTERS, GEORGE 1926
MANURING. *SEE* FERTILIZATION
MAQUET (FRANCE) 1841
MARBLE DRY GOODS PALACE 1848
MARCEL WAVE 1872
MARCHAND (FRANCE) 1678
MARCIONI, ITALO 1896, 1903
MARCONI CO. 1923
MARCUM, M. VAN 1787
MARCUS, HERBERT 1907
MARGARINE 1869, 1871, 1896, 1903, 1944, 1985
MARGGRAF, ANDREAS 1747
MARIA THERESA 1660
MARIOLLES CHEESE 960
MARIUS (PARIS) 1715
MARK I (COMPUTER) 1944
MARKET SQUARE (LAKE FOREST, ILLINOIS) 1916
MARKETS 1634
MARMALADE 1587
MARQUIS, CHARLES 1900
MARS, FRANKLIN 1928, 1930, 1932, 1940, 1954

MARSEILLES, FRANCE 1660, 1671, 1860
MARSH 1851
MARSH, A. L. 1906
MARSHALL, ALBERT T. 1899
MARSHALL FIELD 1852
MARSH SUPERMARKETS 1974
MARTIN, LOUIS 1796, 1797
MARTIN, THOMAS 1715
MARTINIQUE, WEST INDIES 1723, 1806
MARYLAND 1803, 1851, 1879, 1891, 1955
MASON, JOHN 1858
MASON, M. 1851
MASON JAR 1858, 1902, 1914
MASSACHUSETTS 1629, 1639-40, 1642, 1690, 1765, 1778, 1784, 1801, 1825,1827, 1829, 1834, 1840, 1847, 1849, 1851, 1859, 1865, 1868, 1872, 1875,1891, 1895, 1899, 1925, 1938-40, 1943-44, 1947-48, 1985
MASSACHUSETTS INSTITUTE OF TECHNOLOGY 1939-40, 1943, 1948
MASSEY FERGUSON 1958
MASTERS, THOMAS 1850
MATCHES 970, 1270, 1530, 1640, 1680, 1780, 1805, 1810, 1816, 1826-28, 1830-32, 1836, 1842, 1844-45, 1848, 1852, 1855, 1860, 1864, 1872, 1892, 1896, 1898, 1910-11, 1943, 1971, 1985
MATHIEU, GABRIEL 1723
MATIN BROTHERS 1863
MATSUSHITA ELECTRIC 1972
MATTEE, CHARLES 1849
MATTHAI-INGRAM 1890
MATTRESS 1925, 1965, 1970
MATZELIGER, JAN ERNST 1882
MAUGHAM, BENJAMIN 1886
MAXWELL, G. W. 1906
MAY, WILL 1985
MAYTAG 1907, 1913, 1983
MEAN, ROBERT 1800
MEASURING CUPS 1870, 1890, 1896, 1925
MEASURING SPOONS 1900, 1925

PERU -3000, -2000, 1530, 1553, 1602, 1638, 1830, 1840
PESSARY 1838
PESTICIDES 1890, 1920-21, 1924, 1939, 1941-42, 1962, 1969, 1985-86
PET (COMPUTER) 1977
PETER, DANIEL 1875
PETER PAUL MOUNDS 1921
PETRI PAPYRUS -1850
PETROFF, IVAN 1567
PETTIT & SMITH 1856
PEWTER 1384
PHILADELPHIA 1680, 1740, 1748, 1752, 1772, 1796, 1829, 1835, 1857-58, 1860, 1870, 1874, 1876, 1884, 1888, 1902, 1946, 1951
PHILIPPINES 1962, 1964
PHILIPS (HOLLAND) 1937
PHILLIPS, A. D. 1836
PHILLIPS, CHARLES 1873
PHILLIPS' MILK OF MAGNESIA 1873
PHILO OF BYZANTIUM -250
PHOENICIANS -600
PHOSPHORATED HESPIDIN. *SEE* ENVOID
PHOSPHORUS 1680, 1780, 1816, 1830, 1836, 1845, 1861, 1864, 1898, 1910-11, 1914, 1970
PHOSTRIP 1970
PHOTOCHROMATIC GLASSES 1964
PHOTOSYNTHETIC BACTERIA 1985
PHOTOVOLTAIC 1985
PHTALOCYANINE DYES 1934
PICKARD, JAMES 1779
PICKLE -2000
PICK-PROOF LOCK 1818
PIECA 1000
PIERRE, S. F. 1865
PIGGLY WIGGLY 1916
PIGS -6750, 1683, 1769, 1921, 1923, 1937
PILKINGTON, ALISTAIR 1980
PILKINGTON BROTHERS 1959
PILL (CONTRACEPTION) 1952, 1954, 1960, 1986
PILLOWS 1929

PILLSBURY 1949, 1980, 1985
PINCUS, GREGORY 1939
PINEAPPLE 1493, 1616, 1790, 1892
PINEAPPLE CHEESE 1808, 1810
PINS -2001, 1483, 1540, 1543, 1824, 1832, 1838
PIPS (LIGHT BULBS) 1919
PITTSBURGH, PENNSYLVANIA 1825, 1882, 1942
PITTSBURGH CORNING 1942
PITTSBURGH PAINT AND GLASS 1926
PITTSBURGH REDUCTION 1892
PIVOTED SCISSORS -1500, 100, 5TH C
PIXII, HYPOLITE 1832
PIZZA 1000, 1889, 1895, 1958, 1974
PIZZA HUT 1958
PLA, SIMON 1784
PLANTE, R. L. G. 1859
PLANTEX, BALTZAR CARL VON 1926
PLASSE, DINGHEN VAN DER 1564
PLASTICS 1780, 1866, 1914, 1953, 1965, 1976, 1979, 1985. *SEE ALSO* SPECIFIC TYPES
PLASTIC SURGERY 1887, 1905, 1925, 1928, 1930, 1962, 1973
PLATE GLASS 1688
PLATEN, B. VON 1920, 1926, 1929, 1931
PLASTIC BAGS 1962, 1972
PLATINA, BARTHOLOMAEUS 1475
PLATT, HUGH 1594, 1608
PLATZ (GERMANY) 1925
PLINY 77
PLOWS -3000, 1, 890, 990, 1523, 1619, 1627, 1634, 1730, 1797, 1803, 1808, 1819, 1837, 1846-47, 1889-90, 1893
PLUCKNETT, THOMAS 1805
PLUM -1800, 1519, 1825
PLUNKETT, ROY J. 1938
POCHON (FRANCE) 1799
POISON PERFUME 1985
POLA, ITALY 1896
POLAND 1579, 1609
POLAR WARE CO. 1927
POLO, MARCO 1270, 1306

ST. JEROME 400
ST. LOUIS, MISSOURI 1854, 1880, 1884, 1903-4, 1909, 1927, 1937, 1985
SAINT-LOUIS (FRANCE) 1782
SALEM, MASSACHUSETTS 1629, 1856, 1859, 1882
SALICYN 1829
SALINDRES FOUNDRY 1855
SALMON 1825
SALMON (ENGLAND) 1811
SALMON, ISAAC 1861
SALOMONS, DAVID 1888
SALT -6500, -1000, -350, -328, 1260, 1359, 1848
SALTING (PRESERVATION) -1000, 1260, 1359
SALTON-HOTRAY 1965
SALTPETER 1550, 1804
SAMARKAND 751
SAMSON UNITED 1933
SAMSONITE 1969
SAMUELSON, ALEXANDER 1915
SAND FILTERS 1791, 1814, 1827
SANDALS -2000
SANDERS, 1825
SANDERS, BERTEL 1807
SANDFORD, GILBERT 1236
SANDYS, MARCUS SIR 1837
SAN FRANCISCO, CALIFORNIA 1851, 1854, 1876, 1905-6, 1909, 1920, 1953, 1964-65, 1985
SANGER, MARGARET 1916
SANITARY NAPKINS 1921
SANITAS FOOD CO. 1898
SANTORIO (ITALY) 1625
SAPIRSTEIN, JACK 1906
SAPPER, RICHARD 1971
SARA LEE BAKED GOODS 1949
SARATOGA CHIPS 1853
SARDINES 1819
SARGENT, JAMES 1873
SARGONI II 707
SARSAPARILLA 1866
SAUCEPANS 1799, 1890, 1955-56. *SEE ALSO* COOKWARE
SAUNDERS, CLARENCE 1916
SAURIA, CHARLES 1831

SAUSSAUR, NICOLAS THEODORE DE 1804
SAVAGE ARMS CORP. 1924
SAVERY, THOMAS 1698, 1702, 1717
SAVOY HOTEL 1907
SAW, CIRCULAR 1777, 1791, 1810
SAWMILL 300, 1230, 1328, 1592, 1633
SAXON, JOSEPH 1833
SAXONY WHEEL 1533
SCAMP (COMPUTER) 1973
SCANNERS (STORES) 1974, 1987
SCHAFFER, JOHN 1819
SCHAISON, GILBERT 1986
SCHEUTZ, GEORGE 1855
SCHICK, JACOB 1923, 1931
SCHINKEL, HERMAN 1508
SCHLUMBOHM, PETER 1941
SCHNERING, OTTO 1921-22
SCHOLL, DR. 1904
SCHOLL ARCH SUPPORT 1904
SCHROETTER, ANTON 1845
SCHUELLER, EUGENE 1909-10
SCHUTZENBERGER (GERMANY) 1865, 1890
SCHWARZMANN, E. 1930
SCIENTIFIC AMERICAN 1867, 1887
SCISSORS -1500, -3D C, 100, 5TH C, 1761, 1848
SCOPOLAMINE MORPHINE 1902, 1917
SCOTCH TAPE 1937
SCOTLAND -2001, 1696, 1762, 1784, 1792, 1799, 1804-5, 1812, 1815, 1820, 1825, 1830, 1835, 1840, 1847, 1856, 1871, 1873, 1880, 1889-90, 1899, 1927, 1957
SCOTTOWELS 1907, 1984
SCOTT PAPER CO. 1907, 1984
SEA AND SKI 1953
SEALING WAX 1550
SEAMFREE STOCKINGS 1860
SEARLE, G. D. 1952, 1954, 1960, 1965-66, 1981
SEARS 1888, 1894, 1897, 1909, 1951
SEARS, ERNEST ROBERT 1954
SEASONINGS. *SEE* SPECIFIC TYPES (e.g., SALT, SUGAR)
SECRETAIRE À ABATTANT >1700
SECURITY SYSTEM 1972, 1979, 1984

Sources

ACTB   Haas, Albert, Jr. "How to Sell Almost Anything by Direct Mail." *Across the Board* 13, no. 11 (November 1986):45-53.

Mellow, Craig. "Coolest Company in America." *Across the Board* 22, nos. 7-8 (July-August 1985):11-14.

Pick, Grant. "Gerber's Baby Under Stress." *Across the Board* 23, nos. 7-8 (July-August 1980).

Steyer, Robert. "Irradiated Food: A Marketing Hot Potato." *Across the Board* 23, nos. 7-8 (July-August 1986).

Whelan, Elizabeth M. "Sweetener Wars: Taking Aim at Aspartame." *Across the Board* 23, no. 10 (October 1985):

AE   De Camp, L. Sprague. *Ancient Engineers*. New York: Ballantine, 1963. ISBN 0345293479. Note: LC 62 15901.

AHLMA   "Historical Laundry Appliance Display." In *10th National Home Laundry Conference; 1956 Nov. 1-2, Chicago*. Chicago: American Home Laundry Manufacturers' Association, 40.

Snyder, Louis M. "Looking Ahead with the Combination Washer-Dryer." In *11th National Home Laundry Conference, "Wash and Wear--Fact or Fantasy?"; 1957 Nov 1-2; Washington, D.C.* Chicago: American Home Laundry Manufacturers' Association, 40-43.

AMERW   Ecenbarger, William. "Get the Lead Out." *American Way* 14, no. 18 (2 September 1986):51-54.

AO   Oakley, Ann. *Woman's Work: The Housewife, Past and Present*. New York: Vantage, 1976. ISBN 0394719603. Note: LC 75 28281.

BBB   Panati, Charles. *Browser's Book of Beginnings*. Boston: Houghton Mifflin, 1984. ISBN 0395360994. Note: LC 83 26384.

BCT   Schapira, Joel; Schapira, David; and Schapira, Karl. *Book of Coffee and Tea: A Guide to the Appreciation of Fine Coffees, Teas, and Herbal Beverages*. New York: St. Martins, 1975. Note: LC 73-90585.

BHG   "What Aluminum Foil Can Do for You." *Better Homes and Gardens* 26 (May 1948):14.

BP   Butti, Ken, and Perlin, John. *Golden Thread: 2500 Years of Solar Architecture and Technology*. New York: Van Nostrand Reinhold, 1980. ISBN 0-442-24005-8. Note: LC 79-25095.

BR   "Cold-Fighting Tissue Likely to Be Available . . . at Last." *Boardroom Reports* 16, no. 1 (January 1987):2. ISSN 0045-2300.

BUSWK   "America's Supermarket Miracle: New Formats and Technologies Revolutionize the Way We Shop." *Business Week;, 4 May 1987*, 127-131, 134. ISSN 0007-7135.

"And Now There's Even Instant Beer.: *Business Week*, 28 October 1986, 111. ISSN 0007-7135.

Armstrong, Larry, and Dobrzynski, Judith H. "Aprica Kassai: A Fast Ride into the U.S. with Status-Symbol Strollers. *Business Week,* 21 January 1985, 117.

"At 100, the Tuxedo Starts Going Places Again." *Business Week,* 13 October 1986, 177.

Bluestone, Mimi, et al. "Stop--Don't Throw Those Crab Shells Away." *Business Week,* 23 March 1987, 112, 116.

Buell, Barbara. "How P&G Was Brought to a Crawl in Japan's Diaper Market." *Business Week,* 13 October 1986, 71, 74.

Carpenter, Kimberly. "Catalog Showrooms Revamp to Keep Their Identity." *Business Week*, 10 June 1986, 117, 120.

Dreyfack, Kenneth. Quaker Is Feeling Its Oats Again." *Business Week*, 22 September 1986, 80-81.

Dreyfack, Kenneth. "Running Shoes Make a Leap into the Space Age: Kanagroos Is Betting that a Superlight Lining from Astronauts Boots Can Lift It to Stardom." *Business Week,* 19 January 1987, 70.

Dunkin, Amy; Carson, Teresa; Scredon, Scott; and King, Resa W. "Natural Soda: From Health-Food Fad to

Supermarket Staple." *Business Week*, 14 March 1986, 72.

Eklund, Christopher S. "Progresso Cooks Up a Challenge to Campbell." *Business Week*, 21 January 1985, 120-21.

"For Athletes, More Bounce from a 'Space Age' Shoe." *Business Week*, 15 September 1985, 133. ISSN 0007-7135.

"Look, Ma, No Dials: More Major Appliances Go Digital." *Business Week*, 11 November 1984, 97-101.

Mitchell, Russell. "Health Craze Has Kellogg Feeling G-r-r-reat." *Business Week*, 30 March 1987, 52-53.

Pitzer, Mary J. "Obsession Had Minnetonka Smelling Sweet." *Business Week*, 23 June 1986, 114.

Power, Christopher. "How Hodgson Built Its Niche in Prefab Houses." *Business Week*, 8 June 1987, 128b-c. ISSN 0007-7135.

Segal, Troy. "Second Life for Art Deco Dinnerware." *Business Week*, 21 September 1986, 111.

Smith, Emily T. Turning Sewage Sludge into a Squeaky-Clean Fuel." *Business Week,* 20 March 1987, 95. ISSN 0007-7135.

"Stouffer Pre-empts the Kitchen." *Business* Week, 3 June 1961, 85-90.

Therrien, Lori, and Borrus, Amy. "Reeboks: How Far Can a Fad Run." *Business Week*, 24 February 1986, 90-91.

"This Bra Won't Stay Bent out of Shape." *Business Week*, 9 December 1985, 97. ISSN 0007-7135.

BW   Witherspoon, Bruce D. *Great Big Super Giant Book of Trivia*. New York: Galahad Books, 1982. ISBN 0-88365-685-X. Note: LC 83-081093.

CAD   Wright, Lawrence. *Clean and Decent: The Fascinating History of Bathrooms and the Water Closet and of Sundry Habits, Fashions and Accessories of the Toilet Principally in Great Britain, France and America*. London: Routledge & Kegan Paul, 1960.

CB   "No-Frost Refrigerator-Freezer Combination." *Consumer Bulletin* 48 (July 1965):6-12.

CD   Davidson, Caroline. *A Woman's Work Is Never Done: A History of Housework in the British Isles 1650-1950*. London: Chatto & Windus, 1982. ISBN 0701139013.

CDIS

[Advertisement]. *Columbus Dispatch*, 3 December 3, B10.

Andrews, Robert W. "TV Dinner Tray Served to Smithsonian Institution." *Columbus Dispatch*, 1 May 1987, F4.

Aronson, Earl. "Canadians Go Underground to Produce Flowers, Vegetables." *Columbus Dispatch,* 10 February 1985, I13.

Benet, Lorenzo. "Ol' Standbys." *Columbus Dispatch*, 3 February 1987, B1-2.

"'Bugs' Help Clean Navy Sewage Tanks." *Columbus Dispatch*, 29 September 1985, E15.

Carmen, Barbara. "Candy Collector Has a New Relic--35c Bar." *Columbus Dispatch*, 10 January 1986, A1.

"Cold-Defying Home Comes from Sweden." *Columbus Dispatch*, 10 February 1985, I6.

"Contact Lens Makers Trying to Boost Soft Bifocals." *Columbus Dispatch*, 28 December 1986, C1.

Coy, Peter. "Teflon is 25--and Still Slick." *Columbus Dispatch*, 6 April 1986, D4.

Drake, Phyllis, "Life Is Now Just a Bed of Roses for Inventor of Water Wonder." *Columbus Dispatch,* 4 April 1986, D2.

Ellis, Merle. "Irradiation to Be Hot Topic in Meat Industry." *Columbus Dispatch,* 8 January 1986, G7.

Fiely, Dennis. "Campus Caviar: 50 Years of 'Mac and Cheese' Keep Students Going." *Columbus Dispatch*, 26 March 1987, B1.

Fiely, Dennis. "Prize in Every Box: Crackerjack Toy Designer Must Think Small." *Columbus Dispatch*, 25 November 1985, B1.

Gartner, Michael. "Having an Ax to Grind Denotes a Selfish Motive." *Columbus Dispatch*, 11 January 1986, A6.

Gold, Anita. "'Universal Tool' Worked Hard for Its Owner." *Columbus Dispatch*, 19 October 1986, I8.

Gorman, John. "Breakfast Wars Match Fast-Food Burger Wars." *Columbus Dispatch*, 28 May 1985, C7.

Greenwald, Marilyn. "Combibloc Containers Catching On." *Columbus Dispatch*, 14 December 1986, G1.

Hancock, Michelle. "High Heels: A Love-Hate Affair of the Fee." *Columbus Dispatch*, 28 April 1986, D1.

Hausman, Patricia. "Cereal: Just Name Your Grain." *Columbus Dispatch*, 30 January 1983, G2.

Jordan, Janice. "What's New: Bathrooms, Kitchens Given Pizzazz." *Columbus Dispatch*, 3 February 1985, I1.

Keller, Julia. "Treat Goes One: Little Debbie Comes of Age." *Columbus Dispatch*, 2 June 1985, D1.

Lafferty, Michael. B. "Windmills Aided Nation's Growth." *Columbus Dispatch*, 6 July 1986, H2.

Long, Kim. "Instant Yogurt Brews as a Big Hit." *Columbus Dispatch*, 1 January 1987, G3.

Madison, Tanya. "Coca-Cola at Center of Another Dispute." *Columbus Dispatch,* 17 July 1986, D2.

Norman, Bud. "Westerners Getting the Last Laugh." *Columbus Dispatch*, 3 April 1987, C7.

"One Small Sip for Man." *Columbus Dispatch*, 12 May 1985, F10.

Oricchio, Michael. "SPAM at 50: Still No Respect for Popular 'Mystery Meat.'" *Columbus Dispatch*, 29 January 1987, C1.

"Roaming Wild Pigs Raise Cain." *Columbus Dispatch,* 16 December 1986, A8.

"Shuttle's Problem with Potty Is Solved." *Columbus Dispatch*, 12 February 1985, A1.

Starfire, Brian. "Computer Boasts Humble Birth." *Columbus Dispatch*, 10 January 1985, D10.

Swift, Bob. "Moon Pies: The Real Southern Comfort." *Columbus Dispatch*, 2 June 1985, D1.

"Textiles Designers Finding the Computer a Useful Tool." *Columbus Dispatch*, 12 January 1986, I6.

Timberlake, Cotton. "Low-Calorie Sweetener Finds Uses in Booming New Market." *Columbus Dispatch,* 8 March 1985, C12.

Timberlake, Cotton, and Massie, Jim. "Happy Birthday to a Dynamic Duo: Oreos, Elsie the Cow Become Americian Traditions." *Columbus Dispatch*, 9 May 1986, D1.

"Tons of Lightweight Foil." *Columbus Dispatch*, 6 May 1987, G1.

Welzel, Karin A. "Changing Face of Frozen Foods." *Columbus Dispatch*, 6 January 1986, G1.

White, Mary Linn. "Sweet Smell of Success." *Columbus Dispatch*, 19 August 1986, E1-2.

Woods, Mike. "Mother Always Told You Not to Take Long Showers." *Columbus Dispatch,* 2 November 1986, I5.

Yant, Martin. "Hot Dog! Frankfurter's 500 Years Old." *Columbus Dispatch*, 19 July 1987, A11.

Young, Terry, "Murphy Beds Unfold for a New Generation." *Columbus Dispatch*, 26 December 1986, D2.

CJ

[Advertisement]. *Columbus Citizen Journal*, 4 December 1984, 2.

Collins, Susan. "Disposable Contact Lenses May Focus on Fashion. *Columbus Citizen Journal*, 30 January 1985, sec. 1, p. 10.

Collins, Susan. "FDA Approves Oral Medication to Relieve Symptoms of Herpes." *Columbus Citizen Journal*, 30 January 1985, sec. 1, p. 1.

"FDA Proposes Ban on Sulfite Preservative." *Columbus Citizen JOurnal*, 10 August 1985, 1.

"Floating Outhouse: NASA Is Flushed with Success of Simple Toilet." *Columbus Citizen Journal*, 12 February 1985, sec. 1, p.1.

Greene, Bob. "Ounce of Diet Soda Pop Worth Pound of Fudge." *Columbus Citizen Journal*, 23 January 1985, sec. 1, p. 2.

Greenhouse, Steven. "Hyper: Bigg's Is a Supper Market." *Columbus Citizen Journal*, 12 February 1985, sec. 2, pp. 21-22.

Kleinfield, N. R. "HoJo's Loses Its Magic Touch." *Columbus Citizen Journal*, 14 May 1985, 14.

"Potential Gonorrhea Vaccine Developed." *Columbus Citizen Journal*, 30 January 1985, sec. 1, p. 10.

"Something Old . . . Original Coke Will Return to Stores." *Columbus Citizen Journal*, 11 July 1985, 1.

COLLR

Thornley, Betty. "Eve in Action: Current Modes." *Colliers'* 84 (6 July 1929):28.

CP

Melhan, Mrs. Handiwipes, Baggies. Personal communication, 9 April 1985.

CREPT

"Disposable Diapers." *Consumer Reports* 26 (March 1961):1, 51-152.

"Ranges with Self-Cleaning Ovens." *Consumer Reports* 35 (November 1970):691-700.

"Refrigerators: Bottom-Freezer and Top-Freezer No Frost." *Consumer Reports* 33 (September 1968):472-80.

"Two Door Refrigerator-Freezers." *Consumer Reports* 29 (August 1964):368-75.

CT "Shopping Tips: Automatic Filter Drip Coffee Makers." *Changing Times* 27, no. 12 (December 1973):41-43.

DBE De Bono, Edward. *Eureka! An Illustrated History of Inventions from the Wheel to the Computer.* New York: Holt, Rinehart, Winston, 1974. ISBN 003012641x. Note: LC 73-21084.

DH Hunter, Dard. *Papermaking: The History and Technique of an Ancient Craft.* New York: Dover, 1978. ISBN 0-486-23619-6. Note: LC 77-92477.

DID Carter, E. F. *Dictionary of Inventions and Discoveries.* New York: Crane, Russak, 1976. ISBN 0844808679. Note: LC 75 37058.

EA *Encyclopedia Americana.* International edition. Danbury, Conn.: Americana Corp., 1980.

EAA Schapsmeir, Edward L., and Schapsmeir, Frederick. *Encyclopedia of American Agricultural History.* Westport, Conn.: Greenwood Press, 1974. ISBN 0-837179580. Note: LC 74 34563.

EB *Encyclopaedia Britannica.* Chicago: Encyclopaedia Britannica, 1980. ISBN 0852293607. Note: LC 78 75142.

EBM *Medical and Health Annual, 1982.* Chicago: Encyclopaedia Britannica, 1981. ISBN 0-85229-366-0. Note: LC 77-649875.

ECON "Perrier Finds a Zest for America." *Economist,* 15 February 1985, 70.

ECTL *Encyclopedia of Collectibles.* New York: Time-Life Books, 1977. ISBN 0-8094-2764-8. Note: LC 77-99201.

EI Hooper, Meredith. *Everyday Inventions.* New York: Taplinger, 1976. ISBN 0800825446. Note: LC 76 5624.

EL Lifeshey, Early. *The Housewares Story: A History of the American Housewares Industry.* Chicago: National Housewares Assoc., 1973. Note: LC 72 92666.

EMH   McGrew, Roderick E. *Encyclopedia of Medical History*.
             New York: McGraw-Hill, 1985.

EOA   *Encyclopedia of Associations*. 23d ed. Detroit: Gale
             Research, 1989. ISBN 0-8103-2587-X. ISSN 0071-0202.
             Note: LC 76 46129.

EOOET   Panati, Charles. *Extraordinary Origins of Everyday Things*.
             New York: Harper & Row, 1987. ISBN 0-06-096093-0.
             Note: LC 87 213.

EP   [Trademark 18-19 supplement]. *Editor and Publisher* 117,
             no. 48 (1 December 1984):T2, 6, 8.

ER   Hunt, Caroline L. *Life of Ellen H. Richards*. Washington,
             D.C.: American Home Economics Assoc., 1955.
             MacMullan, Jackie. "Shoe Fits." *Eastern Review*, April
             1987, 77-81.

ET   *Encyclopedia of Textiles*. Englewood Cliffs, N.J.: Prentice
             Hall, 1980. ISBN 0132765764. Note: LC 79 26497.

EW   Iversen, Wesley R. "Plugging into the Future." *Electronics
             Week*, 21 January 1985, 18-19.

FD   Dyer, Frank Lewis, and Thomas Commerford Martin.
             Edison: The Man and His Invention. New York:
             Harper, 1910, 2:964.

FF   "Nouvelle Barbecue." *Frequent Flyer*, August 1987, 13.

FFF   Kane, Joseph Nathan. *Famous First Facts*. New York: H.
             W. Wilson, 1981.

FFH   Thwing, Leroy. *Flickering Flames: A History of Domestic
             Lighting through the Ages*. Rutland, Vt.: Charles E.
             Tuttle, 1958. Note: LC 57 12287.

FHTC   Franklin, Linda Campbell. *From Hearth to Cookstove: An
             American Domestic History of Gadgets and Utensils
             Made or Used in America from 1700-1930*. Florence,
             Ala.: House of Collectibles, 1976.

FIH   Tannahill, Reay. *Food in History*. New York: Stein & Day,
             1973. Note: LC 75-160342.

FIIA   Reynolds, Peter J. *Farming in the Iron Age*. London:
             Cambridge University Press, 1976. ISBN 0521210844.
             Note: LC 75 43569.

Sources

1ST   Robertson, Patrick. *Book of Firsts*. New York: Bramhall House, 1982. ISBN 0517216795. Note: LC 82 4309.

500   Maitland, Duck. *500 Years of Tea: A Pictorial Companion*. New York: Gallery Books, 1982. ISBN 0-8317-3335-7.

FORB   Benoit, Ellen. "Raise High the (Precut) Roof Beam." *Forbes*, 137 (10 February 1986):120-24. ISSN 0015-6914.

Gold, Howard. "Maytag Steps Out." *Forbes* 134 (17 December 1984): 96-97.

Mack, Toni. "Elephant Dances." *Forbes* 137 (7 April 1986): 104-6.

McGough, Robert. "Pansies Are Green." *Forbes* 137 (10 February 1986): 89-92. ISSN 0015-6914.

Parr, Jan. "Look out, McDonalds." *Forbes* 136 (30 December 1985): 112. ISSN 0015-6914.

FORT   Baig, Edward C. "General Host Prunes Back to Prosper." *Fortune* 113 (20 January 1986):49. ISSN 0015-8259.

Dumaine, Brian. "Arco's Shortcut to Cheaper Sun Power." *Fortune* 112 (4 February 1985):74.

Flax, Steven. "It's Sugar, All Right, but It's Not Fattening." *Fortune* 112 (9 December 1985):117.

Kessler, Felix. "Tremors from the Tylenol Scare Hit Food Companies." *Fortune* 113 (31 March 1986):54, 62.

"Lettuce and Tomato Wars." *Fortune* 112 (9 December 1985):10-11.

McComas, Maggie. "Quaker Is Feeling Its Oats." *Fortune* 112 (10 June 1985):54-56, 60, 64.

Ramirez, Anthony. "In Hot Pursuit of High-Tech Food." *Fortune* 112 (23 December 1985):89-94.

Sporito, Bill. "Has-been Brands Go Back to Work." *Fortune* 113 (28 April 1986):123-24. ISSN 0015-8259.

FT   Fussell, George Edwin. *Farming Techniques from Prehistoric to Modern Times*. New York: Pergamon, 1966. ISBN 08-0113886. Note: LC 65 28099.

FUTUR   "Tomorrow in Brief." *Futurist* 58 (October 1984): 3.

GABE   GABe, Frances. "GABe Self-cleaning House." In *Technological Woman: Interfacing with Tomorrow*, Edited by Jan Zimmerman, 75-82. New York: Praeger, 1983. ISBN 0-03-062829-6. Note: LC 82 14033.

GBL	Saint-Laurent, Cecil. *Great Book of Lingerie*. New York: Vendome Press: 1986. ISBN 0-86565-072-1.
GH	"Aluminum Foil." *Good Housekeeping*, January 1957, 144-48.
HDI	Irvin, Helen Deiss. "Machine in Utopia: Shaker Women and Technology." In *Women, Technology and Innovation*, edited by Joan Rothchild, 313-19. New York: Pergamon, 1982. ISBN 0-08-028943-6. Note: LC 82 237141.
HEP	Haviland, Willaim A. *Human Evolution and Prehistory*. New York: Holt, Rinehart, & Winston; 1979. ISBN 0-03-044761-5. Note: LC 78 14737.
HFB	Wright, Lawrence. *Home Fires Burning: The History of Domestic Heating and Cooking*. London: Routledge & Kegan Paul, 1964.
HG	Green, Harvey. *Light of the Home*. New York: Pantheon Books, 1983. ISBN 039471329x. Note: LC 82 18867.
HH	Hall, Carl W., and Hendrick, T. I. *Drying of Milk and Milk Products*. 2d ed. Westport, Conn.: AVI, 1971.
HS	Hyams, Jay and Smith, Kathy. *Complete Book of American Trivia*. New York: Rutledge Books, 1983. ISBN 0-87469-041-2. Note: LC 83-11104.
IEEE	"Acoustic Wand May Replace Toothbrush." *IEEE Spectrum* 21 (November 1984):30. "Digital Hearing Aid." *IEEE Spectrum* 22 (December 1985):22. ISSN 0018-9235. "Lasers Assist Shoe Manufacture." *IEEE Spectrum* 22 (May 1985):24. "Whatever Happened to Irradiated Foods." *IEEE Spectrum* 21 (November 1984):28.
INFO	"Nielsen Study Shows 'Electronic Mall' Is an Effective Marketing Tool." *Information Hotline*, May 1985:9.
JHC	Celehar, Jane H. *Kitchens and Gadgets: 1920 to 1950*. Des Moines, Iowa: Wallace Homestead, 1982. ISBN 0870693581. Note: LC 80 53301.

Sources

JM   Munn, Joan. *Fashion in Costume, 1200-1980*. New York: Schocken Books, 1984. ISBN 0-8052-3905-7. Note: LC 83-27089.

KC   Franklin, Linda Campbell. *300 Years of Kitchen Collectibles*. Florence, Ala: Books Americana, 1984.

KM   Kranzberg, Melvin, and Davenport, William H. *Technology and Culture: An Anthology*. New York: Schocken Books, 1972. Note: LC 73-185318.

LB   Boyd, L. M. *Boyd's Book of Odd Facts*. New York: Sterling, 1979. ISBN 0-8069-0166-7. Note: LC 78-66291.

LR   Russell, Loris S. *Handy Things to Have around the House*. Toronto: McGraw-Hill Ryerson, 1979. ISBN 0070827818. Note: LC 79 094523 1.

MB   Benarde, Melvin A. *Chemicals We Eat*. New York: American Heritage Press, 1971. ISBN 07-004422-8. Note: 79-142976.

MH   Harrison, Molley. *Kitchen in History*. New York: Scribner's, 1972. ISBN 850450683.

MHEST   *McGraw-Hill Encyclopedia of Science and Technology*. 6th ed. New York: McGraw-Hill, 1987. ISBN 0-07-079292-5. Note: LC 86 2722.

MI   Coffee, F. K. "Newest Appliance: The Convection Oven." *Mechanics Illustrated* 73 (January 1977): 44-45.

MITU   Dibacco, Thomas V. *Made in the USA: The History of American Business*. New York: Harper & Row, 1987. ISBN 0-06-015524-4. Note: LC 86 45652.

MKTGN   "Consistent Marketing Effort Helps Gatorade Earn Profit and Prize." *Marketing News* 21 (17 July 1987):13.
"Easy Open, Close Bag Gets First Test." *Marketing News* 10 (10 October 1985):24.
"Electronistore Debuts: Donnelley Enters Growing Retail System." *Marketing News* 19 (24 May 1985):1, 42.
"Industrial Marketer Seeks Consumer Niche with a Distinctive New Product." *Marketing News* 19 (22 November 1985):21.

"Innovative Packaging Designs Will Affect Marketing in Food Segments." *Marketing News* 21 (13 March 1987):4.

"Microwave Packaging to Benefit Food Processors." *Marketing News* 20 (9 May 1986):14.

MM   Gimpel, Jean. *Medieval Machine: The Industrial REvolution of the Middle Ages*. 1977, Reprints. New York: Penguin, 1983. ISBN 0140045147. Note: LC 77 6676.

MMGT   Kinder, Faye; Green, Nancy R.; and Harris, Natholyn. *Meal Management*. 6th ed. New York: Macmillan, 1984. ISBN 0-02-354150-9. Note: LC 83-7944.

MONTW   Montgomery Ward & Co. *Catalogue and Buyers Guide*. 1945. Reprint. New York: Dover, 1969. ISBN 486-22377-9. Note: LC 78-86339.

MP   Harris, Leon. *Merchant Princes: An Intimate History of Jewish Families Who Built Great Department Stores*. Hew York: Harper & Row, 1977. ISBN 0-06-011797-4. Note: LC 79-1667.

MR   Louis, J. C. "Cola Wars: Thoughts from the Front." *Management Review* 74 (January 1985):52-58. ISSN 0025-1895.

NASA   Bourland, Charles T.; Rapp, Rita M.; and Smith, Malcolm C., Jr. "Space Shuttle Food System." *Food Technology*, September 1977, 40-45.

"Food for Space Flight." [NASA facts. National Aeronautics and Space Administration: NASA; 1982 Je; NF-133 6-82.]

Siegel, Victor. "Living in the Space Shuttle." [National Aeronautics and Space Administration. About space and aeronautics: NASA; March 1980; ASA 80-4.]

NB   Birmingham, Nan Tillson. *Store*. New York: Putnam, 1978.

NEWSW   Langway, Lynn; Fuller, Tony; and Simons, Pamela Ellis. "Razor Fighting." *Newsweek*, 22 November 1976, 103.

NFFOF   Poole, Gray Johnson. *Nuts from Forest, Orchard, and Field*. New York: Dodd, Mead, 1974. ISBN 0-396-06993-2. Note: LC 74-7662.

NW   Waugh, Norah. *Cut of Women's Clothes: 1600-1930.* New York: Theatre Arts Books, 1968. Note: LC 68-13408.

NY   Bliven, B., Jr. "Annals of Business: Crest Toothpaste." *New Yorker,* 23 March 1963, 83-84ff.

NYT   Claiborne, Craig, and Pierre, Tracy. "20th Century French Revolution." *New York Times Magazine,* 16 March 1975, 82.
Jones, Stacy V. "Fifty Years of Zips." *New York Times Magazine,* 28 April 1963, 108-9.
"Tylenol." *New York Times,* 2 October 1982

OED   Burchfield, R. W. *Supplement to the OED.* Oxford: Clarendon Press, 1972.

OFTAM   Blandford, Percy W. *Old Farm Tools and Machinery: An Illustrated History.* Ft. Lauderdale, Fla: Gale Research, 1976. ISBN 0-8103-2019-3. Note: LC 75 44376.

PA   Wallechensky, David, and Wallace, Irving. *People's Almanac.* Garden City, N.Y.: Doubleday, 1975.

PA78   Wallechensky, David, and Wallace, Irving. *People's Almanac,* Garden City, N.Y.: Doubleday, 1978.

PABL   Wallace, Amy; Wallechensky, David; and Wallace, Irving. *Book of Lists 3.* New York: William Morrow, 1983. ISBN 0688016472. Note: LC 82 20373.

PB   Davis, Myrna. *Potato Book.* New York: Wiliam Morrow, 1972. ISBN 0-688-00186-6. Note: LC 73-7890.

PCW   Parker, Wayne. "From Altair to AT." *PC World,* March 1985, 78-88.

PG   Miller, Sandy, Consumer Services (Procter & Gamble). Pampers. Telephone conversation, April 1985.

PM   Prisco, Dorothy D., and Moore, Harold W. *Fashion Merchandise Information: Textiles and Non-Textiles.* New York: Wiley, 1986. ISBN 0-471-89577-6. Note: LC 86-13337.

POPSC   Miller, Ann. "Aluminum Foil Goes to Work in Home and Shop." *Popular Science* 154 (January 1949):218-19.

PW   Wynne, Peter. *Apples.* New York: Hawthorn, 1975. ISBN 0-8015-0340-X. Note: LC 74-15646.

RC   Cowan, Ruth Schwartz. *More Work for Mother: The Ironies of Household Technology from the Open Hearth to the Microwave*. New York: Basic Books, 1983. ISBN 0465047319. Note: LC 83 70759.

RD   "Bugs with Iron Stomachs." *Readers Digest*, August 1985, 154. ISSN 0034-0475.
Ecenbarger, William. "Our New National Dish." *Reader's Digest*, May 1986, 127-30.
"Notes from All Over." *Reader's Digest*, March 1985, 162. ISSN 0034-0375.
"What a Plant Can Do." *Reader's Digest*, August 1985, 154. ISSN 0034-0375.

RJ   Forbes, R. J. *Studies in Ancient Technology*. 2d ed. Leiden, Holland: E. J. Brill, 1964.

RM   Manchester, Richard B. *Mammoth Book of Fascinating Information*. New York: A. & W. Visual Library, 1980. ISSN 0891041915. Note: LC 80 80792.

RMRB   Iverson, Nick. *Record Makers and Record Breakers*. Middlevillage, N. Y., 1979.

SA   Reynolds, Terry S. "Medieval Roots of the Industrial Revolution." *Scientific American* 251 (July 1984):122-30.

SBI   *Smithsonian Book of Inventions*. New York: Norton, 1978. ISBN 0895990024. Note: LC 78 62960.

SBR   "Executive Health--Fighting Plaque." *Small Business Report* 11 (November 1986):88.

SCI   Caporael, L. R. "Ergotism: The Satan Loosed in Salem?" *Science* 192 (2 April 1976):21.
Davis, Richard S. "Social Structure in the Pleistocene." *Science* 232 (23 May 1986):1024-26. ISSN 0036-8075.
Karban, Richard, and Carey, James R. "Induces Resistance of Cotton Seedlings to Mites." *Science* 225 (6 July 1984):53-54.
Norman, Colin. "No Panacea for Firewood Crises." *Science* 226 (11 November 1984):676.
Vietmeyer, Noel D. "Lesser-known Plants of Potential Use in Agriculture and Forestry." *Science* 232 (13 June 1986):1379-84. ISSN 0036-8075.

SCI85   "The Newest Way to Make a Baby." *Science 85* (8 March 1985):8.

SG   Giedion, Siegfried. *Mechanization Takes Command*. New York, 1948.

SI   Jewkes, John; Sawers, David; and Stillerman, Richard. *Sources of Invention*. 2d ed. New York: W. W. Norton, 1969. ISBN 39300502X. Note: LC 79 90986.

SKY   Shaw, Russell. "Bring on the Brie." *Sky* 14 (September 1985):71-81.

SM   Mintz, Sidney W. *Sweetness and Power: The Place of Sugar in Modern History*. New York: Viking, 1985.

SMM   *1978 Portfolio Sales & Marketing Plans*. New York: Sales and Marketing Management, 1977.

SS   Strasser, Susan. *Never Done: A History of American Housework*. New York: Pantheon Books, 1982. ISBN 039451024. Note: LC 81 48234.

SSVF   Friedlander, Barbara. *Secrets of the Seed Vegetables, Fruits, and Nuts*. New York: Grosset & Dunlap, 1974. ISBN 0-448-01368-1. Note: LC 72-90854.

TAC   Busch, Jane. "Cooking Competition: Technology on the Domestic Market in the 1930s." *Technology and Culture* 24 (April 1983):222-45.

        Cowan, Ruth Schwartz. "From Virginia Dare to Virginia Slims: Women and Technology in American Life." *Technology and Culture* 17 (January 1976): 1-23.

        Rose, MArk H. Urban Environments and Technological Innovation: Energy Choices in Denver and Kansas City, 1900-1940." *Technology and Culture* 25 (July 1984):503-39. ISSN 0040-165X.

        Schroeder, Fred E. H. "More 'Small Things Forgotten': Domestic Electrical Plugs and Receptacles, 1881-1931." *Technology and Culture* 27 (July 1986):525-43.

        Thrall, Charles A. "The Conservative Use of Modern Household Technology." *Technology and Culture* 23 (July 1982):175-95.

        Vanek, Joann. "Household Technology and Social Status: Rising Living Standards and Status and Residence Differences in Housework." *Technology and Culture* 19 (January 1978):361-75.

TH   Grun, Bernard. *Timetables of History*. New York: Simon
            & Schuster, 1979. ISBN 0671249886. Note: LC 79
            9952.

TIME   "Baby's New Bottle." *Time*, 7 June 1963, 52. ISSN 0040-
            791X.
            Cocks, Jay. "Black Tie Still Required: The Tuxedo
            Celebrates Its 100th Birthday." *Time*, 7 October 1986,
            99.
            Murphy, Jamie. "Morning after Pill." *Time*, 29 December
            1986, 62.
            "No. 32164: Necchi Sewing Machine." *Time*, 21 April 1452,
            100ff. ISSN 0040-781X.
            Shearton, Mimi. "Sweet Smell of Success: Lofty Cinnamon
            Rolls Are the Latest Snack Fad." *Time*, 21 April 1986,
            79. ISSN 0041-781X .
            "Udder Excitement." *Time*, 29 September 1986, 59. ISSN
            0040-781X.
            "Waterbeds: A Rising Tide." *Time*, 8 February 1971, 71.
            "Winning Spirit." *Time*, 19 May 1986, 49-60. ISSN 0040-
            781X.
            Zinti, Robert. "New Heat over an Old Issue: Renewing the
            Abortion Fight." *Time*, 4 February 1985, 17. ISSN
            0040-781X.

TJ   Trager, James. *Enriched, Fortified, Concentrated, Country
            Fresh, Lip-Smacking, Finger-Licking, International, Oral,
            and Unexpurgated Food Book*. New York: Grossman,
            1980.

TODHE   "Feeding Baby through the Ages." *Today's Health* 42 (April
            1964):76-79.

TR   Connell, Elizabeth B. "Crisis in Contraception. *Technology
            Review*, May-June 1987, 47-55.
            Kahn, Robert D. "Harvesting the Wind." *Technology
            Review*, November-December 1984, 55-61.

TT   *Timetable of Technology*. New York: Hearst Books, 1982.
            ISBN 0878512098. Note: LC 82 11899.

TW   Luker, Kristin. "Abortion: A Domestic Technology." In
            Technological Woman: Interfacing with Technology,
            edited by Jan Zimmerman, 128-350 New York:
            Praeger, 1983. ISBN 0-03-062829-6. Note: LC 82
            14033.

Sources

TWA   Blodgett, Bonnie. "Dirty Linen." *TWA Ambassador*, November 1984, 34-38.

UNITD   Fichter, George. "Birth of a Notion." *United* 31 (October 1896):87-96.

USA   "Bacteria Could Give the Space Shuttle a Boost." *USA Today*, 4 March 1985, D4.
"Office Building Brings Fuel Cell down to Earth." *USA Today*, 4 March 1985, B5.

USAIR   Beneson, Esther. "Delights of Broccoli," *USAIR* 9 (April 1987):18-22.
"English Potter in America." *USAIR* 8 (July 1986):28.
Jaffe, Walter W. "Romance of Tea." *USAIR* 9 (February 1987):16-21.
Karoff, Barbara. "In Praise of Peanuts." *USAIR* 8 (January 1986):14-20.
McBee, Julit. "Summer Nectar." *USAIR* 9 (August 1987):12-17.
Orwoll, Mark. "In Praise of Rolltop Desks." *USAIR* 8 (December 1986):52-58.
Schrambling, Regina. "All about Avocados." *USAIR* 8 (March 1986):68-72.
Taylor, Patricia A. "Time for Tomatoes. *USAIR* 8 (July 1986):12-19.
Van Benthuysen, Patricia. "No Matter How You Slice It, Spam Is Here to Stay." *USAIR* 9 (September 1987):86-93.

UZ   Eco, Umberto, and Zorzoli, G. B. *The Picture History of Invention: From Plough to Polaris*. New York: Macmillan, 1963.

VIS   O'Brien, George. "On the Home Front: The Big Little Halogen Light." *Vis à-vis* 1 (May 1987):380.

WA   Editors of World Almanac. *The Good Hosekeeping Woman's Almanac*. New York: Newspaper Enterprise Associates, 1977. ISBN 0385133774. Note: LC 77 75353.

WABI   Giscard d'Estaing, Valerie-Anne. *World Almanac Book of Inventions*. New York: World Almanac, 1985. ISBN 0-911816-96-0.

WAS   Wright, Lawrence. *Warm and Snug: The History of the Bed*. London: Routledge & Kegan Paul; 1962.

WB   *World Book Encyclopedia*. Chicago: Field Enterprises, 1975. ISBN 0716600757. Note: LC 73 93660.

WBW   Weinberger, Bernhard W. *An Introduction to the History of Dentristry*. St. Louis: C. Vimoshy, 1948.

WC   Willett, C., and Cunnington, Phyllis. *History of Underclothes*. Revised by A. D. Mansfield and Valerie Mansfield. London: Faber & Faber, 1981. ISBN 0-571-11747-3.

WG   Garrison, Webb. *How It Started*. Nashville: Abingdon Press, 1972. ISBN 0687176050. Note: LC 72 173951.

WIGR   Lefkowitz, Mary R., and Fant, Maureen B. *Women's Life in Greece and Rome: A Source Book in Translation*. Baltimore: Johns Hopkins University Press, 1982. ISBN 0800182866X. Note: LC 82 7756.

WJ   Abrams, Bill. "Sport Boss: Adidas Makes Friends, Then Strikes Deals That Move Sneakers." *Wall Street Journal*, 28 January 1986, 1, 25.

"Adieu, Beurre." *Wall Street Journal*, 4 April 1985, 1, 31.

Alsop, Ronald. "Dannon Co. Stirs into Action as Yogurt Competition Grows." *Wall Street Journal*, 12 December 1985, 2, 35.

Alsop, Ronald. "Old Chewing-Gum Favorites Find Life after Death." *Wall Street Journal*, 11 September 1986, 2, 37.

Alsop, Ronald. "Restaurant Battle Heating Up as Breakfast Trade Mushrooms." *Wall Street Journal*, 4 April 1985, 2, 31.

Bishop, Jerry E. "If the Shoe Fits, It's Probably Been Mostly Luck until Now." *Wall Street Journal*, 1 March 1985, 2, 17.

Bishop, Jerry E. "Sex Lives of Shellfish Reveal Clues to Increased Production." *Wall Street Journal*, 1 February 1985 , 2, 19.

Brooks, Geraldine. "Use Beer Scum, Salt to Make Vegemite: Australians Eat 4,500 Tons Annually." *Wall Street Journal,* 13 October 1986, 1, 15.

Bulkeley, William M. *"Who Built the First PC? Hint: His Name Isn't Wozniak or Jobs." Wall Street Journal*, 14 May 1986, 2, 27.

Bussey, John. "Procter & Gamble Is Laying Groundwork to Introduce Its Next Wave of Products." *Wall Street Journal*, 11 December 1984, 2, 31.

Carroll, Paul B., and William M. Bulkley. "Leaders of Personal Computer Firm Proclaim the Industry's Slump Is Over." *Wall Street Journal*, 24 February 1987, 12.

Days, Michael. "Scott Narrowing Its Paper Towel in Wipe-up Bid." *Wall Street Journal*, 1 April 1985, 2, 25.

Duke, Paul, Jr. "Compaq to Introduce Speedier Version of Portables, Desktops, Using 8386 Chip." *Wall Street Journal*, 24 September 1987, 10.

Freedman, Alix M. "Sticking Points: Teflon Is Versatile, but It Is Hell on Skis." *Wall Street Journal*, 4 October 1984, 1.

"Fungus Eats Metal." *Wall Street Journal*, 9 April 1987, 2, 29.

Galante, Steven P. "Fresh-Cookie Stores Feel Bit as Snack-Food Market Shifts." *Wall Street Journal, 24 February 1986, 2, 33.*

"Gillette Says Lubra-Smooth Will Expand Shaving Line." *Wall Street Journal*, 8 March 1985, 2, 48.

Hall, Trish. "And New Foods Tap Yogurt Boom." *Wall Street Journal*, 8 May 1986, 2, 29.

Hall, Trish. "New Packaging May Soon Lead to Food that Tastes Better and Is More Convenient." *Wall Street Journal*, 21 April 1986, 2, 21.

Hudson, Richard L. "French Entrepreneur Labors in Obscurity despite His Big Feat." *Wall Street Journal, 18 September 1985, 1, 18.*

James, Frank E. "If Tonight's Party Is Black Tie, Give Credit to Grizzy Lorillard." *Wall Street Journal*, 31 December 1985, 1, 4.

Johnson, Robert. "Frozen Custard Is Soft Ice Cream, but Hard to Find." *Wall Street Journal*, 19 June 1986, 1, 21.

Kilman, Scott. "Coca-Cola Co. to Bring Back Its Old Coke." *Wall Street Journal*, 11 July 1985, 1, 2.

Kilman, Scott. "Coca-Cola to Begin Market Test of Can Made from Plastic." *Wall Street Journal*, 14 October 1985, 2, 32.

Koenig, Richard. "Wendy's Battles Fast-Food Competition." *Wall Street Journal*, 12 November 1986, 1, 6.

Lowenstein, Roger. "Aqueduct Carries Hope to Colombia Town." *Wall Street Journal*, 19 February 1985, 2, 32.

MacDonald, Stephen. "Competing Designs of Toothpaste Pumps are Vying for Supremacy in Marketplace." *Wall Street Journal*, 13 November 1984, 2, 35.

McNish, Jacquie. "They Make Pickles Old-Fashioned Way, and It's Disgusting." *Wall Street Journal, 15 April 1986, 1, 28.*

Mufson, Steve. "Ostriches Don't Bury Their Heads in Sand; They're Just Hungry." *Wall street Journal,* 15 May 1985, Sec. 1, p. 1.

Murray, Alan. "Mail-Order Homes Sears Sold in 1909-37 are Suddenly Chic." *Wall Street Journal*, 11 February 1985, sec. 1, pp. 1, 18.

"Ovulation Predictors Find a Hot New Market despite a Steep Price." *Wall Street Journal*, 7 November 1985, 1.

"P&G Is Test Marketing Form of Tide Detergent." *Wall Street Journal*, 12 February 1985, 1, 14.

Pasztor, Andy. "EPA Bans Use of 2 Herbicides Containing Dioxin." *Wall Street Journal*, 19 March 1985, 1, 19.

Rundle, Rhonda L. "Vipont Labs Is Taking on Formidable Foes in Marketing Anti-Plaque Dental Products." *Wall Street Journal*, 28 January 1986, 2, 56.

Schacter, Mark. "As Long as They Really Schmeck Who Cares if They're Patented." *Wall Street Journal*, 11 February 1986, 2, 35.

Schwadel, Francine. "Campbell Soup Co. Plans Dry-mix Line to Rival Lipton's." *Wall Street Journal*, 12 July 1985, 2, 23.

Schwadel, Francine. "Revised Recipe: Burned by Mistakes, Campbell Soup Co. Is in the Throes of Change." *Wall Street Journal*, 14 August 1985, 1, 15.

Solomon, Jolie B. "P&G Introduces a Crest Toothpaste to Combat Tartar." *Wall Street Journal*, 13 March 1985, 1, 2.

Stipp, David. "Toxic Shock Linked to Certain Fibers Used in Tampons." *Wall Street Journal*, 6 May 1985, 1, 10.

Watkins, Linda M. "Bar Codes Are Black-and-White Stripes and Soon They Will Be Read All Over." *Wall Street Journal*, 8 January 1985, 2, 17.

Wessel, David. "To New Englanders, It's Never Too Cold for Eating Ice Cream." *Wall Street Journal*, 28 January 1985, sec. 1, pp. 1, 22.

"What's New: Solar Fan, Fungi and Nose Drops." *Wall Street Journal*, 13 November 1986, 31.

Zehr, Leonard. "Chester's Cold Ears: The Authentic Story of a Vital Invention." *Wall Street Journal*, 10 February 1986, 1, 8.

WK   Kowinkski, William Severini. *Malling of America: An Inside Look At the Great Consumers Paradise.* New York: William Morrow, 1985. ISBN 0-688-04180-9. Note: LC 84-22597.

WL   Weiser, Marjorie P. K., and Arbeiterer, Jean S. *Womanlist.* New York: Athenaeum, 1981. ISBN 0689111134. Note: LC 80 65983.

WT   Hirshorn, Paul, and Izenau, Steven. *White Towers.* Cambridge, Mass.: MIT Press, 1979. ISBN 0262080966.

WU   "LINC, First Personal Computer." *Washington University Magazine*, 1984 Fall; 54: 4; ISSN 0162-7570.